Composites Science and Technology

Series Editor

Mohammad Jawaid, Lab of Biocomposite Technology, Universiti Putra Malaysia, INTROP, Serdang, Malaysia

Composites Science and Technology (CST) book series publishes the latest developments in the field of composite science and technology. It aims to publish cutting edge research monographs (both edited and authored volumes) comprehensively covering topics shown below:

- Composites from agricultural biomass/natural fibres include conventional composites-Plywood/MDF/Fiberboard
- Fabrication of Composites/conventional composites from biomass and natural fibers
- Utilization of biomass in polymer composites
- Wood, and Wood based materials
- Chemistry and biology of Composites and Biocomposites
- Modelling of damage of Composites and Biocomposites
- Failure Analysis of Composites and Biocomposites
- Structural Health Monitoring of Composites and Biocomposites
- Durability of Composites and Biocomposites
- Biodegradability of Composites and Biocomposites
- Thermal properties of Composites and Biocomposites
- Flammability of Composites and Biocomposites
- Tribology of Composites and Biocomposites
- Bionanocomposites and Nanocomposites
- Applications of Composites, and Biocomposites

To submit a proposal for a research monograph or have further inquries, please contact springer editor, Ramesh Premnath (ramesh.premnath@springer.com).

More information about this series at http://www.springer.com/series/16333

Shamsher Bahadur Singh ·
Madappa V. R. Sivasubramanian ·
Himanshu Chawla
Editors

Emerging Trends of Advanced Composite Materials in Structural Applications

Springer

Editors
Shamsher Bahadur Singh
Department of Civil Engineering
Birla Institute of Technology and Science
Pilani, India

Himanshu Chawla
Department of Civil Engineering
Thapar Institute of Engineering
and Technology
Patiala, India

Madappa V. R. Sivasubramanian
Department of Civil Engineering
National Institute of Technology
Puducherry
Puducherry, India

ISSN 2662-1819 ISSN 2662-1827 (electronic)
Composites Science and Technology
ISBN 978-981-16-1687-7 ISBN 978-981-16-1688-4 (eBook)
https://doi.org/10.1007/978-981-16-1688-4

This Springer imprint is published by the registered company Springer Nature Singapore Pte Ltd.
The registered company address is: 152 Beach Road, #21-01/04 Gateway East, Singapore 189721, Singapore

Contents

Composite Behaviour of Thin Precast Concrete Sandwich Panels

Roger P. West and Oliver Kinnane

Abstract This chapter discusses the composite behaviour of thin wythe precast concrete sandwich panels comprising sustainable mixes, hybrid fibres and non-conductive shear connectors. Following a review of available high performance concretes, insulation and shear connector types, it focuses on the individual and collective load-deflection behaviour of insulation, single wythes, unconnected and connected wythes with short and long spans under flexural loading where both grid and pin fibre reinforced polymer shear connectors are used to assist in developing better composite action. It concludes that, depending on insulation thickness and stiffness and connector type, partial composite action can be achieved in enhancing the ultimate load capacity of precast concrete sandwich panels.

Keywords Facades · High performance concrete · Precast concrete · Sandwich panels

1 Introduction

In the context of the absolute imperative to reduce the carbon footprints associated with the construction and operation of buildings, which accounts for about 40% of carbon dioxide emissions worldwide (Directorate General for Energy 2017), the criticality of the manufacture, construction and structural and thermal performance of façades in use is acute. The significant majority of domestic and commercial buildings are energy inefficient in terms of heating or cooling demands in delivering occupant comfort and the use of conventional concrete in the building envelope

R. P. West (✉)
Department of Civil, Structural and Environmental Engineering, Trinity College, University of Dublin, Dublin 2, Ireland
e-mail: rwest@tcd.ie

O. Kinnane
Architecture, Planning and Environmental Policy, University College Dublin, Belfield, Dublin 2, Ireland
e-mail: oliver.kinnane@ucd.ie

S. B. Singh et al. (eds.), *Emerging Trends of Advanced Composite Materials in Structural Applications*, Composites Science and Technology,
https://doi.org/10.1007/978-981-16-1688-4_1

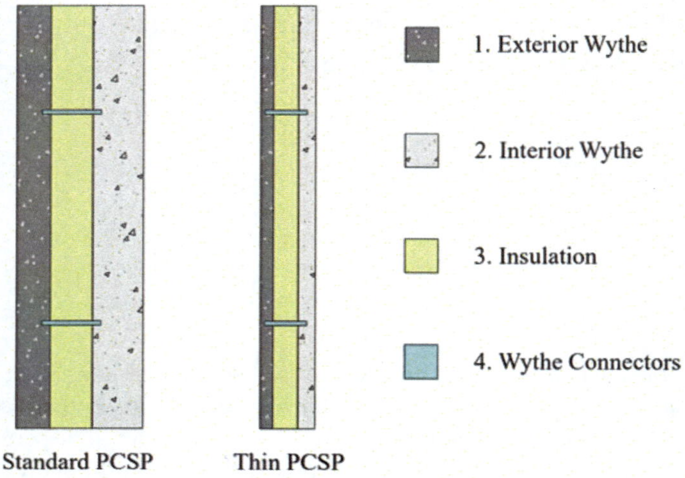

Fig. 1 Typical thick and thin PCSPs (O'Hegarty et al. 2019)

is not sustainable (Kinnane et al. 2014). However, as shall be discussed here, the evolution of more sustainable high performance thin lightweight precast concrete sandwich panels (PCSP) has the potential to have a significant impact on the cladding, re-cladding and over-cladding of new and existing building stock (FIB 2017).

Well-established thick heavy sandwich panels comprise an inner leaf, or wythe, and a thinner water-resistant outer rain screen wythe, infilled with insulation and connected by some form of shear connector, usually concrete webs or steel plates (Fig. 1). Panels are traditionally steel reinforced and structurally load-bearing, resisting wind, dead and live loads including self-weight, often with steel shear connectors (Kinnane et al. 2020). Typically 300–400 mm wide and weighing of the order of 500 kg/m^2, they impose considerable loads on lower floor columns and foundations.

However, in the provision of non-load bearing over-cladding or re-cladding facades, there has been considerable progress in reducing the inner and outer wythe thicknesses through the utilisation of high performance and more sustainable concrete, more efficient and/or greater thicknesses of insulation and non-conductive shear connectors to reduce the self-weight loads while aiming to provide the requisite structural strength and improved thermal performance (O'Hegarty and Kinnane 2019). Herein lies the main challenge—how to provide sufficient composite action between two thin high strength concrete wythes which accommodate larger insulation thicknesses using non-conductive connectors which transfer shear adequately between the wythes. In essence the aim is to maximise structural efficiency through composite action in conjunction with low thermal transmittance and minimal thermal bridging.

This is unobtainable without using innovative concrete composites containing alternative non-corrosive reinforcement, low thermal conductivity insulation and effective mechanical connectors such that partial composite action is achieved. This chapter will discuss how this might be successfully accomplished.

2 PCSP Elements

2.1 Concrete Wythes

In thick PCSPs, made using normal strength concrete (with water to cement ratios (w/c) of the order of 0.4–0.6), durability issues and the need for cover to the reinforcing steel bars drive up the minimum wythe thickness. For thin wythes, where the strength must be much higher, alternative concrete types are needed and there are many choices, some of which, with their abbreviations, are listed in Table 1. For more sustainable mixes, supplementary cementitious materials (SCMs) either partially or fully replace the less sustainable normal Portland cement (NPC), particularly the pozzolans: fly ash (FA), ground granulated blast-furnace slag (GGBS) and silica fume (SF). The latter is particularly used for high performance or ultrahigh performance concrete (HPC or UHPC) as it is very fine, chemically reactive and helps fill the natural voids in the cement matrix.

To have a viable HPC or UHPC with the high flexural strength needed by PCSPs, the w/c ratio must be low (circa 0.25–0.3) which, with a high cement content, means that there will be a much lower quantity of capillary voids (those water or air filled voids formed by virtue of surplus water not needed to hydrate all of the cement present). The higher strength of the cement matrix also bonds more effectively with the shear connectors. Often small diameter coarse aggregates only are used, with a void filler to particle pack the mix, minimising the void content, reducing porosity and improving compressive and flexural strength and durability. To maintain workability, a superplasticiser is essential, with long mixing times. Assuming good mechanical compaction and curing, the high strength thus attained creates a more brittle concrete which is also vulnerable to shrinkage cracking, both of which can be overcome by adding fibres. The type, size and dosage of fibre depends on the required properties, but principally they are added to control cracking and add flexural toughness post-cracking, when the elastic limit of the concrete is exceeded. In sufficient dosage, the fibres bind the concrete together, distribute the cracks and prevent them from opening under increasing load and/or displacement in a standard prism flexural test and in the panels themselves.

A series of typical mixes are given in Table 2 for illustration. Mix A is a mix with limestone aggregate which could be used for a thick wythe, where, for example, an NPC (or European CEM I) mix with a w/c ratio of 0.53 delivers cube compressive and beam flexural strengths of 37 and 6.9 MPa, respectively at 28 days. By increasing the cement content from 400 to 610 kg/m^3, with SCMs of GGBS and SF and 400 kg/m^3 of a calcium carbonate filler, the w/c of 0.33 delivers 89 and 13.3 MPa compressive and flexural strengths, respectively. By adding a high dosage of hybrid fibres, these

Table 1 Some concrete types for thin wythe PCSPs

Abbrev.	Name	Description
OPC/NPC	Ordinary/normal Portland cement concrete	Concrete containing traditional Portland cement as binder
FC	Foamed concrete	Low strength with no stone and highly voided to improve thermal properties
SCC	Self-compacting concrete	Highly fluid but sufficiently viscous to avoid segregation and requires no compaction
FRC	Fibre reinforced concrete	Contains one or two types (hybrid) of fibres to enhance crack control and post-cracking toughness
GRC	Glass-fibre reinforced concrete	Glass fibres added to mix, short or long, for shrinkage cracking resistance
SFRC	Steel fibre reinforced concrete	Steel fibres of different shapes and aspects ratios added to mix for shrinkage, cracking control and toughness
RCA	Recycled concrete aggregate	A more sustainable mix where old concrete has been crushed to a given size
RPC	Reactive powder concrete	High quantity of very fine cement without stone
GC	Geopolymer concrete	NPC is replaced by supplementary cements, usually pozzolans, activated by chemicals
PCMC	Phase-change material concrete	Solid granules added to mix to enhance the thermal storage capacity, while reducing strength
TRC	Textile reinforced concrete	A fabric or textile of different composition added to act as reinforcing
HPC	High performance concrete	Any concrete with compressive cube strength above 80 MPa
UHPC	Ultra-high performance concrete	Any concrete with compressive cube strength above 125 MPa. May or may not include fibres
UHPFRC	Ultra-high performance fibre reinforced concrete	A combination of any UHPC with FRC

strengths rise to 97 and 15.5 MPa, respectively. To make the concrete more sustainable still, supposing a recycled concrete aggregate (RCA) replaces the virgin limestone aggregate in full. It would be expected that the strengths would drop because the surfaces of the RCA contain porous adhered cement paste which is normally weaker and more absorbent than the parent limestone aggregate and so it is the case that the strengths drop to, for example, 80 and 11.0 MPa, respectively. To overcome this, a lower w/c of 0.25 is used together with additional SF to pre-coat (and thus block) the voids on the RCA, whereupon the strengths rise to 90 MPa and 13.5 MPa,

Table 2 Typical wythe concrete mixes: A: control (limestone); B: HPC limestone; C: HPFRC limestone; D: HPFRC RCA; E: HPFRC RCA + SF (Lipszynska et al. 2020)

Mass in kg/m^3	A	B	C	D	E
CEM I cement	400	254	254	254	254
GGBS	0	254	254	254	254
SF	0	102	102	102	109
10 mm crushed limestone	1050	760	760	0	0
10 mm recycled concrete aggregate	0	0	0	760	760
Fine aggregate	700	715	715	715	715
Calcium carbonate filler	0	400	400	400	400
Water	210	200	200	200	154
Superplasticiser	0	18	18	18	18
Steel fibres (30 mm)	0	0	39	39	39
Polypropylene fibres (24 mm)	0	0	18	18	18
w/c ratio	0.53	0.33	0.33	0.33	0.25
Cube strength (MPa)	37	89	97	80	90
Flexural strength (MPa)	6.9	13.3	15.5	11.0	13.5

respectively, representing a viable and more sustainable fibre reinforced HPC in the manufacture of thin wythes for use in PCSPs.

To observe the effect of fibres in term of enhancing strength and post-cracking toughness, Fig. 2 demonstrates that a HPC without fibres has a brittle response with flexural (tensile) and compressive strengths of 9.4 and 99.6 MPa, respectively, while a mix with 2.9% of 24 mm alkali resistant glass fibres added has a flexural strength of 15.3 MPa, with a compressive strength of 96 MPa (with a modulus of elasticity of 49.6 GPa). The considerable toughness post cracking (at circa 9.2 MPa) and residual strength post peak are notable. These properties are essential for PCSPs when subject to extreme wind or impact loading.

Fig. 2 Flexural test results on a 160 × 40 × 40 prism of HPC with and without 2.9% of 24 mm glass fibres (O'Hegarty et al. 2020)

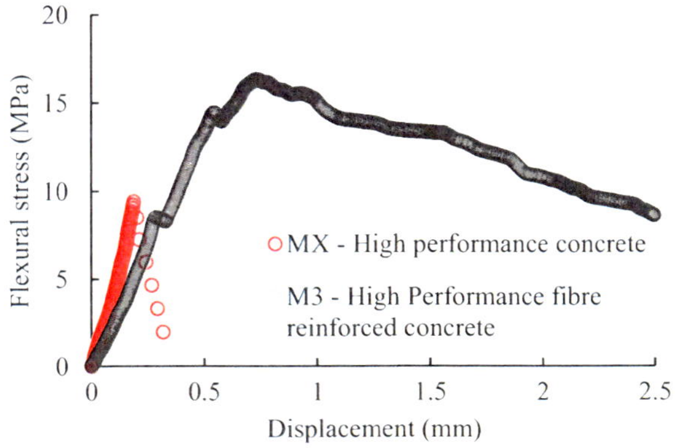

2.2 Insulation

The primary purpose of insulation in a PCSP is to provide thermal resistance across the façade element. Usually building standards specify a maximum thermal transmittance or U-value, which is the inverse of the sum of the resistances of each of the components of the façade through the thickness. There is a trade-off between the coefficient of thermal conductivity, k, and the thickness of the insulation layer needed to meet the standard. Table 3 lists some of the more common insulation types, from the more conductive (concrete) to the least conductive (VIP), with typical k values. While the VIP insulation is highly thermally efficient, it is also prohibitively expensive. The mineral wool insulations are highly compressible and so are unsuitable for PCSPs.

Figure 3 shows the different panel thicknesses which arise when VIP (30 mm thick), PF (110 mm) and EPS (180 mm) are used to deliver a U-value of 0.18 W/m K with 80 mm inner and outer concrete wythes. However, in the context of optimising the composite structural action between two wythes, the further apart the wythes become, as necessitated by the insulation choice for a given thermal performance, the more work the shear connectors have to do, recognising also that the compressive and flexural resistance of the insulation can also contribute to the load resistance of a PCSP.

The insulation and shear connectors can both offer compression resistance under lateral wind loads to promote load share between the two wythes and the concrete-insulation bond in longitudinal shear under flexure can also assist in enhancing the load carrying capacity to some degree. Figure 4, for example, shows the efficacy of XPS and EPS insulation in resisting compressive loads in a displacement control axial compression test (Fig. 5). It may be observed that after the short linear elastic region, continued crushing resistance is offered by the insulations such that a residual load capacity exists even under large displacements. The plots tail upwards as the bottom and top platens of the test machine engage.

Table 3 Different insulation types with indicative thermal conductivity coefficients, k

Abbrev.	Name	Thermal conductivity, k (W/m K)
–	Concrete	1.800
FC	Foamed concrete	0.500
MW	Mineral wool	0.038
EPS	Expanded polystyrene	0.037
XPS	Extruded polystyrene	0.030
PUR	Polyurethane foam	0.028
PIR	Polyisocyanurate	0.023
PF	Phenolic foam	0.021
VIP	Vacuum insulation panel	0.007

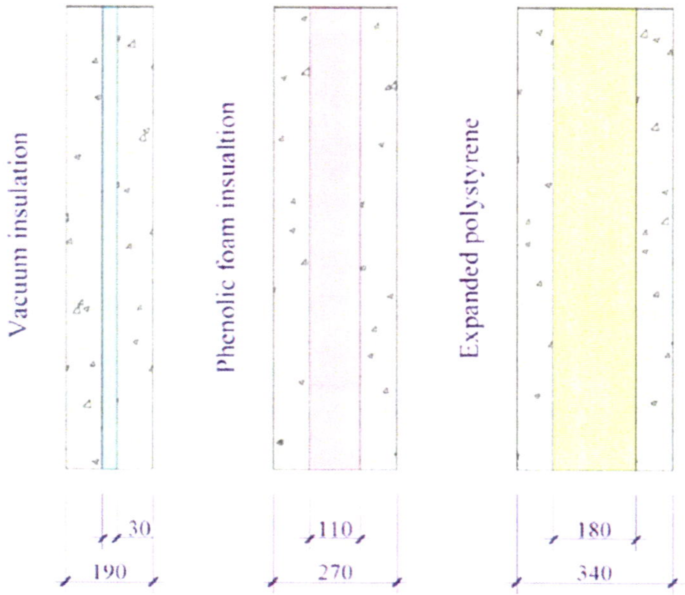

Fig. 3 Insulation thickness for a PCSP to satisfy a U-value of 0.18 W/m K using different insulating materials (O'Hegarty and Kinnane 2019)

Fig. 4 Compression test on EPS (left) and XPS (right) insulation sheets (Shukla 2019)

Similarly, Fig. 6 shows, from a flexural test on 5 sheets of plain XPS insulation siliconed together, that the insulation offers linearly increasing load capacity up to about 0.5 kN at about 30 mm displacement, indicating excellent flexibility in flexure.

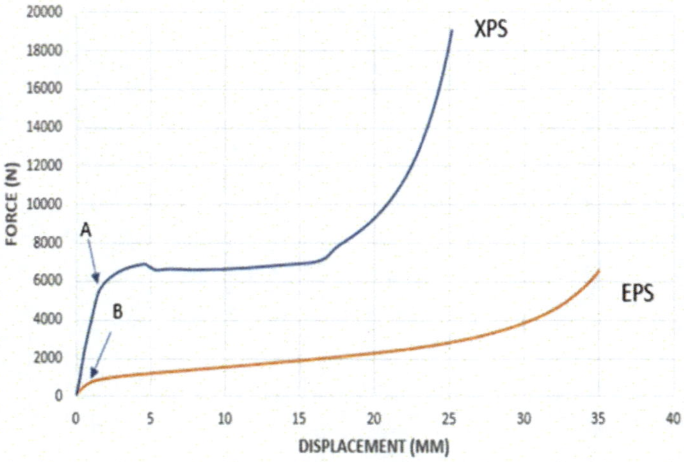

Fig. 5 Force-displacement plot for XPS and EPS insulation sheets (Shukla 2019)

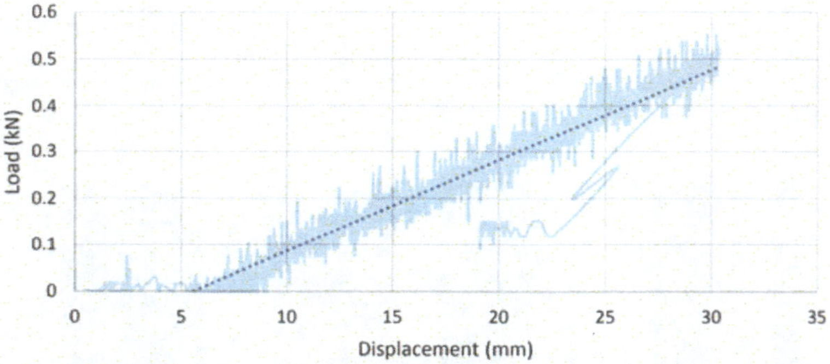

Fig. 6 Load–displacement plot for the flexural testing of 5 sheets of XPS insulation (180 mm thick) glued together using silicone (Lipszynska et al. 2020)

2.3 *Connectors*

The nature of the connection between the two wythes determines not only the structural but also the thermal performance of the PCSP. These connectors can be continuous, such as a vertical web of concrete or an FRP grid/truss, or can be discrete entities placed at the required centres in rows, such as steel plates or FRP pins (Fig. 7). The more conventional connectors, such as concrete or steel trussed bars or flat plates can be effective at developing good composite action, but have a significant drawback in their ability to create a thermal bridge between the outside and inside of a building due to their thickness (Fig. 7) and high thermal conductivity (see Table 4).

Furthermore, while most connectors have good to excellent tensile strength capacity (Table 4), their connector geometries also determine their susceptibility

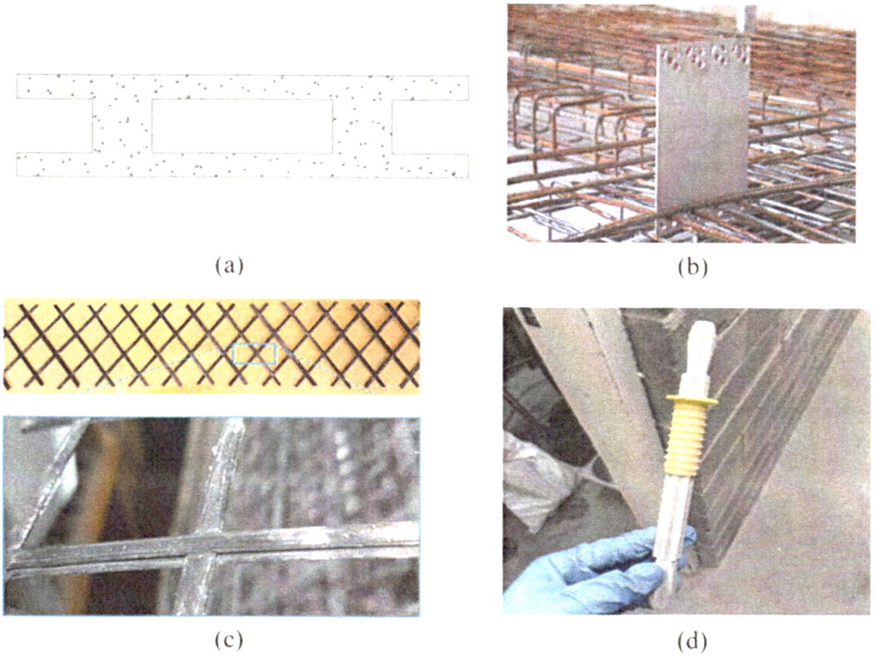

Fig. 7 Examples of **a** concrete web (O'Hegarty et al. 2020), **b** steel plate (Kinnane et al. 2020), **c** CFRP truss (O'Hegarty et al. 2020) and **d** FRP pin connectors (Lipszynska et al. 2020)

Table 4 Indicative properties of different connector types (O'Hegarty and Kinnane 2019)

Abbrev.	Name	Tensile strength (MPa)	Young's modulus (GPa)	Thermal conductivity, k (W/m K)
–	Ultra high strength concrete	12 MPa	40	2
–	Steel plates	450–700	200	50
CFRP	Carbon fibre reinforced polymer	600–3700	120–580	5–8
GFRP	Glass fibre reinforced polymer	480–1600	35–50	0.3–1
BFRP	Basalt fibre reinforced polymer	1100	70	1

to buckle under compression, such as in thin plates or CFRP grids or trusses in lateral shear. For example, the compressive and tensile responses of CFRP grids when embedded in a rigid foam is exemplified in Fig. 8 in which the spacing of the grid matrix affects tensile strength while the compressive strength is negligible, noting the change of axis scale in the compressive zone in the graph. What this

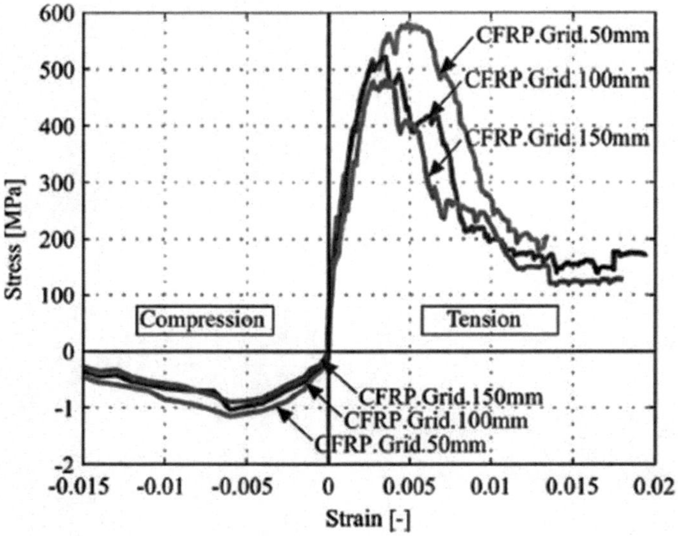

Fig. 8 Behaviour of CFRP grids within rigid foam subject to pure tension and compression (O'Hegarty and Kinnane 2019)

implies in practice is that the insulation becomes a key factor in determining the elastic composite behaviour of PCSPs made with such connectors—the lateral shear is resisted by a virtual truss in which alternate members of a CFRP truss act in tension while the insulation acts in compression due to the inability of the CFRP to resist compressive forces without buckling. Therefore, in addition to the spacing and orientation of the connectors, the insulation stiffness and width can also affect the composite behaviour of the PCSP.

On the other hand, the FRP pins (Fig. 7d), while much more effective at transferring compressive loads from an outer to an inner wythe in the transverse direction, are poor at resisting lateral shear forces parallel to the plane of the wythes. These matters will be explored further in due course using case studies.

3 Determining the Extent of Composite Action in PCSPs

Considering a single wythe of a PCSP, the thermal efficiency is determined by the thermal resistance (= 1/U value) which is proportional to the thickness, t, while the structural efficiency is determined by t^3 where t is the wythe thickness. If there are two wythes which are joined by connectors which are incapable of transferring lateral shear but can transfer applied load, such as through non-bonded insulation or isolated FRP pins, then the relative flexural stiffnesses of the wythes will determine the load share, as illustrated in Eqs. 1–3 in Table 5 and Fig. 9. This scenario represents

Table 5 Equations to calculate the second moment of area, I, of fully composite and non-composite PCSPs, with dimensions defined in Fig. 9 (O'Hegarty et al. 2019)

Non-composite panel		Fully composite panel	
$I_1 = \frac{b_1 t_1^3}{12}$	(1)	$A = b(t_1 + t_2)$	(4)
$I_1 = \frac{b_2 t_2^3}{12}$	(2)	$c_1 = \frac{[0.5bt_1^2 + bt_2(h - 0.5t_2)]}{A}$	(5)
$I_{nc} = I_1 + I_2$	(3)	$I_c = \frac{bt_1^3}{12} + bt_1 y_1^2 + \frac{bt_2^3}{12} + bt_2 y_2^2$	(6)

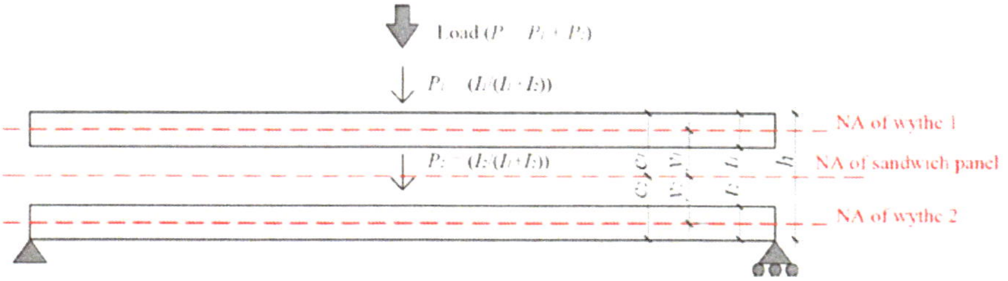

Fig. 9 Relevant dimensions and neutral axes for a PCSP for composite and non-composite cases, showing load share depending on the second moment of area of wythes if non-composite (O'Hegarty and Kinnane 2019)

non-composite action, as illustrated in Fig. 10b in which the individual wythes bend about their own neutral axes.

If, in contrast, in full composite action (Fig. 10a), there is complete transfer of shear between the two wythes, then the second moment of area, I_c, increases very substantially, as illustrated in Eqs. 4–6 in Table 5 and stresses and deflections are considerably smaller under identical loads. The additional terms involving the distance of the centroid of each wythe from the neutral axis (y_1 and y_2) contain a square term and so dwarf the t^3 term due to bending about the thin individual wythe's neutral axes. Therefore, the further apart the wythes are, the greater the structural efficiency if

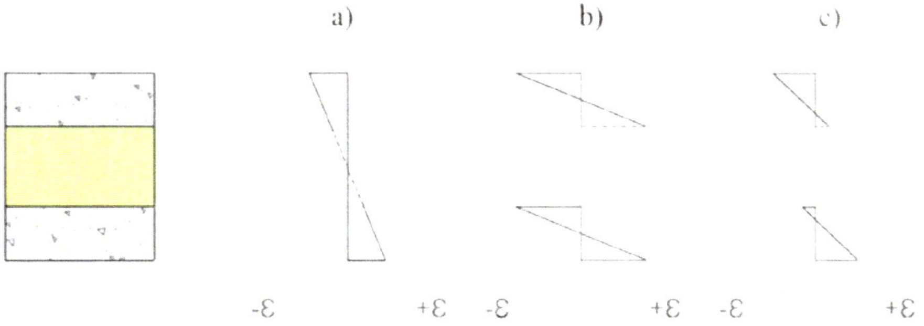

Fig. 10 Degree of composite action displaying approximate strain plots for **a** fully composite, **b** non-composite and **c** partially composite action (O'Hegarty and Kinnane 2019)

there is composite action. However, the further apart the wythes, the harder it will be for the shear connectors to develop full composite action and partial composite action (Fig. 10c) is more likely. For a given insulation/wythe/connector type, the wider the cavity the insulation fills, the better the thermal efficiency but only in proportion to the cavity width. So there is an incentive to increase the cavity width to improve the U value, but at a potential cost of structural efficiency and thus a balance must be found, especially if the connectors become less effective at transferring lateral shear.

The degree of partial composite action can be determined by assessing the elastic load-deflection response in a flexural test to establish the actual second moment of area being achieved in practice, comparing this to the full composite and non-composite theoretical values, as determined by Eq. (7) (O'Hegarty et al. 2020). The effective second moment of area is calculated from a three point flexural test value for deflection using Eq. (8) (O'Hegarty et al. 2020).

$$k(\%) = \frac{I_{exp} - I_{nc}}{I_c - I_{nc}} \tag{7}$$

$$EI_{exp} = \frac{PL^3}{48\delta} \tag{8}$$

In terms of testing a PCSP in flexure, although four point loading is often used, given there is a particular interest in lateral shear, the bending moment diagram for four point loading (Fig. 11) has two disadvantages: there is no shear in the central third of the beam and the moment in the central third is a constant, leading to more variable results due to the wider area over which flexural tensile cracking is likely to occur. Therefore, in the tests described here, they will all be conducted using three point loading, for which Eq. (8) applies. In practice, a uniformly distributed wind load is a common loading condition on a PCSP, so if, say, the wind pressure is 1.6 kPa, then on a span of 3.4 m, a PCSP has to resist a moment of 2.3 kN m to remain elastic. Therefore, the PCSP being tested must have a first cracking capacity of at least this magnitude.

Furthermore, it is normal when testing fibre reinforced concrete beams to use displacement control—that is, a displacement is imposed on the beam at a given

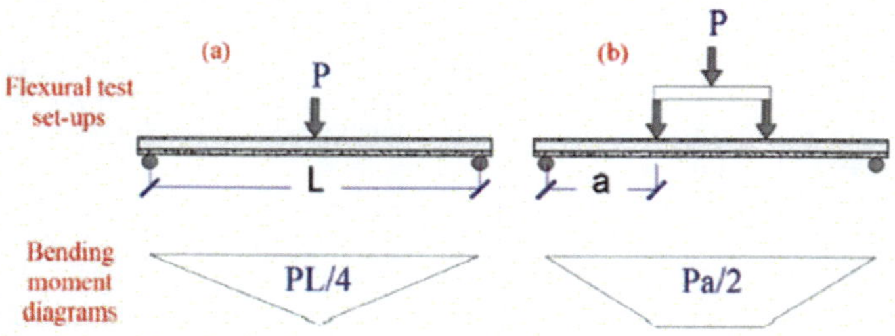

Fig. 11 Bending moment diagrams for **a** three and **b** four point flexural tests (O'Hegarty and Kinnane 2019)

rate (in mm per minute) so that the post-cracking toughness of the composite can be observed as the fibres pull-out of the cement matrix, such as observed in Fig. 2. The recorded load in such tests is the load resistance offered by the beam to that imposed displacement. If a load control test were to be used instead, then when the peak load is reached, the actuator would attempt to apply the next increment in load (at a given rate in kN/min), the beam would not have the capacity to resist such a load and failure would be sudden as the actuator head extension accelerates in an attempt to reach the next load increment. Therefore, there would be no post-cracking response in such cases. Hence, all the flexural results reported on here will be for displacement control tests.

Finally, there is a difference in the response of short or long span PCSPs because, under a given central point load, the transverse shear forces will not vary with span, whereas the bending moment and deflection will depend on the span (in proportion to L and L^3, respectively). Hence, in testing short span panels, the shear effects are exaggerated compared to the deflections and so present a sterner test of the shear force transfer capacity of the connectors.

4 Flexural Tests on Short Span PCSPs

4.1 Single Wythe Response

Using Mix C from Table 2 to create a single 900 × 600 × 20mm wythe in a simply supported flexural test with a central line load arrangement (Fig. 12), the load displacement graph in Fig. 13 demonstrates the initial linear response (to point A) followed by post-cracking ductility as the steel fibres gradually pull out (at B and C in particular), up to almost 40 mm central displacement, whereupon the slab rests on the lower platen causing an upward surge in the load. Strain gauge G2 (on the bottom of the wythe, see Figs. 12 and 14) also confirms the ability of the wythe to increase strain while sustaining load resistance capacity as the displacement is imposed incrementally. A further test was undertaken on a slab with two 35 × 35 mm ribs (used

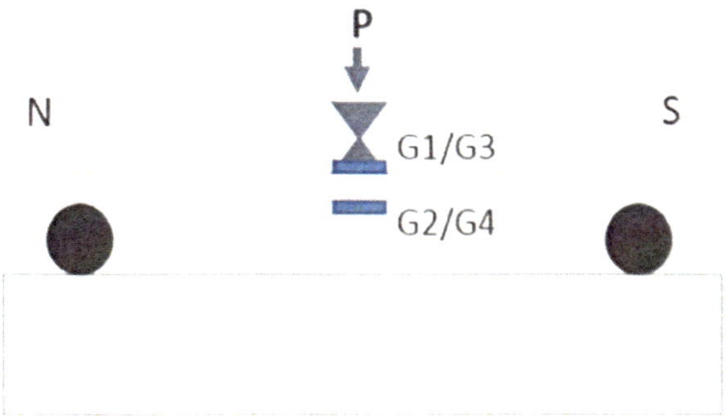

Fig. 12 Single wythe flexural test (Lipszynska et al. 2020)

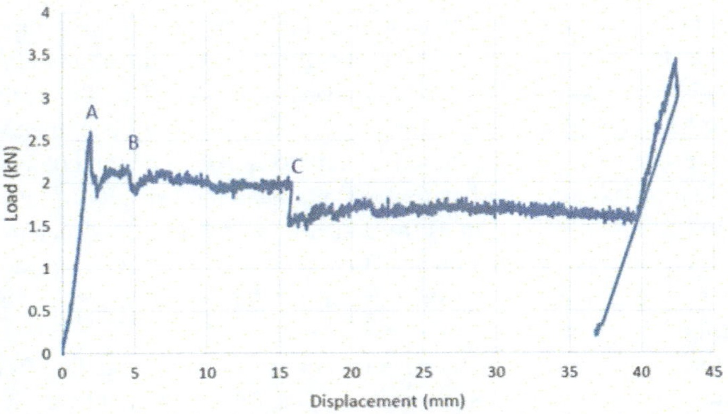

Fig. 13 Load versus mid-span displacement for single unribbed wythe in flexure (Lipszynska et al. 2020)

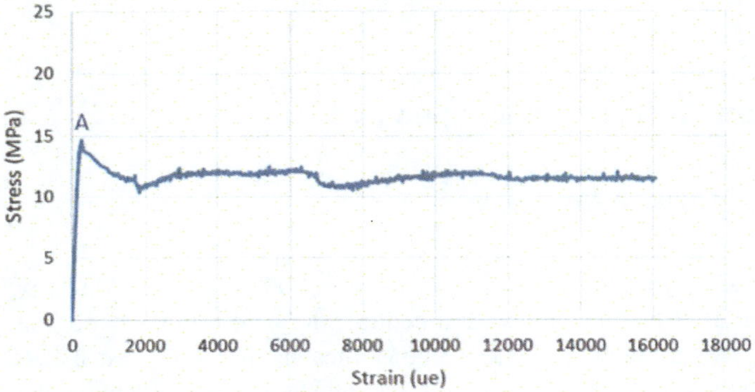

Fig. 14 Stress versus microstrain (G2) for single unribbed wythe in flexure (Lipszynska et al. 2020)

to embed connectors in the wythes more deeply) running longitudinally, which had a slightly higher capacity but very similar toughness response. Even with a fully developed crack, the wythe did not collapse due to the binding nature of the steel fibres (Fig. 15).

4.2 *Unconnected Panel Short Span Response*

By way of example, the flexural response of a 900 × 600 mm plan by 180 or 195 mm deep PCSP with two different insulation types will be discussed. As Fig. 16 shows, there is a lower thick wythe, of depth 120 mm, which is lightly reinforced with 4 no. 12 mm diameter high tensile steel bars. The upper thin UHPFRC wythe of 30 mm has a mix similar to Mix C in Table 2. The wythes are not connected by shear connectors.

Fig. 15 Fully developed crack post-peak load for a ribbed single wythe (Lipszynska et al. 2020)

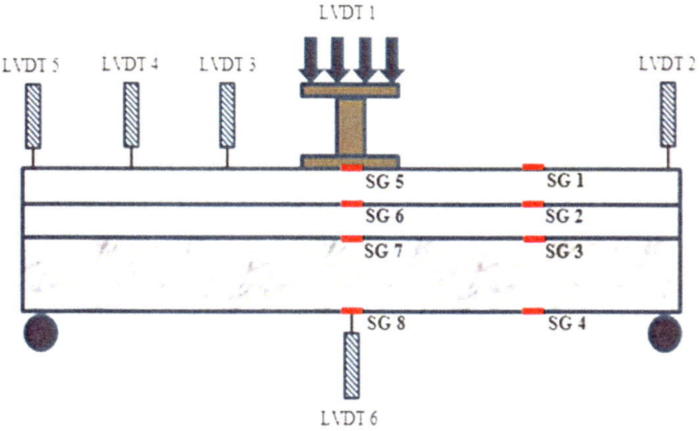

Fig. 16 Experimental set-up with strain gauges (SG) and displacement transducers (LVDT) locations (Shukla 2019)

The insulation is either 30 mm thick XPS or 45 mm of EPS, where the XPS initial stiffness is about 6 times that of the EPS (Fig. 5).

The influence of the insulation in the displacement response (over time, at a rate of 1 mm/min) of the unconnected PCSP may be observed from Fig. 17. Initially, the top wythe bends like a slab on an elastic foundation, where the lower wythe, which is much stiffer, provides relatively rigid support to the insulation, which helps to spread the load over the surface of the lower wythe. The upper wythe experiences its first crack at about point C in both XPS and EPS plots and, between points C and D, the ends of that wythe begin to lift, the crack widths under the load point increase, restrained by the fibres and a progressively smaller area of the outer edges of the insulation is engaged by the wythe as its edges lift progressively. From point D onwards, the insulation, under the V shaped rotated top wythe, is extensively compressed and the lower wythe now sustains almost all of the load in a simply supported slab

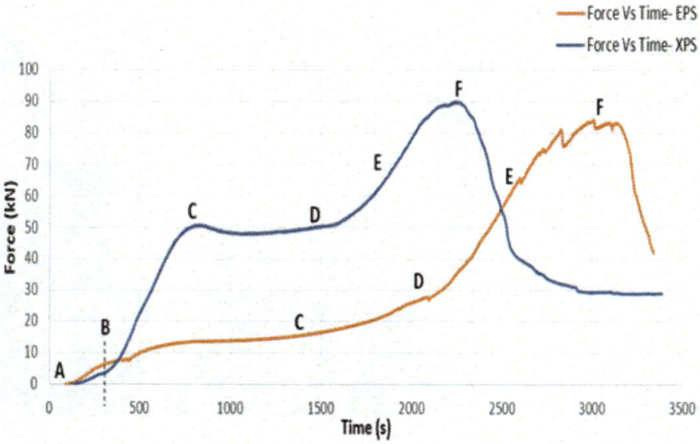

Fig. 17 Force versus time response in a flexural test on unconnected XPS and EPS insulated PCSPs (Shukla 2019)

arrangement. At point F, the reinforced lower wythe fails and the wythe's residual load capacity drops off substantially with continuing displacement. It may be observed that the insulation type plays a significant role in the responses, though peak load is determined by the bottom, stronger, wythe, at point F, as illustrated in Fig. 18. More importantly, from a serviceability viewpoint, there is a significant difference in the first crack load in the top wythe for the two insulation types due to the degree of flexure which arises in the top wythe, depending on the stiffness of the insulation.

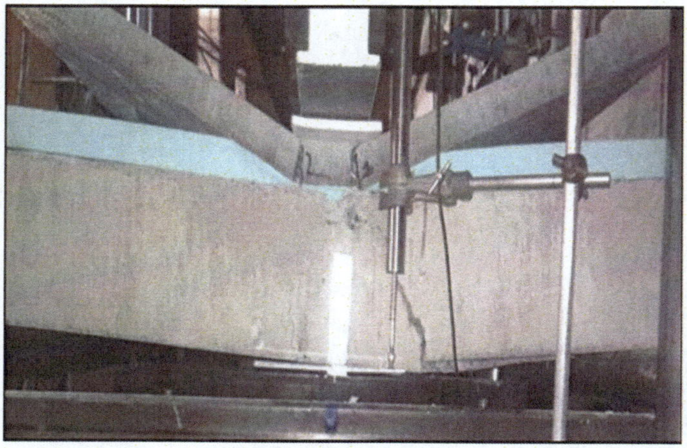

Fig. 18 Failure point of XPS insulated PCSP with top wythe disengaged (Shukla 2019)

4.3 Connected Short Span Thin Sandwich Panel Response

Consider a similar case but where a trussed CFRP grid (Fig. 7c) or discrete FRP pins (Fig. 7d) are used to connect the top and bottom wythes and with two sheets of siliconed XPS insulation.

As the load is applied, the early response for the trussed grid connector case is similar to the one above, where the grid provided very little compression resistance (Fig. 8) and a first crack appears on the bottom of the top wythe on either side of the load spreader (see Fig. 19a and the discontinuity in the curves in Fig. 20

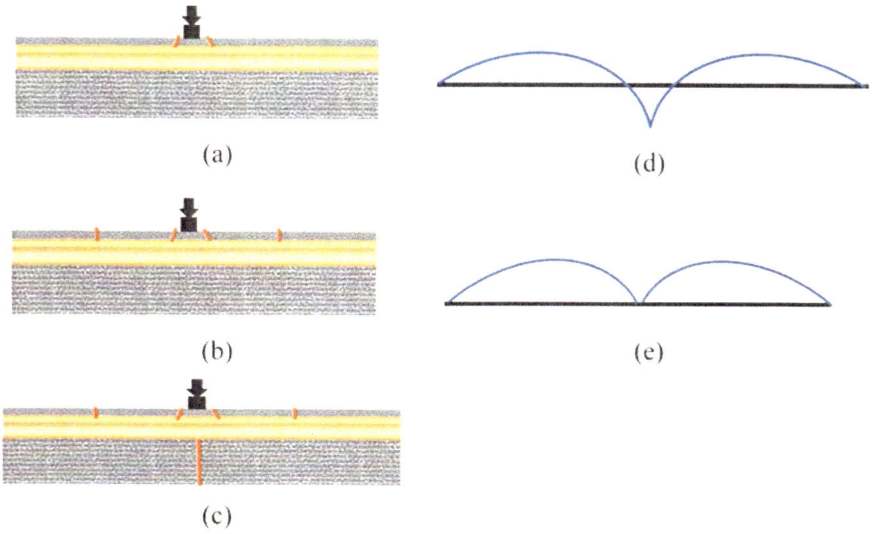

Fig. 19 Schematic of connected PCSP crack development (**a**–**c**) and the relevant bending moment diagrams for the top wythe (**d**, **e**) (Sexton et al. 2019)

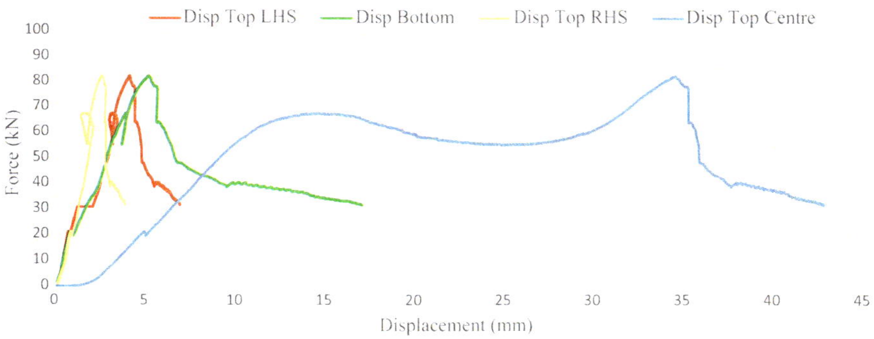

Fig. 20 Load displacement plot for the PCSP with XPS insulation and a trussed CFRP shear connector

at about 20 kN). This is because the bending moment diagram for a plate on an elastic foundation has peak sagging moment at the centre of the beam (Fig. 19d). The previous case showed that the edges now have a tendency to lift but the trussed connector resists this by developing a tensile force in response (see the top left hand (LH) and right hand (RH) curves in Fig. 20). At no point did the shear connector pull out of the 30 mm top wythe, suggesting that the bond with the UHPFRC was sufficient to embed the connector satisfactorily. An enhanced hogging bending moment ensues (Fig. 19e) which leads to tension on the top surface some distance from the load spreader, leading to a second pair of cracks propagating from the top surface of the top wythe (Fig. 19b) some distance either side of the load spreader. The top wythe continues to compress the insulation substantially as it moves downwards (see the top centre curve in Fig. 20). Ultimately, as further displacement is imposed in this displacement control test, the top wythe has little fibre pull-out resistance left and the lower wythe now takes all of the load until it too fails, at approximately 80 kN, by cracking in the centre of the bottom wythe (Fig. 19c). It should also be noted in Fig. 20 that the displacement at the centre of the bottom wythe (in green) is substantially less than the equivalent top wythe (in blue) under a given load because it is much stiffer as it is supported at its ends by a rigid vertical simply supported restraint, unlike the top wythe which is only supported by the highly compressible insulation. The resulting local crushing in the insulation in the vicinity of the load spreader at the point of failure of the bottom wythe may be observed in Fig. 21 in which the wythe ends may be observed as not having lifted due to the connector restraint.

In contrast, when a FRP pin shear connector is used, the pin can transfer the load more effectively to the bottom wythe as its axial stiffness is much higher than the insulation's. Therefore, as may be observed from Fig. 22, the top wythe displacements are considerably lower than the previous case and the local deformations near the spreader at the point of failure confirm this in Fig. 23. Figures 22 and 23 also confirm that the FRP pins are capable of resisting the uplift loads when the top wythe cracks (again at approximately 20 kN), with failure at approximately 80 kN as in the previous case.

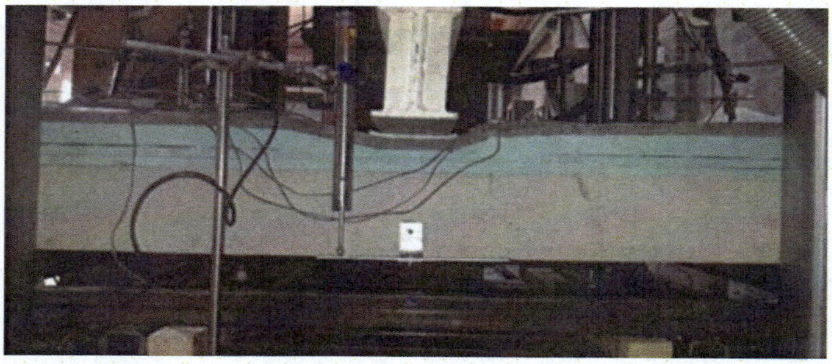

Fig. 21 The left and right wythe ends being restrained against uplift by the trussed CFRP connector in tension and insulation crushing under the load spreader due to local buckling of connector

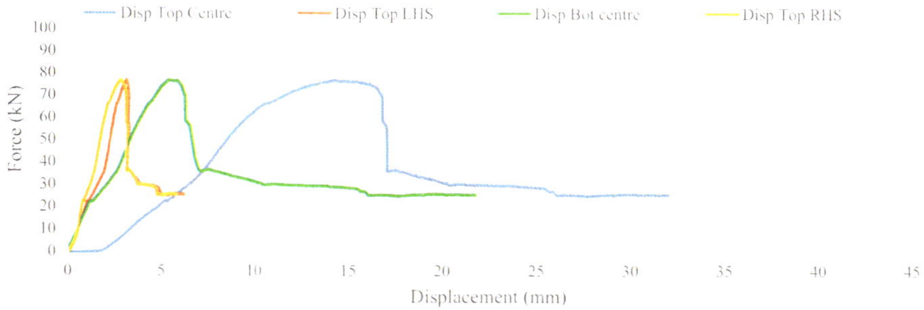

Fig. 22 Load displacement plot for the PCSP with XPS insulation and a pin FRP shear connector

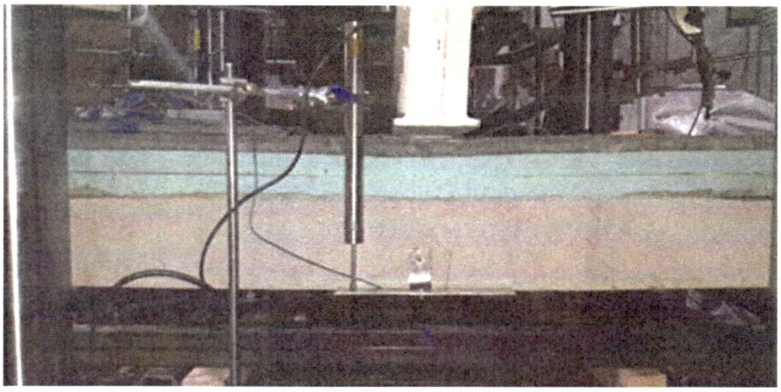

Fig. 23 The left and right wythe ends being restrained from lifting and insulation largely uncrushed under load spreader due to FRP pin connector

4.4 Connected Short Span Deep Sandwich Panel Response

With the desire for lighter, more thermally efficient PCSPs, the possibility of using a panel with two thin UHPFRC wythes and considerably greater insulation thicknesses pose additional challenges in developing composite action. Consider the case of a 900 × 600 mm panel with two 20 mm thick wythes with two 35 × 35 mm ribs along their length (Fig. 24) to accommodate the two rows of FRP pin shear connectors. The pins are 200 mm long so would only have 10 mm of embedment in each wythe were it not for the ribs. 180 mm of XPS insulation, in 5 siliconed sheets, was used, making a panel depth of 220 mm in total. The HPFRC concrete mix was highly sustainable with 50% GGBS, 100% recycled concrete as aggregate, coated with SF, and with hybrid fibres (mix E in Table 2). With a w/c of 0.25, the compressive and flexural strengths were 90 and 13.5 MPa, respectively (Table 2).

A finite element analysis of thick and thin bottom wythe panels under, say, a 50 kN point load with similar insulation thicknesses (Fig. 25) shows clearly that, in the linear elastic range, there is a significant difference in behaviour between the two cases. In case (a) the bottom wythe is so stiff that the top wythe flexes relatively

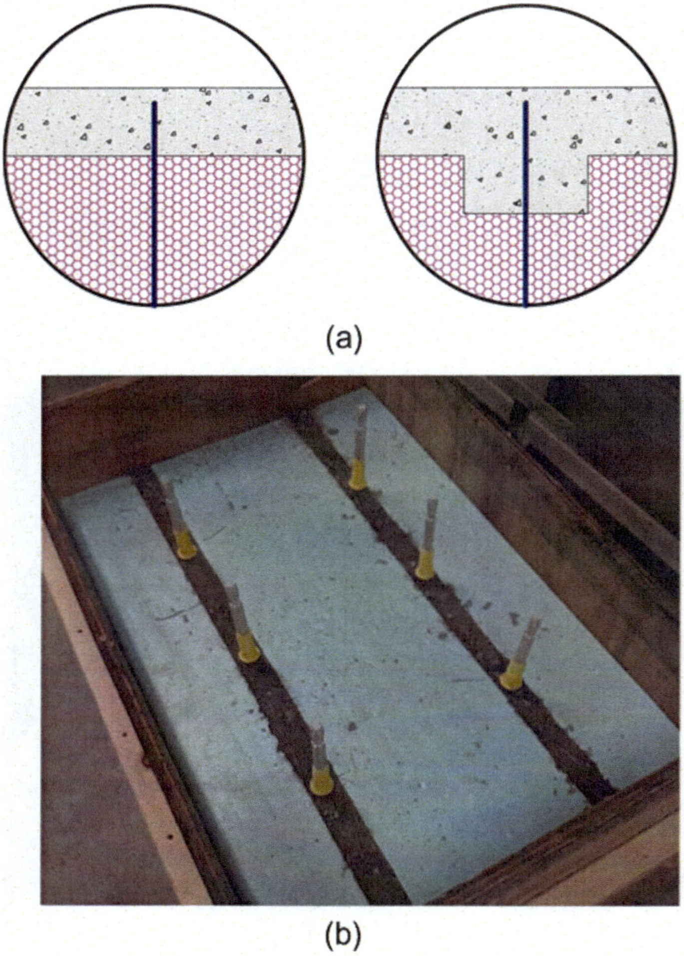

Fig. 24 **a** Graphic of 20 mm wythes with and without 35 mm ribs to enhance the anchorage of the shear connectors, used both top and bottom (O'Hegarty et al. 2020) and **b** the embedment of the FRP pin connectors at 200 mm centres between the two concrete panel pours in practice (Lipszynska et al. 2020)

so much more, giving rise to early cracking on the bottom face of that wythe, as observed previously (Fig. 19a). In case (b), with an equally thin bottom wythe, the stress on the bottom wythe can be greater (depending on the insulation and shear connector used) and so this lower wythe can crack first. One potential drawback of this is that if the panel is a vertical façade element, it is possible that the (unseen) inner wythe can crack without the outer wythe exhibiting any visible sign of distress, making the panel less safe under future loading. In practice, therefore, it is advisable to make the inner wythe slightly thicker than the outer one.

Fig. 25 Finite element elastic analysis leading to stress contours indicating a 50 kN load share for **a** thick and **b** thin bottom wythe PCSPs (O'Hegarty et al. 2018)

If two 220 mm thick ribbed panels are made up as described, one with and one without FRP pin connectors, with a simply supported test arrangement as before (Fig. 26), then the resulting load-deflection plots are as shown in Fig. 27a and b, respectively. In the unconnected case, as the panel is so deep, the top wythe and insulation leaves act as load spreaders and the bottom wythe cracks first at approximately the same load (2.5 kN) as a single wythe (Figs. 13 and 27a). Thereafter, with load resistance from the uncracked top wythe and the pull-out resistance of the bottom wythe, with a minor contribution from the insulation, the load resistance capacity increases to a second peak when fibre pull-out occurs in the lower wythe as the lower wythe cracks open up. The peak load (at D), when the top wythe develops a crack at about 3.5 kN, is somewhat less than the sum of the capacity of the two wythes (if they were load sharing equally). Thereafter, to point E, both fibre pull-out and shear slippage of the insulation leaves occur, with as much as 2 kN residual load capacity at failure. The physical state of the panel at this point may be observed in

Fig. 26 Test set-up for unconnected and connected thin wythe deep PCSPs (Lipszynska et al. 2020)

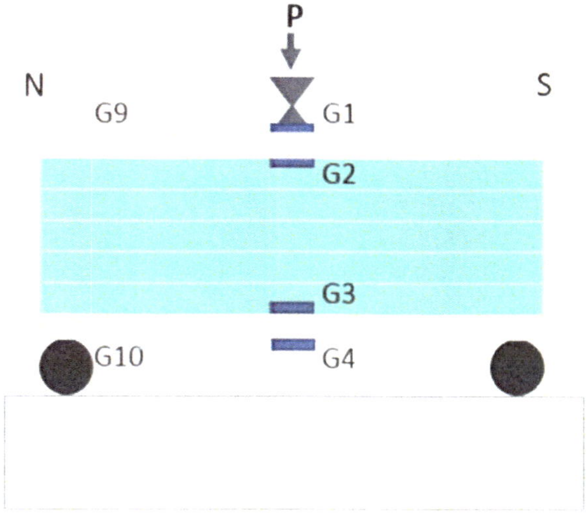

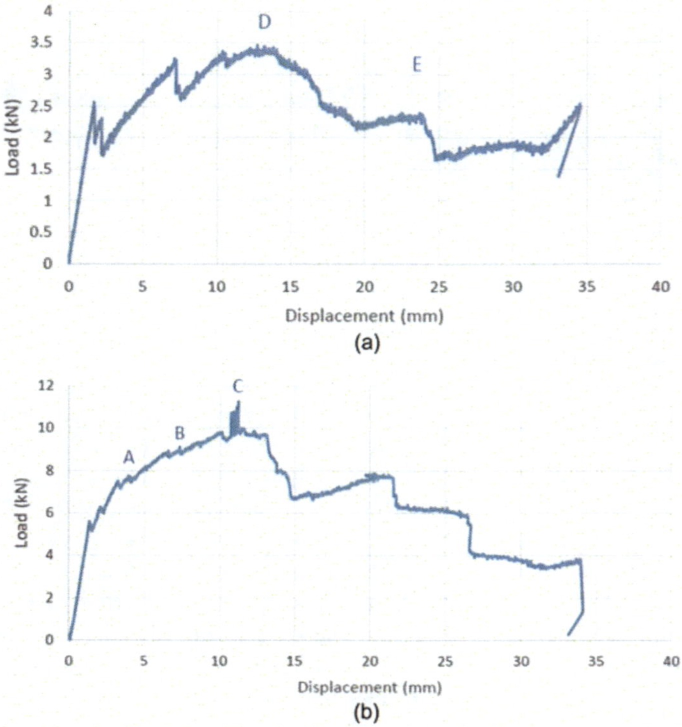

Fig. 27 Load mid-span displacement plots for **a** unconnected and **b** FRP pin shear connected PCSP made with sustainable UHPFRC panels (Lipszynska et al. 2020)

Fig. 28 in which the lower wythe is seen to be disengaged, the insulation sheets have slid horizontally (as may be observed from the staggered black vertical line on the insulation near the centre of the panel in the figure) and the top wythe has cracked but is holding the panel in place through the pull-out capacity of the fibres and the residual strength of the insulation (see also Fig. 6) which has not failed in tension despite the extent of sliding horizontally due to shear.

The connected panel has evidence of almost equal load share between panels at the point of the first crack at 5.5 kN (Fig. 27b) slightly more than twice the uncracked capacity of a single wythe.

The theoretical second moment of area of unconnected and fully composite arrangements are found to be 8×10^5 and 240×10^6 mm^4, respectively using Eqs. (1)–(6) from Table 5, the fully composite value being some 300 times the unconnected case. Using the single wythe test results (Fig. 13 and Eq. (8)) to estimate the modulus value, E, of the HPFRC as 35 GPa, the experimental composite value, I_e (Fig. 27b) can be estimated as 8.4×10^5 mm^4 from Eq. (8). Therefore the degree of partial composite action in the linear elastic range is calculated to be negligible (less than 0.1%), using Eq. (7), within experimental error.

Fig. 28 Unconnected thin UHPFRC wythe PCSP close to failure (Lipszynska et al. 2020)

Thus, there is no strong evidence here of composite action up to the elastic limit but, significantly, without evidence of a drop off in load due to wythe cracking, there is progression of load resistance from A to B until a peak at C of about 11.3 kN. It is reassuring that the post-cracking toughness during this displacement control test is such that the displacement at ultimate load is approximately 6 times that at the serviceability (first crack) limit.

The maximum load capacities of the insulation, single wythes with and without ribs, and the unconnected and connected panels are given in Table 6. It may be noted that the 11.3 kN capacity of the connected panel is approximately 50% higher than the sum of the individual peak capacities of the two wythes and insulation acting independently, indicating good composite action. The only difference between the two cases is the presence of the FRP shear pins which provides for this enhanced strength and the subsequent post-cracking toughness, with evidence of crack development and fibre pull-out in Fig. 27b. However, as the illustration of the final physical state of the panel shows in Fig. 29, the bottom wythe is again disengaged, the top has residual strength with restraint from the pull-out resistance of the steel fibres while the insulation shows much lower signs of slippage and subsequent tensile tearing

Table 6 Maximum load capacity of individual elements in thin UHPFRC wythes PCSPs (Lipszynska et al. 2020)

Element	Maximum load (kN)
Insulation	0.50
Wythe without ribs	3.44
Whythe with ribs	3.68
Unconnected thin wythe PCSP	3.46
Connected thin wythe PCSP	11.22

Fig. 29 Connected thin UHPFRC wythe PCSP close to failure (Lipszynska et al. 2020)

instead due to the binding and composite nature of the FRP pins shear connectors, which showed no sign of being pulled out of the HPFRC wythes.

5 Long Span PCSPs

A series of displacement controlled three point flexural tests were conducted on 3000 mm span by 750 mm wide panels with 25 or 40 mm upper wythes and 40 mm lower wythes, using a HPFRC concrete similar to mix C in Table 2, except either 10 or 60 kg/m^3 of steel fibres were used in conjunction with 3% of 24 mm long alkali resistant glass fibres. The concrete compressive strengths at 28 days was 96 MPa. Either FRP grid or pin connectors were used and there was 90 mm of foam insulation between the wythes. The protrusion of the FRP grid connectors above the insulation prior to the top wythe concrete being poured may be observed in Fig. 30a. The test arrangements for the two wythe depths are shown in Figs. 30b and 31.

A comparison of the deflection behaviour of the two shear connector types (Fig. 32) shows that initially the pin connectors have greater stiffness and better load-sharing than the grid connectors (with k values of approximately 13% and 8%, respectively), as was the case with the short span. But post first crack, the composite behaviour of the grid connector is better because as the lateral shear develops with increasing deflection in a slender panel, the insulation provides some stiffness in compression while the grid resists the lateral tension. This is evidenced by an examination of the failed panel (Fig. 33) in which the trussed grid is seen in many cases to have snapped not pulled out of the HPFRC matrix, suggesting that the extra bond of the grid with the concrete matrix is sufficient to facilitate the tensile failure in the truss elements. There is significant load capacity in the trussed grid post cracking which is reflective of the composite load sharing between the wythes, as may be observed through two additional lines on Fig. 32 in which the theoretical non-composite and fully composite linear elastic responses are shown.

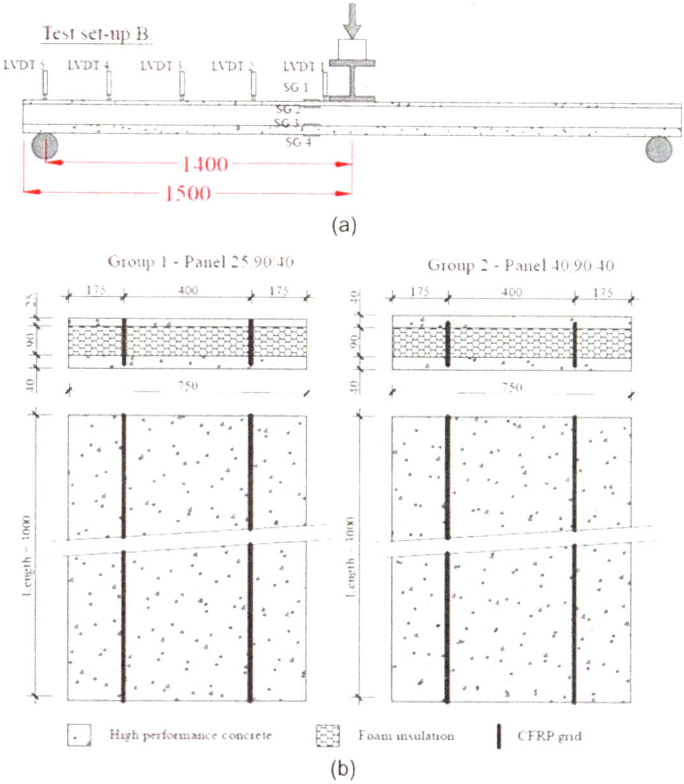

Fig. 30 Test set-up with **a** front sectional elevation and **b** end elevation and plan (O'Hegarty et al. 2019)

Other factors that influence the response can also be examined. In Fig. 34, the influence of the upper wythe thickness may be observed. The initial stiffness and the ultimate strength are both considerably enhanced through a relatively small change (some 15 mm) in the thickness of one wythe—for less than a 25% increase in the weight, the ultimate load capacity increases by about 80%.

In relation to the benefit of the presence of the steel fibres in the HPRFC mix, Fig. 35 indicates that changing from 10 to 60 kg/m^3 will significantly improve the post-cracking toughness, where for the former, some residual pull-out capacity exists on a softening load-displacement curve while a higher elastic capacity exists as well as a hardening curve leads to more than a two-fold increase in load capacity for the latter.

Finally, the importance of the insulation's contribution with and without CFRP grid connectors can be seen from Fig. 36. The additional benefit of the insulation in developing composite action leading to lower displacements can be seen by the divergent nature of the curves with increasing insulation stiffness.

Fig. 31 FRP grid protruding above the insulation prior to the second concrete pour and the flexural test arrangement for long span panels (O'Hegarty et al. 2019)

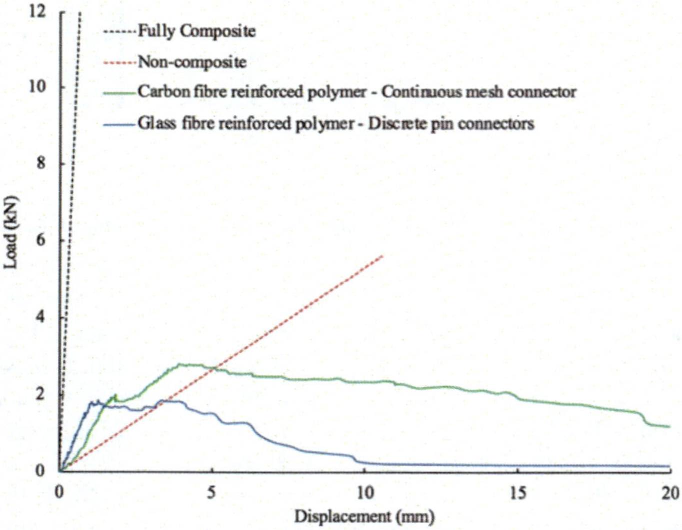

Fig. 32 Flexural test comparison of CFRP grid connector with FRP pin connectors (O'Hegarty et al. 2020)

6 Concluding Remarks

While the use of steel or concrete as shear connectors results in a high degree of composite action between wythes in PCSPs, the lower thermal performance due to significant bridging has led to an examination of alternative non-conductive shear

Fig. 33 The condition of the trussed grid connected when the top wythe is lifted after failure of the panel (O'Hegarty et al. 2019)

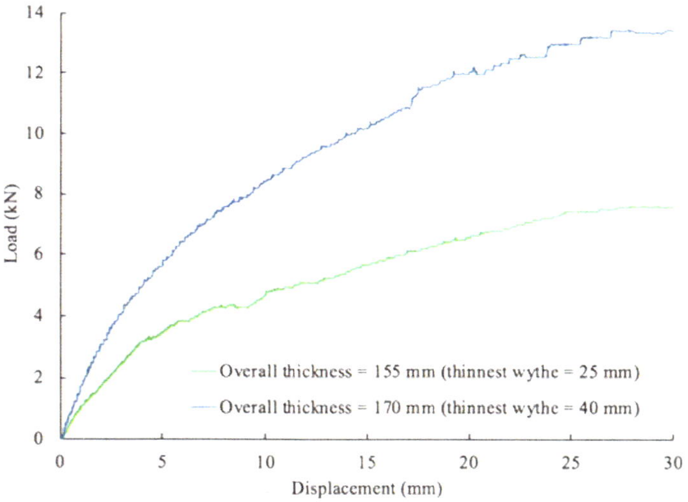

Fig. 34 Flexural test comparison of panel wythe thicknesses (O'Hegarty et al. 2020)

connectors. In particular, grid and pin FRP connectors have been examined by testing short and long panels in flexure. It may be concluded that for shorter spans the pin connectors allow excellent load-sharing between wythes by virtue of their axial stiffness but they perform poorly in more slender long span panels in which their lateral shear resistance is less effective. In contrast, in short beams, grid CFRPs have very poor initial load transfer and rely on the insulation stiffness, due to the local buckling of the grid under a compressive force. In longer span panels, the grids form a tension-compression truss with the insulation in resisting the lateral shear necessary for partial composite action and so, post-peak, there is hardening in the response and ultimate failure loads are significantly enhanced. The presence of fibres

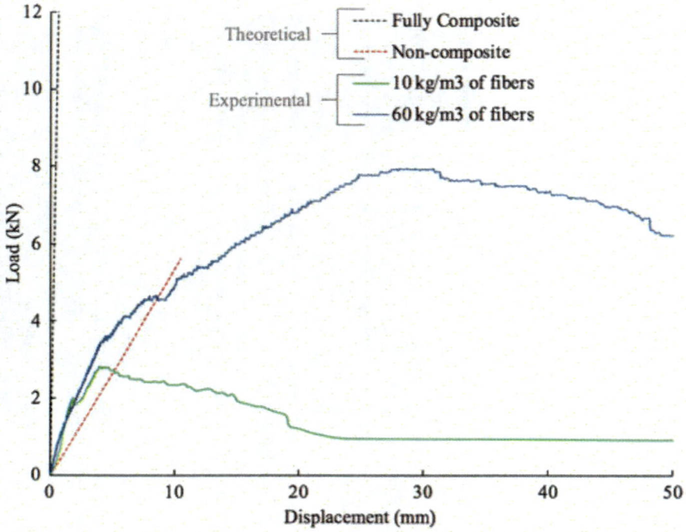

Fig. 35 Comparison of PCSPs encompassing concrete wythes of varying fibre content (O'Hegarty et al. 2020)

Fig. 36 Comparison of flexural rigidities with increasing insulation modulus for PCSPs with and without CFRP grid connectors (O'Hegarty et al. 2019)

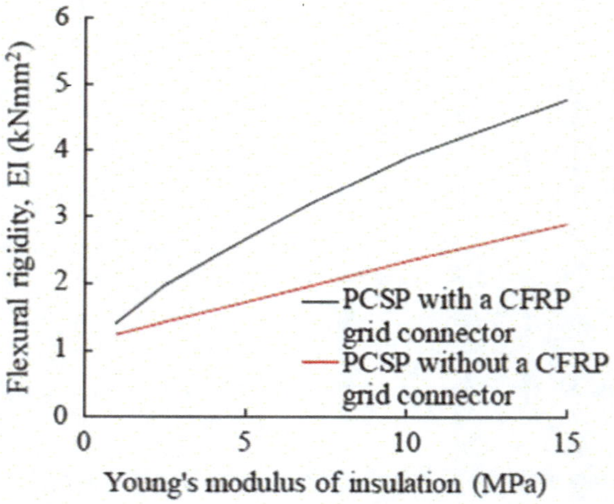

and high compressive/tensile/flexural stress capacities in the HPFRC wythes enables effective pull-out behaviour to be observed for the hybrid fibres and both types of the embedded FRP connectors, making such thin elements viable as an over-cladding panel. With the need for axial load-bearing capacity in the facades of buildings, such as in re-cladding projects, the benefits of a thicker inner wythe have been observed and an example of its use in practice may be seen in Fig. 37, where a model of a single precast panel and its execution may be observed.

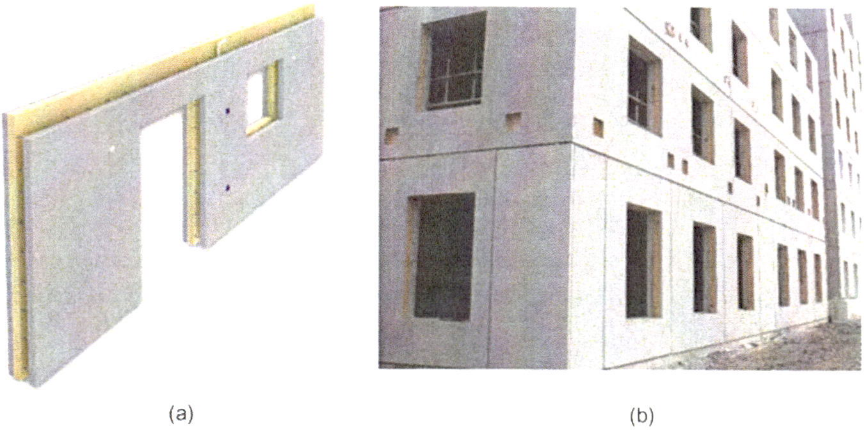

Fig. 37 **a** Computer model of an actual designed single PCSP with FRP pin connectors and PIR insulation and in situ concrete infill and **b** as assembled in practice on site (Keegan Precast 2020)

References

Directorate General for Energy, EU Energy in Figures, European Commission (2017). https://op.europa.eu/en/publication-detail/-/publication/2e046bd0-b542-11e7-837e-01aa75ed71a1

FIB, Precast Insulated Sandwich Panels, Federation Internationale du Beton Bulletin 814 (2017)

Keegan Precast (2020) Precast KC Wall. Retrieved from: http://www.keeganprecast.com/products/precast-walls/kc-wall

Kinnane O, Dyer M, Grey T (2014) Energy and environmental forensic analysis of public buildings. Eng Sustain Proc Inst Civ Eng 167(4):143–156

Kinnane O, West RP, O'Hegarty R (2020) Structural shear performance of insulated precast concrete sandwich panels with steel plate connectors. Eng Struct 215:10. https://doi.org/10.1016/j.engstruct.2020.110691

Lipszynska J, West RP, Grimes M, Niall D, Kinnane O, O'Hegarty R (2020) Composite behaviour of wide sandwich panels with high performance concrete thin wythes with and without Thermomass shear connectors. In: Proceedings of civil engineering research in Ireland (CERI20) conference, Cork, pp 125–130

O'Hegarty R, Kinnane K (2019) Review of precast concrete sandwich panels and their innovations. Constr Build Mater 233:19. https://doi.org/10.1016/j.conbuildmat.2019.117145

O'Hegarty R, Kinnane O, Grimes M, Newell J, Clifford M, West RP (2020) Development of thin precast sandwich panels: challenges and outcomes. Constr Build Mater 26. https://doi.org/10.1016/j.conbuildmat.2020.12098

O'Hegarty R, West RP, Reilly A, Kinnane O (2019) Composite behaviour of fibre-reinforced concrete sandwich panels with FRP shear connectors. Eng Struct 198:15. https://doi.org/10.1016/j.engstruct.2019.109475

O'Hegarty R, Kinnane O, West RP (2018) Finite element analysis of thin precast concrete sandwich panels. In: Proceedings of civil engineering research in Ireland (CERI18) conference, Dublin, pp 146–151

Sexton E, Kaur G, West RP, Niall D, O'Hegarty R, Kinnane O (2019) Design of UHPC and PCM concrete for thin lightweight sandwich panels. In: UKIERI congress, NIT Jalandhar, p 10

Shukla R (2019) Unconnected insulated concrete sandwich panels. MSc Dissertation, Trinity College Dublin and Birla Institute of Technology and Science, Pilani, p 108

Masonry-Infilled RC Frames Strengthened with Fabric-Reinforced Cementitious Matrix

Durgesh C. Rai, Bhushan Raj Selvaraj, and Lalit Sagar

Abstract Fabric reinforced cementitious matrix (FRCM) has shown significant potential to reduce seismic vulnerabilities of unreinforced masonry infills, especially the out-of-plane collapse when cracked under the action of in-plane forces. Recent experimental and numerical studies have demonstrated the effectiveness of the direct application of the fabric to masonry substrate embedded in the mortar with adequate anchorage to the surrounding RC frame. Experimental studies are concerned with characterization of FRCM strengthened masonry and performance verification of strengthened specimens of masonry-infilled RC frames tested under bidirectional loading of slow cyclic drifts for in-plane loading and shake table-generated motions for out-of-plane loading. Numerical models were developed to simulate the observed behavior up to failure and used to develop fragility curves for evaluation and design purposes.

Keywords Fabric reinforced cementitious matrix · Infilled masonry · Seismic loads · Out-of-plane response

1 Introduction

Construction of reinforced concrete (RC) frame buildings with masonry-infilled walls is widely practiced throughout the world. Although infills are helpful in increasing the lateral stiffness of structures and attaining higher energy dissipation, the brittle nature and poor bond of masonry with the surrounding RC frame makes them vulnerable to collapse during seismic events. During earthquakes, infill walls undergo significant damage due to their high in-plane stiffness. Past earthquake damage surveys (Scawthorn and Johnson 2000; Braga et al. 2011; Li et al. 2008; Rai et al. 2012, 2015) have exposed the vulnerability of masonry infill panels due to the poor connection between the frame and infill and interaction of in-plane damage on the out-of-plane response. Several experimental and analytical studies have addressed the reduction

D. C. Rai (✉) · B. R. Selvaraj · L. Sagar
Department of Civil Engineering, Indian Institute of Technology Kanpur, Kanpur, India
e-mail: dcrai@iitk.ac.in

S. B. Singh et al. (eds.), *Emerging Trends of Advanced Composite Materials in Structural Applications*, Composites Science and Technology,
https://doi.org/10.1007/978-981-16-1688-4_2

in out-of-plane capacity of infills with increasing in-plane damage (Angel et al. 1994; Calvi and Bolognini 2001; Komaraneni et al. 2011a, b; Agnihotri et al. 2013; Sagar et al. 2019). Strengthening schemes for such masonry should not only provide better flexural strength and ductility but also maintain the structural integrity under in-plane forces.

Research on strengthening masonry by epoxy-based polymer has shown improved performance (Triantafillou 1998; Stratford et al. 2004; Tumialan and Nanni 2001, 2003; Milani et al. 2006; Mosallam 2007; El-Dakhakhni et al. 2006; Kalali and Kabir 2012; Bischof et al. 2014). However, its undesirable properties such as poor performance at high temperatures, difficulty in applying on wet/uneven surfaces, incompatibility with masonry material, health hazards related to its application and high cost have led to the replacement of organic matrices with inorganic matrices. Such mortar-based polymers are commonly known by the terms Fabric Reinforced Cementitious Matrix (FRCM), Engineered Cementitious Composites (ECC), Cement based Matrix Grid (CMG), Inorganic Matrix composite Grid (IMG) and Textile Reinforced Mortar (TRM). In this chapter, the term Fabric Reinforced Cementitious Matrix (FRCM) is used to refer such reinforcement scheme.

Experimental studies have shown that FRCM-strengthened masonry walletes tested under in-plane shear and out-of-plane flexural loads exhibit good performance (Prota et al. 2006; Aldea et al. 2007; Tomaževič et al. 2011; Özkan et al. 2012; Parisi et al. 2013; Sagar et al. 2017). Also, FRCM-strengthened walletes have been able to match their fiber-reinforced polymer (FRP)–strengthened counterparts in terms of ultimate strength and deformability (Papanicolaou et al. 2007, 2008; Babaeidarabad et al. 2014a, b). However, most of the tests conducted so far have been on FRCM strengthened walletes. Tests on RC frames with infilled walls, especially under dynamic loads, are limited. Koutas et al. (2014a, b) tested single-bay, 3-storey infilled RC frames under in-plane cyclic loading. The FRCM-strengthened specimen exhibited superior strength, deformability, and energy dissipation; also, anchors helped in better utilization of the tensile capacity of the fabric resulting in increased load capacity and deformability. The unanchored specimens tested by Sagar et al. (2017, 2019) showed poor energy dissipation due to detachment of the infill panel from the bottom beam, which resulted in the formation of plastic hinges at the column bases. Overall, the anchored specimens demonstrated better load sharing between the infill and the RC frame.

During earthquakes, infilled walls are subjected to simultaneous in-plane and out-of-plane loading. When the infills are damaged under in-plane loading, out-of-plane strength is drastically reduced, endangering their stability. Therefore, to understand the behavior of FRCM strengthened infills under seismic forces, a unique loading protocol has been developed which simulates a more realistic behavior of infilled walls under bidirectional loading, consisting of slow cyclic drifts in the in-plane direction and shake table-generated ground motion in the out-of-plane direction (Komaraneni et al. 2011a, b). Experimental studies were conducted to evaluate the two modes of fabric application, which differ in the position of fabric placement, and the contribution of mechanical anchors in enhancing the efficiency of the strengthening system under bidirectional loading. Also, the effect of fabric orientation on the performance of FRCM-strengthened infilled RC frames has been investigated (Sagar et al. 2017, 2019).

Further, there are not many analytical models which can be readily used for the analysis of FRCM strengthened masonry infills. Finite element model of FRCM reinforced masonry has been developed to study the behavior of masonry under both in-plane and out-of-plane forces. The improvement in interaction behavior of in-plane damage on the out-of-plane capacity of FRCM reinforced infills was studied by developing interaction curves, and vulnerability assessment was performed by integrating the out-of-plane strength with in-plane damage (Bhushan 2015).

2 FRCM Strengthened Masonry Walletes

An experimental program was conducted to evaluate the in-plane diagonal shear and out-of-plane flexural strength of masonry walletes strengthened using FRCM. The experimental program consisted of 45 walletes, with 15 specimens each for diagonal tension tests and flexure tests along parallel and perpendicular to the bed joint. Two sets of specimens were each strengthened via direct (D) and sandwich (S) modes, of which mechanical anchors (A) were provided for one set (D-A) and (S-A). The walletes were constructed by an experienced mason using half-scale solid burnt clay bricks laid in running bond with a joint thickness of approximately 6 mm. It has been shown that such half-scaled masonry could adequately capture prototypical behavior (Singhal and Rai 2014a). Two modes of wet lay-up application namely direct and sandwich, differing in the sequence of placing the fabric, were studied (Fig. 1). In the direct mode of application, the fabric was placed directly on the surface of the masonry and then coated with a single cement mortar layer of 6-mm thickness. In the sandwich mode of application, the fabric was placed between two cement mortar layers with thicknesses of 2 mm (first layer) and 4 mm (second layer). Also, the contribution of mechanical anchors was investigated in enhancing the performance of masonry subassemblages.

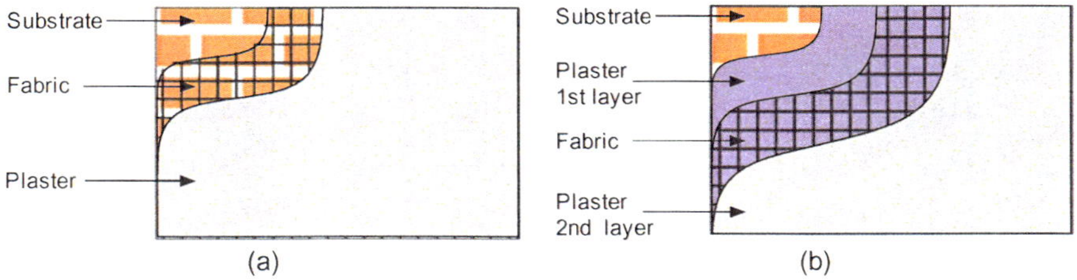

Fig. 1 Modes of application of the fabric: **a** direct application, and **b** sandwich application

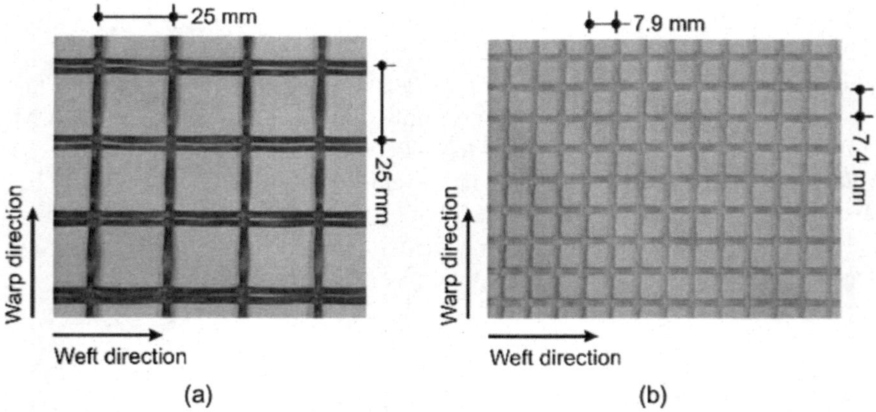

Fig. 2 Fabrics used for FRCM strengthening of wallete: **a** main fabric, and **b** edge fabric

2.1 Material Characterization

Specially made half-scale burnt clay bricks (130 × 64 × 43 mm) and a Type-M1 (BIS 1987) lime-cement mortar mix of 1:1:6 proportions of cement, lime, and sand were used for preparing the masonry walletes. The physical and mechanical properties of the bricks correlated well with those of the full-scale prototype bricks (Singhal and Rai 2014a). The average compressive strength of the bricks was 14.4 MPa with average water absorption of 13.2% by weight. A water-binder ratio of 0.80 was used to obtain workable mortar in a hot and dry climate. The average compressive strength of mortar obtained 8.6 MPa. The average compressive strength of the masonry prism was 7.3 MPa with coefficient of variation (COV) of 7.5%. Cement mortar of mix proportion 1:4 (cement to sand) was used for plastering, and its average compressive strength was 19.7 MPa. The fabrics used in this study consisted of glass fiber strands woven along two orthogonal directions at equal spacing, forming approximately square grids. Grid sizes were about 25 and 8 mm for the main and edge fabrics, respectively (Fig. 2a and b).

2.2 Tensile Properties of FRCM Coupons

The tensile properties of the FRCM were determined by performing coupon tests. The dimension of coupons was 520 × 100 × 10 mm and 520 × 50 × 10 mm for the main and edge fabric, respectively. Ten coupons were tested for both the main and edge fabrics, of which five specimens each were tested to determine the tensile properties of the fabric along the primary (warp) and secondary (weft) directions. The dimensions were fixed based on the guidelines of AC 434 (ICC 2013), as shown in Fig. 3a. The thickness of the coupon was taken as 10 mm and the width of the

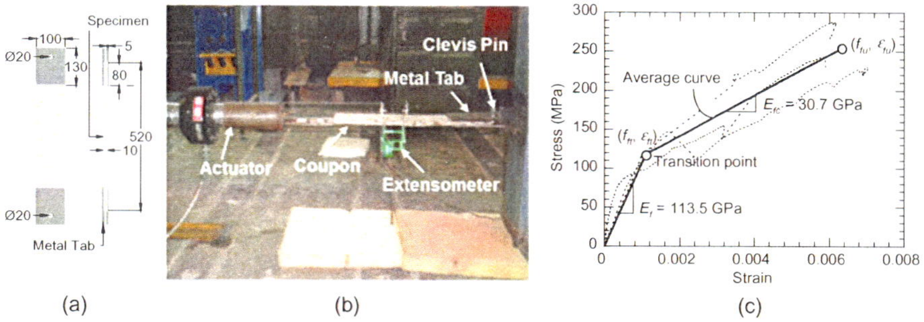

Fig. 3 Direct tension test of FRCM coupon: **a** specimen and metal tab dimensions, **b** test setup, and **c** stress versus strain curve along warp direction

Table 1 Mechanical properties of FRCM coupon

Mechanical property	Main fabric		Edge fabric	
	Warp dir.	Weft dir.	Warp dir.	Weft dir.
Area per unit width (mm^2/mm), A_f	0.104	0.096	0.117	0.079
Uncracked modulus of elasticity (GPa), E_f	114 (26)[a]	101 (25)	181 (38)	144 (45)
Cracked modulus of elasticity (GPa), E_{fc}	30.7 (10)	26.7 (29)	59.4 (13)	29.4 (41)
Ultimate tensile strength (MPa), f_{fu}	254 (12)	262 (12)	286 (19)	208 (13)
Ultimate tensile strain, ε_{fu} ($\times 10^{-3}$)	6.33 (9)	6.65 (21)	3.64 (19)	5.05 (32)

[a]*Note* Values in the parentheses indicate percentage coefficient of variation (COV)

coupon was chosen such that it includes at least 3 strands, and not less than four times the coupon thickness.

From the stress versus strain curve for the coupon (Fig. 3c), it is evident that FRCM exhibited bilinear behavior under tension, consisting of the pre-cracked and post-cracked phases. Initially when the coupons were loaded in the undamaged state, the stresses were resisted by the cement matrix. After few cracks were formed, the stresses were transferred effectively from the mortar to the fabric and ultimately the specimen failed due to the rupture of fabric. The average tensile properties of FRCM coupons for the main and edge fabric along the warp and weft directions are presented in Table 1. The main fabric showed similar tensile properties along both the warp and weft directions, but for the edge fabric higher strength and modulus were observed along the warp direction.

2.3 *Bond Strength of FRCM*

The bond test was performed as per ASTM C1583 (ASTM 2013) to determine the bond strength of the FRCM overlay with the masonry substrate. For main fabric

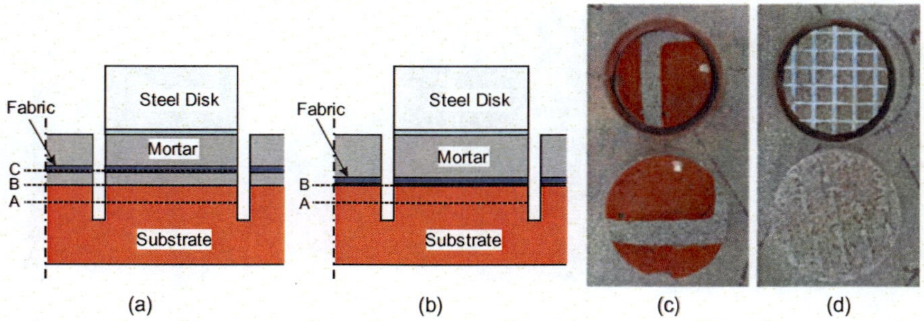

Fig. 4 Bond test: **a** possible failure mode for sandwich application, **b** possible failure mode for direct application, **c** failure at substrate level (mode A), and **d** interlaminar failure (mode C)

strengthened walletes, all the cores failed at the level of the substrate (cohesive failure) as shown in Fig. 4b, and the bond strength for direct and sandwich applications were 0.83 MPa and 0.63 MPa, respectively. For the edge fabric strengthened walletes, some cores failed at the level of the substrate and some failed at the level of the fabric-mortar interface (interlaminar failure) as shown in Fig. 4c, and the bond strength for direct and sandwich applications were 0.86 MPa and 0.51 MPa, respectively. Higher bond strength was attained for direct application compared to sandwich application for both main and edge fabric strengthened specimen. The inter-laminar failure observed when the edge fabric was used shows that the bond of the edge fabric with the mortar was weaker than the main fabric.

2.4 *Diagonal Tension/Shear Test*

The diagonal tension/shear test was performed as per ASTM E519 (ASTM 2010) to determine the in-plane shear strength of masonry. The specimens were loaded in compression along one of the diagonals to cause diagonal tension failure, with the specimen splitting apart parallel to the direction of loading. The results of diagonal tension test are shown in Table 2 and the shear stress versus shear strain plots for the walletes is presented in Fig. 5. The specimen without FRCM suffered brittle failure after reaching peak load, whereas the strengthened specimens sustained large deformation after reaching the peak load (Fig. 5b and c). The average shear strength of the unstrengthened specimen was 1.78 MPa and for the different strengthening schemes, strength varied between 1.54 MPa to 1.86 MPa. The strengthening of the walletes on the single face apparently did not help in improving the shear strength since the specimen failed due to formation of cracks on the unstrengthened face of the wallete, but the deformability of the specimen was significantly improved.

The performance of the different strengthening schemes were almost comparable, however the anchored specimens showed higher pseudo-ductility and shear modulus, and exhibited good post-peak behaviour with lower rate of strength degradation

Table 2 Results of diagonal tension/shear test

Spec.	Peak load (kN)	Shear strength (τ_u)	Shear strain at cracking (γ_{cr})	Ultimate shear strain (γ_u)	Modulus of rigidity G_s (MPa)	Pseudo-ductility (μ)
C	64.60 (5)[a]	1.78 (8)	0.0017 (3)	0.0017 (3)	1067.26 (7)	1.00
D-N	66.25 (15)	1.75 (15)	0.0014 (8)	0.0115 (5)	839.61 (3)	8.06 (5)
D-A	57.72 (6)	1.54 (6)	0.0014 (14)	0.0125 (45)	943.26 (15)	9.15 (51)
S-N	68.17 (2)	1.72 (3)	0.0014 (48)	0.0093 (2)	905.29 (23)	8.02 (48)
S-A	74.09 (3)	1.86 (2)	0.0018 (15)	0.0154 (21)	930.52 (30)	10.92 (26)

[a]Figures in brackets () indicate percentage coefficient of variation (COV)

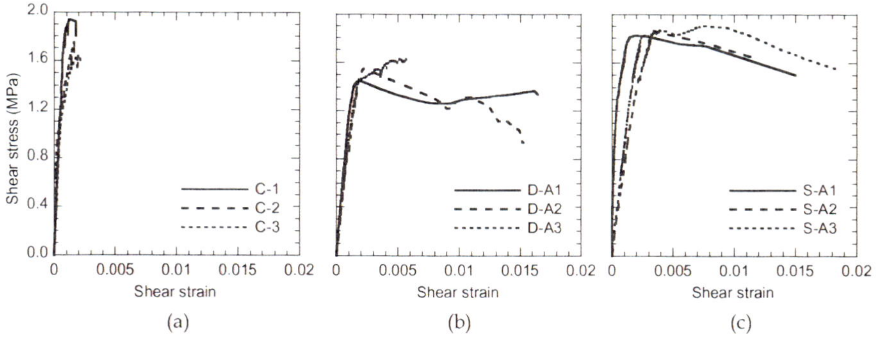

Fig. 5 Shear stress versus shear-strain curve for masonry wallettes: **a** unstrengthened, **b** strengthened with direct application and anchored and **c** strengthened with sandwich application and anchored

(Fig. 5b and c). The value of ultimate strain was increased by six to nine times compared to the control specimen, and the average pseudo-ductility was 10.04 for anchored specimens (D-A and S-A). Specimen strengthened using direct application with anchors (D-A) showed an unusually lower shear strength of 1.54 MPa.

The experimental and analytical results for in-plane shear strength are compared in Table 3. The average experimental peak load for FRCM strengthened specimens was 66.56 kN. However, the in-plane capacity obtained from the different analytical models ranges from 40.86 to 44.90 kN, with the analytical models of ACI 549 (2013) and CNR DT200 (2004) providing a better prediction of the experimental results. The analytical models predicted shear friction failure, which occur because of bond failure followed by step-type crack propagation through the horizontal and vertical joints. This prediction was consistent with experimental observations, where the specimen failed due to the formation of diagonal tension and shear friction cracks.

Table 3 Diagonal tension test: comparison of experimental and analytical results

Reference	Analytical results				Experimental results		Experimental/Analytical Peak load (V_{exp}/V_n)
	Peak load (kN)			Failure mode			
	Masonry contribution (V_m)	Composite contribution (V_f)	Total strength (V_n)		Peak load, V_{exp} (kN)	Failure mode	
ACI 549.4R	27.63	17.27	44.90	SF	66.56	SF + DT	1.48
CNR DT200		16.24	43.87				1.52
Lin et al. (2014)		13.23	40.86				1.63

SF Shear friction; *DT* Diagonal tension

2.5 Flexure Test

During a seismic event, masonry walls are subjected to out-of-plane loads. In the past earthquakes, out-of-plane collapse of walls due to flexural failure has been identified as one of the predominant modes of failure. Therefore, evaluation of flexural strength of masonry is helpful in selecting a suitable retrofitting scheme for masonry structures. Flexure tests were performed along directions parallel and perpendicular to the bed joint as per BS EN 1052-2 (BIS 1999). The flexure test results for the failure planes parallel and perpendicular to the bed joint are given in Tables 4 and 5 and the flexural stress versus deflection plots are presented in Fig. 6a–c and Fig. 7a–c, respectively.

As shown in Figs. 6a and 7a, the unstrengthened specimens exhibited brittle failure after reaching the peak load at ultimate mid-span displacement of less than 1 mm, and the average flexural strength of the control specimens for failure plane parallel and perpendicular to the bed joint was 1.07 MPa and 1.46 MPa, respectively. However, the flexural capacities for FRCM strengthened specimens were significantly enhanced due to the redistribution of tensile stresses in the fabric after the formation of cracks in the masonry (Figs. 6b, c and 7b, c), and the ultimate flexural strength was about 2.0 times and 1.4 times the control specimen for failure plane parallel and perpendicular to the bed joint, respectively.

The ultimate flexural strength for failure plane parallel to bed joint of strengthened specimen was 2.22 MPa for sandwich application and 1.66 MPa for direct application. For failure plane perpendicular to bed joint, the flexural strength for sandwich application was 2.25 and 2.01 MPa for direct application. Unstrengthened wallettes

Table 4 Results of flexure test parallel to bed joint

Spec.	Peak load (kN)	Deflection at cracking, δ_{cr} (mm)	Stress at cracking, f_{cr} (MPa)	Ultimate deflection, δ_u (mm)	Ultimate stress, f_u (MPa)	Ductility index, I
C	6.20 (8)	0.82 (10)	1.07 (17)	0.82 (10)	1.07 (17)	1.0
D-A	10.10 (12)	0.83 (2)	1.10 (13)	5.49 (3)	1.66 (15)	7.55 (7)
S-A	14.10 (6)	1.02 (26)	1.36 (9)	6.09 (36)	2.22 (12)	7.79 (27)

Figures in brackets () indicate percentage coefficient of variation (COV)

Table 5 Results of flexure test perpendicular to bed joint

Spec.	Peak load (kN)	Deflection at cracking, δ_{cr} (mm)	Stress at cracking, f_{cr} (MPa)	Ultimate deflection, δ_u (mm)	Ultimate stress, f_u (MPa)	Ductility index, I
C	4.04 (6)	0.47 (29)	1.46 (2)	0.47 (29)	1.46 (2)	1.0
D-A	6.65 (14)	0.84 (6)	1.43 (11)	6.05 (7)	2.01 (12)	7.28 (4)
S-A	6.66 (23)	1.07 (19)	1.87 (2)	5.35 (14)	2.25 (20)	5.03 (6)

Figures in brackets () indicate percentage coefficient of variation (COV)

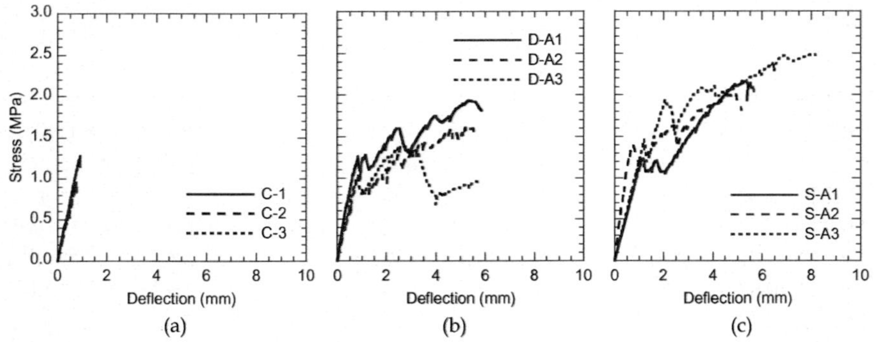

Fig. 6 Flexure test parallel to the bed joint: comparison of flexural stress versus deflection curve for various specimens

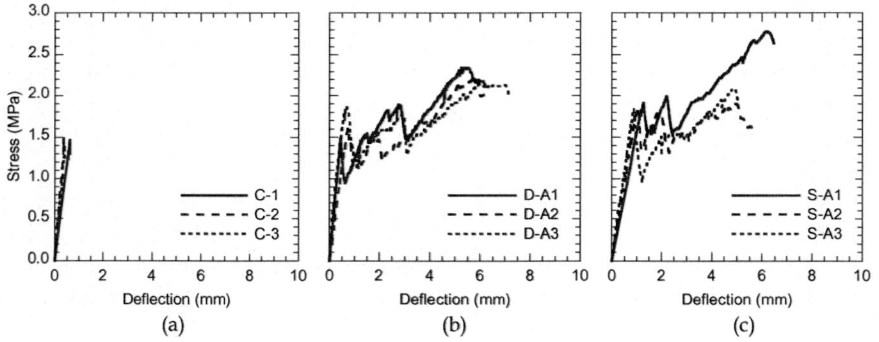

Fig. 7 Flexure test perpendicular to the bed joint: Comparison of flexural stress versus deflection curve for various specimens

exhibit higher flexural strength for failure perpendicular to the bed joint than failure parallel to bed joint. However, in FRCM strengthened specimens, similar strength was attained along both the failure planes since the mode of failure was through the rupture of fabric and the tensile strength of FRCM overlay governs the ultimate flexural strength of the wallete.

The ductility of masonry is improved significantly in FRCM strengthened specimens due to the contribution of fabric in resisting flexural tensile stress. The ultimate mid-span deflection for strengthened specimens was about 7.3 times and 12.7 times the control specimen for failure plane parallel and perpendicular to the bed joint, respectively. The ductility index gives an estimate of the energy dissipation of the specimen after reaching its elastic limit. Ductility index of 7.7 and 6.2 was attained for strengthened specimens having failure plane parallel to bed joint and perpendicular to bed joint, respectively. The anchored specimen with sandwich application had unusually lower value since it reached a higher pre-cracking stress compared to the other strengthened specimen Fig. 7c).

Table 6 Flexure test: comparison of experimental and analytical results

Reference	Analytical moment capacity (kN m)		Experimental moment capacity (kN m)		Experimental/analytical moment capacity	
	Parallel	Normal	Parallel	Normal	Parallel	Normal
Hamoush et al. (2001)	0.46	0.33	0.60	0.44	1.31	1.33
Tan and Patoary (2004)	0.50	0.36			1.21	1.23
Harajli et al. (2010)	0.47	0.34			1.28	1.30
ACI 549.4R (2013)	0.46	0.33			1.31	1.34

The average moment capacity of the strengthened specimen for failure plane parallel and perpendicular to bed joints was 0.60 kN m and 0.44 kN m, respectively. The analytical models provided consistent prediction of flexural strength values, ranging from 0.46 to 0.50 kN m and 0.36 to 0.40 kN m for failure plane parallel and perpendicular to the bed joint as shown in Table 6. However, the equations proposed by Tan and Patoary (2004) provided better prediction when compared to other analytical models.

3 FRCM Strengthened Infills

Single-bay, single-storey RC frame infilled walls, measuring 2.3 × 1.3 m, were constructed from specially made half-scale bricks to represent an interior bay of a multi-storied RC building as shown in Fig. 8. The model's frame members were 200 × 125 mm. Four 12-mm bars were provided for longitudinal reinforcement, and 6-mm stirrups, at a spacing of 150 mm, were provided for shear reinforcement. The overall thickness of the masonry infill, including the cement mortar coat, was 76 mm.

In the test matrix, the first specimen was unstrengthened (control) and the remaining specimens were strengthened with different FRCM configurations. Specimen DA_{0-90} was strengthened using direct application and an orthogonal fabric orientation with respect to bed joints, and with anchors. Specimens SA_{0-90} was similar to DA_{0-90} with the sandwich application of fabric. Specimen DA_{45} was strengthened using direct application with anchors and a fabric orientation of 45° with respect to bed joints. Two types of fabrics, main and edge, were used for FRCM strengthening. The main fabric was placed on the entire surface of the infill with an overlap of 150 mm on the frame; the edge fabric, in the form of 300-mm-wide strips, was placed along the frame–infill interface (Fig. 9).

Mechanical anchors, consisting of bolt, nut, and washer assembly, were provided. The bolts had a diameter of 6 mm, and the washers had an external diameter of 25 mm.

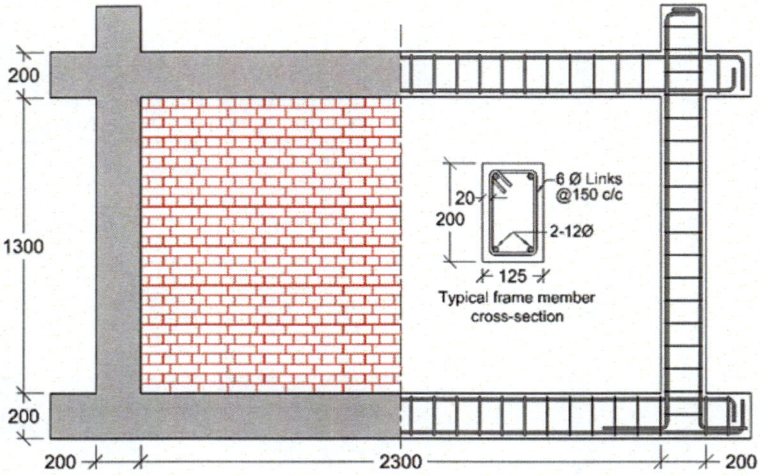

Fig. 8 Geometric and reinforcement details of the test specimen (all dimensions are in mm)

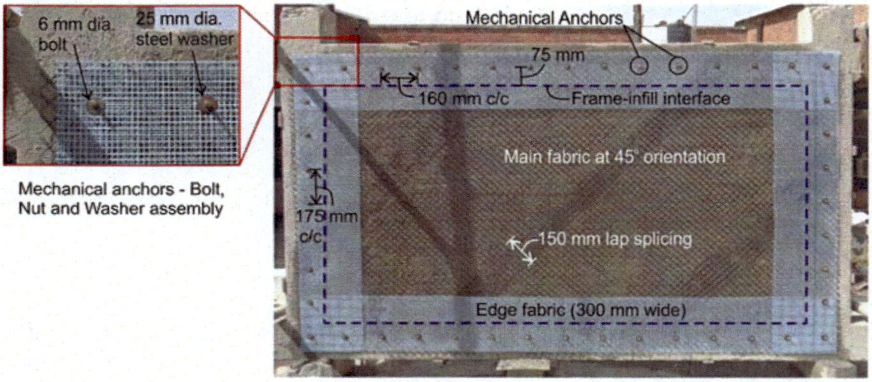

Fig. 9 Layout of FRCM strengthening of specimen DA_{45}

The external washer diameter was chosen to accommodate at least three strands of edge fabric. The anchors were used to prevent fabric debonding and better utilize the tensile capacity of the FRCM.

3.1 Material Characterization

Specially made half-scale burnt clay bricks (130 × 64 × 43 mm) and a Type-M1 (BIS 1987) lime-cement mortar mix of 1:1:6 proportion (cement:lime:sand) were used for building the infills. A Type-H1 (BIS 1987) cement mortar mix of proportion

Table 7 Average properties of the materials used in the wall specimen

Specimen	Properties (MPa)				
	Plastering mortar compressive strength	Masonry prism		Concrete	
		Compressive strength	Elastic modulus	Compressive strength	Tensile strength
CS	14.8 [2][a]	8.1 [12]	2630 [43]	31.3 [7]	2.5 [16]
DA_{0-90}	13.7 [19]	8.0 [6]	1970 [25]	33.5 [1]	2.3 [14]
SA_{0-90}	18.5 [8]	9.6 [9]	3200 [26]	36.7 [3]	2.7 [16]
DA_{45}	18.1 [6]	10.4 [11]	2560 [10]		

[a]*Note* Values in the parentheses indicate percentage coefficient of variation (COV)

1:4 (cement: sand) was used for plastering the walls and for the FRCM strengthening. The average properties of the materials and prisms for all six specimens are provided in Table 7.

3.2 *Test Setup*

A unique loading protocol was adopted involving a predefined sequence of slow cyclic drifts for in-plane loading and shake table-generated motions for out-of-plane loading (Komaraneni et al. 2011a, b; Singhal and Rai 2014b). The proposed loading program addressed reduction in out-of-plane resistance of a wall due to prior in-plane damage under a realistic out-of-plane shake table motion corresponding to a given hazard level. The test setups for the out-of-plane and in-plane loading are shown in Fig. 10. A servo-hydraulic uniaxial shake table was used for out-of-plane loading, and a 500 kN actuator was used for in-plane loading. To simulate realistic boundary conditions for the infilled RC frame, an adequate number of lateral supports was provided on both sides of the specimen (Fig. 10a and b). The lateral supports were braced at the top to provide torsional restraint to the infilled wall during in-plane and out-of-plane loading. The in-plane supports (Fig. 10b) were attached to the strong-reaction floor to transfer overturning loads generated during in-plane loading without overstressing the shake table bearings.

3.3 *Gravity Load Simulation*

In order to simulate gravity load, precompression was maintained on the wall and columns using a flexible wire rope arrangement (Fig. 10a and b). A precompression load of 0.3 times Dead Load (DL) was planned; however, a load of only 25% of 0.3 DL was applied due to the limitations of the test setup. A load of 25 kN was

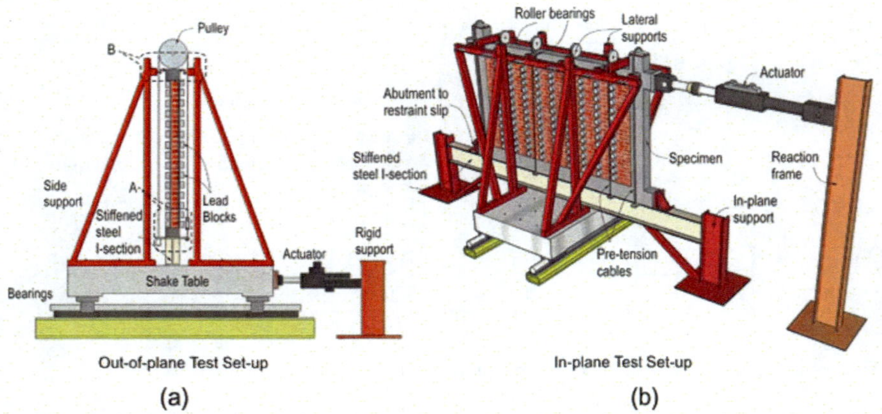

Fig. 10 Schematic showing various components of the test setup: **a** out-of-plane loading, and **b** in-plane loading

applied on each column, and the wall was subjected to precompression of 0.1 MPa corresponding to the weight of the infill from the storey above.

3.4 *Artificial Mass Simulation*

Structurally ineffective masses (artificial masses) need to be added to reduced-scale models for proper simulation of both gravitational and inertial forces (Singhal and Rai 2014b). For the half-scale model bricks considered, the additional weight to be added was approximately equal to the mass of each brick. These additional masses were attached in the form of cylindrical lead blocks (60-mm diameter, 28-mm height, 0.865-kg weight), which were arranged in a regular grid pattern on both faces of the wall to simulate the inertial forces generated due to out-of-plane ground motions.

For the out-of-plane tests, 17 accelerometers and 8 wire potentiometers were used for recording out-of-plane acceleration and displacement of the masonry infill, respectively. For in-plane testing, 5 LVDTs and 2 wire potentiometers were used to record in-plane displacements and diagonal deformation of the infill. Six load cells were used to monitor precompression load on the specimens.

3.5 *Loading History and Test Procedure*

The specimens were subjected to simulated earthquake ground motions generated by a shake table in the out-of-plane direction. The N21 E component of the 1952 Taft earthquake with a peak ground acceleration (PGA) of 0.156 g and duration of 54.16 s was chosen as the target ground motion, as it corresponds well with Zone V

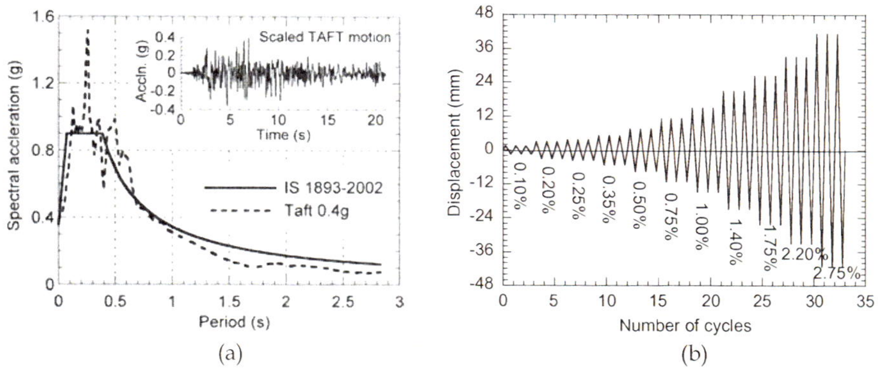

Fig. 11 **a** Comparison of scaled response spectra of Design Basis Earthquake as per IS 1893-2002 with original TAFT motion up scaled to 0.4 g, and **b** in-plane loading history

design earthquake spectra of IS 1893 (BIS 2002a, b) when scaled to 0.4 g with 5% damping. The selected Taft ground motion was scaled to the PGA values of Zones II, III, IV, and V and then referred to as Levels II, III, IV, and V, respectively (Fig. 11a). In addition, the Taft motion scaled to a PGA of 0.055 g was defined as Level I.

In-plane loading consisted of displacement-controlled slow cyclic drifts, which were selected following the guidelines of ACI 374.1 (ACI 2005). The loading history consisted of gradually increasing storey drifts of 0.10, 0.20, 0.25, 0.35, 0.50, 0.75, 1.00, 1.40, 1.75, 2.20, and 2.75% (Fig. 10b).

3.6 Results and Discussion

3.6.1 Physical Observations

Failure patterns for all specimens at the end of testing are shown in Fig. 12a–d. Physical observations included cracks formed in the infills, concrete spalling, and damage to the reinforcement in the columns; also observed were the behavior of the FRCM strengthening system and the corresponding variation in load response. The infilled wall of the unstrengthened specimen experienced several cracks under in-plane loads and disintegrated owing to the brittle nature of the masonry. Subsequently, the loosened masonry fragments showed poor response with excessive displacements under out-of-plane loading (Fig. 12a). The poor out-of-plane stability of the infill was further aggravated by the lack of an adequate bond between the infill and the RC frame. However, the extent of cracking under in-plane loads was much less in the FRCM-strengthened specimens. As a result, the strengthened walls preserved structural integrity and showed a superior out-of-plane response.

The mechanical anchors restricted the tendency of the infills to separate from the RC frame. With a large contact area between the frame and the infill, stress

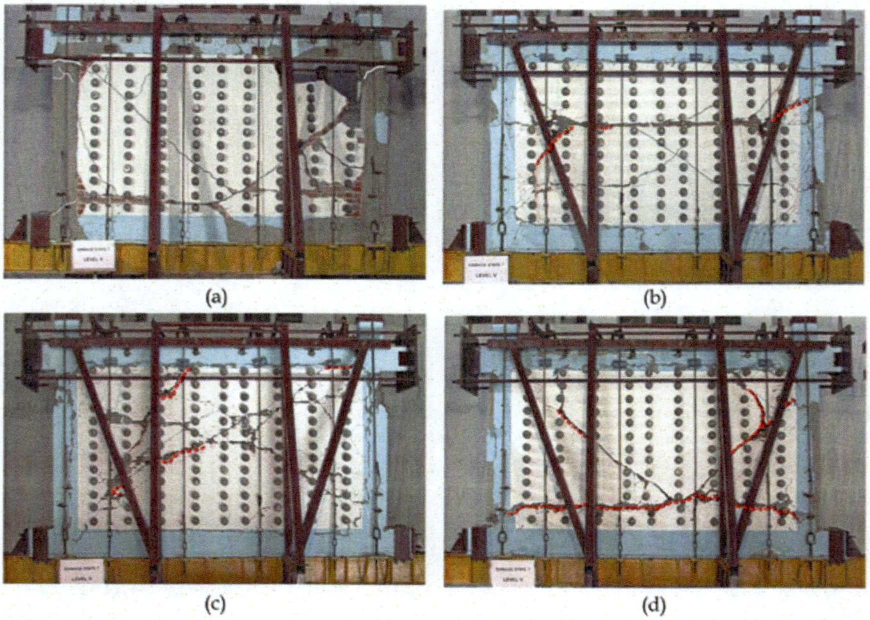

Fig. 12 Damage pattern of the specimens at the end of the test: **a** CS, **b** $DA_{0\text{-}90}$, **c** $SA_{0\text{-}90}$, and **d** DA_{45} (*Note* broken lines indicate rupture of fabric)

concentration at the corners of the infill panel and the column ends was avoided. The fabrics ruptured along the diagonal cracks in $DA_{0\text{–}90}$ and $SA_{0\text{–}90}$ during the 2.2% drift cycles as shown by broken lines in Fig. 12b and c; however, their out-of-plane stability was preserved due to the superior connection along the frame-infill interface.

In Specimen DA_{45}, the fabric experienced overstressing due to its oblique orientation, which was evident from the early appearance of well-defined cracks around the anchors during the 0.5% drift cycle (Fig. 12d). The FRCM in Wall DA_{45} experienced an extensive rupture of fibers along the horizontal bed-joint cracks and thus was not so effective in resisting stresses generated by shear sliding cracks(Fig. 12d).

3.7 Out-of-Plane Behavior of Damaged Infill Panels

Uniform out-of-plane pressure was calculated from the value of acceleration experienced by the infill. The variation in equivalent uniform pressure of the infills with respect to different in-plane drift levels is shown in Fig. 13a. Unstrengthened specimens (CS) experienced a drop in uniform out-of-plane pressure at the 0.5% drift level; however, in all other specimens pressure remained almost constant up to the 1.4% drift level. The uniform pressure in Specimens CS, and DA_{45} again showed a sharp decline after the 1.75 and 2.2% in-plane drift cycles (Fig. 13a). The stiffness

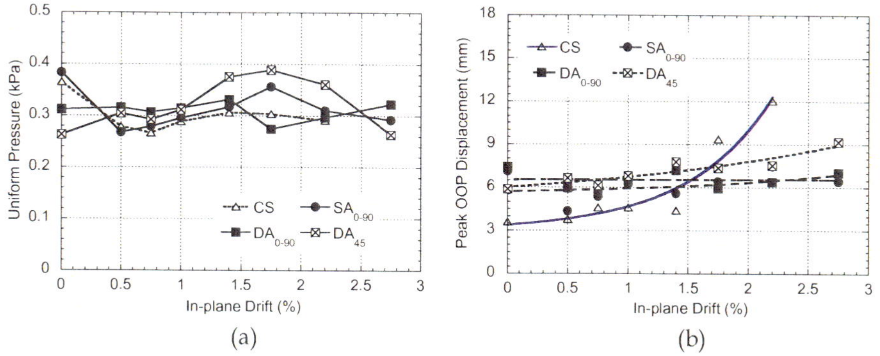

Fig. 13 **a** Variation of peak uniform acceleration and with different in-plane drifts levels; **b** variation of peak out-of-plane displacements with different in-plane drifts levels

degradation of these specimens due to the accumulation of damage under in-plane loads was responsible for the drop in observed out-of-plane pressure. However, Specimens DA_{0-90}, and SA_{0-90} preserved structural integrity without significant change in in-plane stiffness until the end of the test; subsequently, out-of-plane pressure remained almost constant.

Variations in peak out-of-plane displacements of the infills at different in-plane drift levels and corresponding trends in displacement are shown in Fig. 13b. Infill displacement profiles at different damage states are shown in Fig. 14a–d. Unstrengthened specimens (CS) showed excessive out-of-plane deflections of about 12 mm due to disintegration of the infill panel (Figs. 13b and 14a).

The out-of-plane behavior of the anchored specimens was considerably superior due to a better bond at the frame-infill interface. Despite damage to the infills in the anchored specimens—DA_{0-90}, SA_{0-90}, and DA_{45}—out-of-plane displacements were restricted to 8 mm (i.e., about one-eighth of the wall thickness) and the specimens exhibited superior out-of-plane response (Fig. 14b–d). However, the out-of-plane performance of these specimens was qualitatively different at the higher damage

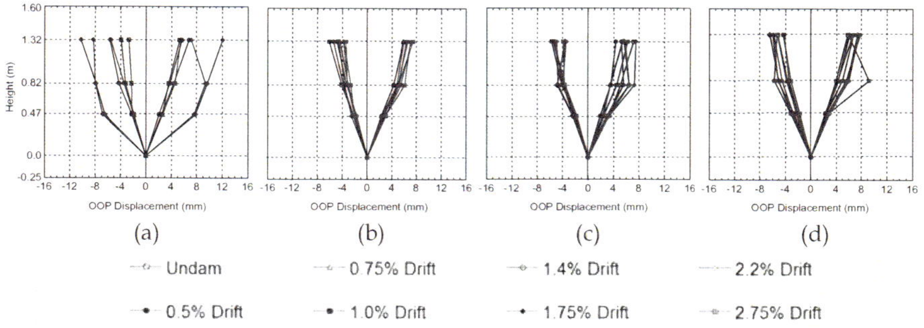

Fig. 14 Out-of-plane displacements profile along the height of specimen after in-plane drifts cycles: **a** CS, **b** DA_{0-90}, **c** SA_{0-90}, and **d** DA_{45}

states. Specimens $DA_{0–90}$ and $SA_{0–90}$ showed similar out-of-plane performance at lower drift levels, but, after debonding of the fabric from the wall, Specimen $SA_{0–90}$ showed greater variability in displacements and hence lesser out-of-plane stability than $DA_{0–90}$ (Fig. 14c). The performance of Specimen $DA_{0–90}$ was more consistent due to the absence of fabric debonding. Specimen DA_{45}'s extensive fabric rupture was responsible for its relatively poor out-of-plane performance. The specimen's inability to resist stresses due to shear sliding in the infill led to out-of-plane walking, and it experienced comparatively higher out-of-plane displacement during the load test corresponding to the final damage state (Fig. 14d).

3.8 In-Plane Load Displacement Response

The hysteresis curves with failure modes are given in Fig. 15a–d and comparison of backbone curves for all specimens, along with the numerical bare-frame response, is shown in Fig. 16. The in-plane behavior of these specimens was evaluated in

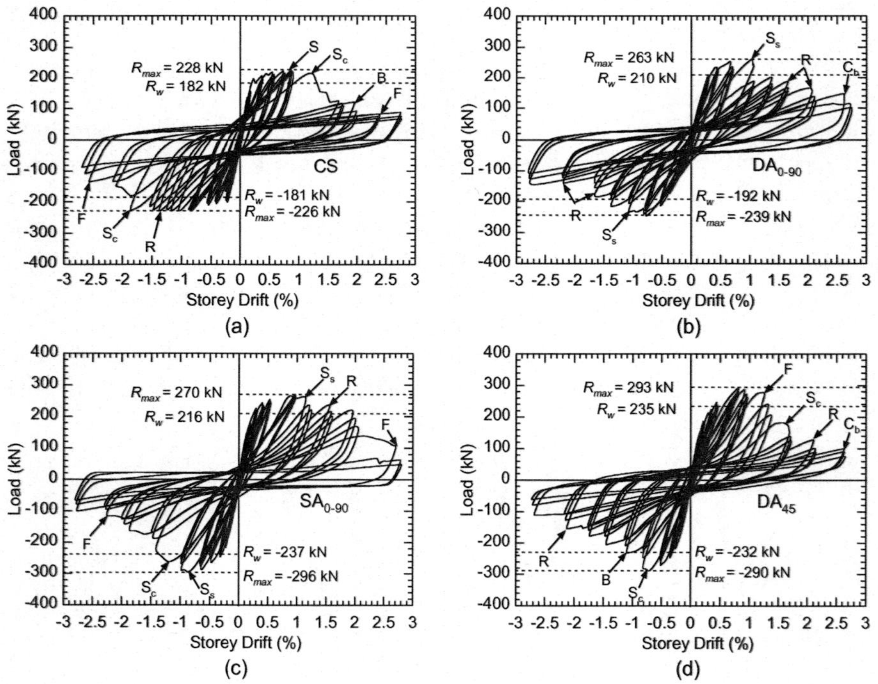

Fig. 15 Comparison of hysteretic behavior of the specimens: **a** CS; **b** $DA_{0–90}$; **c** $SA_{0–90}$; **d** DA_{45} (S = separation at wall to column interface, S_c = shear crack in the column, S_s = shear sliding crack in the wall, B = buckling of column rebars, F = fracture of lateral tie in column, R = rocking of masonry panel and, P_H = plastic hinging at the column ends and C_b = crushing of brick)

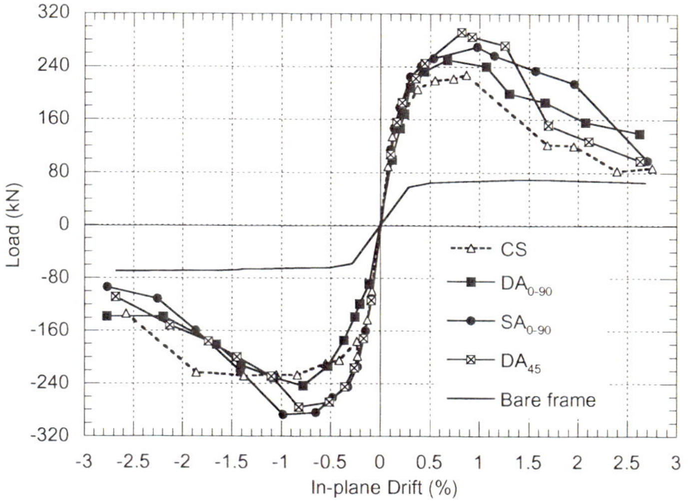

Fig. 16 Comparison of observed envelope value of load versus storey drift

terms of peak load capacity, R_{max}, and the corresponding displacement, δ_{Rmax}; safe working load, R_w (60% of R_{max}), and the corresponding displacement, δ_u; displacement ductility, μ, given by the ratio, δ_u/δ_{Rmax}; effective stiffness, K_e; strength degradation, C_{sd} and energy dissipation, $\sum E_D$. Elastic stiffness, K_e, was approximated as the secant stiffness at $0.6R_{max}$ prior to reaching peak load (Rai et al. 2014). To indicate strength deterioration in the specimen after reaching peak load capacity, a strength degradation factor, C_{sd}, was used, given by the ratio of residual strength (strength at the end of the test), R_r, to peak load capacity, R_{max}. Global energy dissipation and infill energy dissipation values were computed to assess the deformability of the entire specimen and the infill, respectively. In-plane results for all specimens are summarized in Table 8.

The brittle failure observed in unstrengthened Specimen (CS) resulted in considerable strength degradation, and the response showed convergence with that of the bare frame at approximately 2% drift level (Fig. 16). Strength degradation was more gradual in the strengthened specimens. Strength degradation in the anchored specimen, DA_{0-90}, was gradual and uniform up to the end of the test (Fig. 15b). However, the Specimens SA_{0-90} and DA_{45}, showed an abrupt drop in in-plane load capacity at higher drift levels due to fabric debonding and separation of the infill from the RC frame (Fig. 15c and d). Specimen DA_{0-90} showed the highest average strength degradation factor, 0.57 and highest average displacement ductility, 3.02 (Table 8). The FRCM strengthened specimens showed a superior connection at the frame–infill interface and their infill panel load-displacement hysteresis loops were symmetric and commensurate with the global behavior.

As depicted in Table 8, energy dissipation capacity was lower for unstrengthened specimen than for the FRCM strengthened specimens. Specimen CS showed the lowest infill energy dissipation, 3.1 kN m and lowest global energy dissipation

Table 8 Summary of observed response for all specimens

Wall	Ultimate load, $R_{\max}$ (kN)	$\delta^{a}_{R\max}$ (mm)	δ^{b}_{u} (mm)	Displ. ductility, μ ($\delta_u/\delta_{R\max}$)	Effective stiffness, K_e (kN/mm)	Strength degradation, C_{sd}	Cumulative energy dissipated, $\sum E_D$ (kN m)	
							Global	Infill
CS	+229 −227	+13.0 −20.7	+23.6 −38.3	1.83	58.1 (+) 68.9 (−)	0.38 (+) 0.59 (−)	17.0	3.1
DA_{0-90}	+251 −244	+10.1 −11.7	+33.9 −31.5	3.02	47.8 (+) 32.5 (−)	0.56 (+) 0.57 (−)	33.7	9.1
SA_{0-90}	+270 −287	+14.7 −14.7	+34.4 −26.5	2.07	60.4 (+) 65.3 (−)	0.36 (+) 0.33 (−)	24.3	9.2
DA_{45}	+292 −276	+12.3 −12.3	+24.2 −28.6	2.15	50.7 (+) 62.5 (−)	0.33 (+) 0.39 (−)	29.5	15.0

[a] $\delta_{R\max}$ = displacement corresponding to peak load
[b] δ_u = post-peak displacement corresponding to R_w, 60% of the peak load

value, 17 kN m. Among the strengthened specimens, Specimens DA_{0-90} and DA_{45} showed the highest global and infill energy dissipation values, 33.7 and 15.0 kN m, respectively. Although the in-plane performance of DA_{45} was comparable to that of DA_{0-90} in terms of in-plane and energy dissipation capacity, due to extensive fabric rupture the out-of-plane performance of this specimen was inferior to that of DA_{0-90} with orthogonal fiber orientation.

4 Finite Element Modeling of FRCM Strengthened Masonry

Developing a complete interaction curve by experiments requires testing multiple number of specimens which is generally very expensive. However, finite element models validated with experimental results can be used to develop the complete interaction curve. Hence, the experimental model was simulated in finite element analysis to study the effect of in-plane damage on the out-of-plane strength of the wall. Masonry was modeled using macro-modeling technique in ABAQUS (Simulia 2011) for its simplicity and less computational effort to simulate the behavior under in-plane and out-of-plane forces.

4.1 Material Models

Concrete damaged plasticity (CDP) model available in ABAQUS (Simulia 2011) was used for simulating the non-linear behavior of masonry, concrete, and cement plaster due to the quasi-brittle nature of these materials. For masonry, Kaushik et al. (2007b) proposed a simple piece-wise linear stress-strain curve to input into the finite element programs for structural analysis. The tensile behaviour of masonry was assumed to be linearly elastic till it reaches its tensile strength followed by a linear softening curve. The Kent and Park (1971) model of concrete under compression without confinement was used for modeling the compression behavior of concrete. Tensile response of concrete was modeled based on the tension softening law proposed by Gopalaratnam and Shah (1985). The modulus of elasticity and the peak tensile strength were assumed according to Indian code IS 456 (BIS 2000). As per the experimental results of compression tests on cement mortar cubes by Kaushik et al. (2007a, b), the modulus of elasticity for the cement plaster was assumed as 185 times the cube compressive strength of the cement mortar. The strain at the peak and 50% of peak compressive strength was assumed as 0.01 and 0.03, respectively.

The fabric was modeled as two-node truss elements possessing elastic-perfectly plastic material property with limited ductility (Carozzi et al. 2014). The modulus of elasticity and ultimate strength of the fabric were obtained from the cracked phase of direct tension coupon tests. The failure strain of the yarns was assumed to be

3%, as given in the manufacturer's specifications. Stress-strain behavior proposed by Kunnath et al. (2009) for steel incorporating the strain hardening was used for longitudinal reinforcement in RC members. However, the stirrups were modeled as elastic-perfectly plastic material to reduce the computational effort.

4.2 Interfaces

Due to the various materials used, interfaces were also introduced in the model viz. masonry-plaster, masonry-bounding frame, and frame-plaster. Coulomb friction model is widely acceptable to simulate the interface of masonry with the surrounding RC frame (Crisafulli et al. 2000). Hence, *Hard Contact* was defined in the normal direction and isotropic friction with coefficient of 0.7 was defined for the tangential direction. *Tie Constraint* was defined between masonry and cement mortar, as there was no debonding observed in the experimental study. Poor bond between the cement plaster and concrete was observed in the experimental study due to the smoothness of concrete surface. Hence in the normal and tangential direction, *Hard Contact* and *Friction behavior* with a very low coefficient of 0.2 was defined. The cohesion interaction properties were calibrated such that the plaster has very low bond strength with concrete surface.

When mechanical anchors were used for the FRCM reinforcement, the bond between frame and plaster had enhanced significantly, due to which the spalling of plaster was very less than that observed in unanchored specimens. Moreover, the cracking pattern in the experimental study revealed that the masonry infill and frame acted together as a composite prolonging the shear cracking of column ends. Hence, for anchored specimens, the bond between plaster and frame was assumed to be perfect by defining *Tie constraint* between them for 150 mm width around the periphery of the infill.

4.3 Element

Linear hexahedral 3D solid continuum elements with reduced integration from the Explicit element library (C3D8R) were used for the concrete frame, masonry, and cement plaster. The reinforcing bars were modeled as 2 node 3D truss elements (T3D2). The element size for the solid elements was chosen as 40 mm after several iterations balancing the accuracy and the computational cost of the analyses. The coarse mesh masonry was prone to Hourglassing effect at high drift levels, hence *Enhanced Hourglass control* and finer mesh was used for masonry to avoid the zero energy deformation.

4.4 Model Verification

The finite element modeling technique and the material properties used for the elements were validated with the experimental flexure tests of masonry assemblages. In FE analysis there was no distinction between the simulation of direct and sandwich application due to the assumption of perfect bond between plaster and masonry and perfect embedment of fabric in matrix. The numerical model of the strengthened wallets was able to simulate the experimental results in terms of ductility, redistribution of stresses in the fabric and ultimate strength as observed in Fig. 17.

5 Interaction of in-Plane Damage on Out-of-Plane Strength of FRCM Reinforced Infill Frames

The numerical study was targeted on understanding the effect of in-plane damage on out-of-plane strength of FRCM strengthened infills. Interaction curves are failure envelopes, which help in identifying the vulnerability of infill wall under bidirectional loading. In the analytical study, CS denotes control or unstrengthened specimen, SA_{0-90} denotes FRCM strengthened with anchoring and fabric oriented along the direction of bed joints, and SA_{45} is similar to SA_{0-90} with 45° fabric orientation. These strengthened specimens are compared with the experimental results of direct application specimens as there was perfect bond assumed between FRCM and masonry in the analytical study.

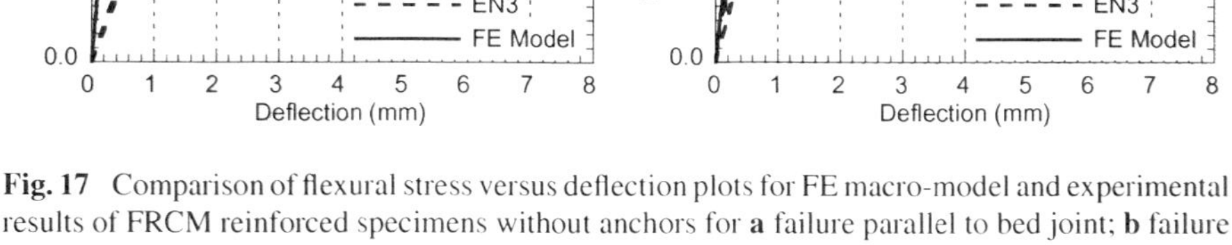

Fig. 17 Comparison of flexural stress versus deflection plots for FE macro-model and experimental results of FRCM reinforced specimens without anchors for **a** failure parallel to bed joint; **b** failure perpendicular to bed joint

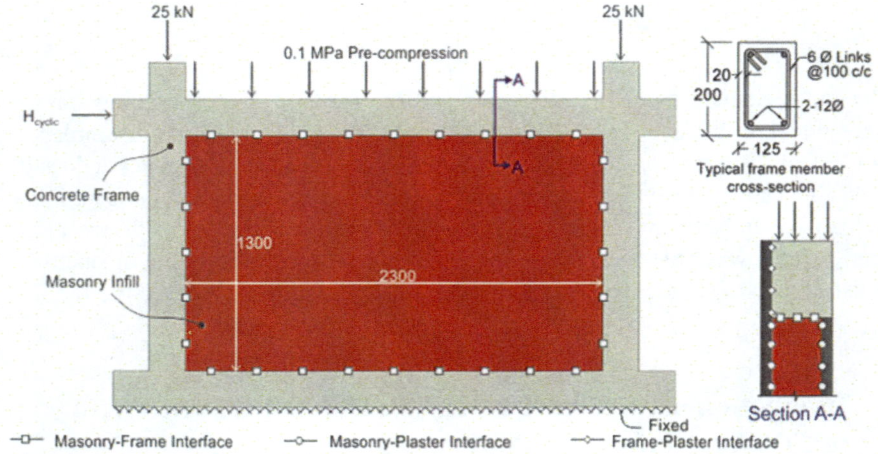

Fig. 18 Boundary, frame-infill interaction and initial loading in the FE model

5.1 Boundary and Loading Conditions

The bottom beam was assumed to be fixed using the *encastre* boundary condition. The out-of-plane restraint was simulated by restraining only the displacement normal to the plane of the frame at the top. A schematic representation of the boundary conditions and loads applied in the finite element model is shown in Fig. 18. To study the interaction between in-plane damage and the out-of-plane strength of the infill walls, the damage was applied by in-plane cyclic loading and followed by increasing out-of-plane pressure till the specimen reached its maximum capacity. The time period of loading was fixed as 0.4 s for which the slow loading conditions were satisfied evident with the ratio of kinetic energy to the internal energy. The masonry panel was loaded in the out-of-plane direction after each damage state (drift level) to develop the interaction curve. The out-of-plane load was applied by *pressure* loading with monotonic linear *smooth* increase in amplitude.

5.2 In-Plane Behavior

The backbone curve of the hysteresis curves obtained in the FE analysis is compared with the experimental results in Fig. 19a–c. The RC frame being a non-ductile moment resisting frame failed predominantly by shear failure at column ends. However, when the fabric was anchored, due to the development of composite behavior of infill and frame, the failure in the column ends was prolonged to higher drifts as the forces were distributed to both the frame and infill. This was evident from the ductile failure of the specimen after reaching the peak strength in $SA_{0\text{-}90}$ and SA_{45}. Anchored FRCM reinforcement was able to develop stress due to perfect bond

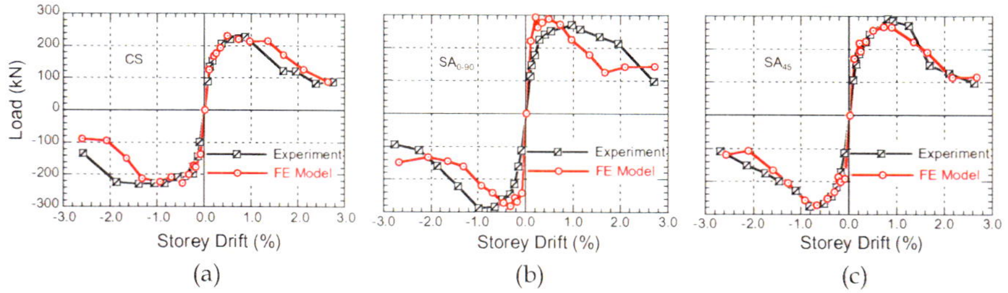

Fig. 19 Comparison of backbone curves obtained from experiment and FE analysis for **a** unstrengthened infill specimen (CS); **b** FRCM (0–90° orientation) strengthened infill with mechanical anchors (SA_{0-90}); **d** FRCM (45° orientation) strengthened infill with mechanical anchors (SA_{45})

with the frame resulting in a peak strength of 293.5 kN at 0.5% drift level. However, there was no significant increase in the residual strength due to FRCM reinforcement as it had reached its ultimate strain. As observed in Fig. 19b and c, the backbone curves obtained from the FE analysis for all schemes show that the peak strength, drift level at peak capacity and degradation of strength are well in agreement with the experimental results.

5.3 *Out-of-Plane Behavior*

5.3.1 Unstrengthened Masonry Infill (CS)

Out-of-plane capacity (OP) obtained from finite element analysis was found to be in good agreement with the analytical predictions by Dawe and Seah (1989), Angel et al. (1994), Flanagan and Bennett (1999), and Bashandy et al. (1995) Fig. 20a). For the undamaged wall, the out-of-plane pressure was applied without subjecting the infill to in-plane cyclic loading. For simulating the damage, in-plane cyclic loading was applied till the desired drift level i.e. including the cycles of lower drift level as per the loading history before applying out-of-plane pressure. The OP capacity vs OP displacement curves are plotted for undamaged infill and damaged infill with 0.50, 0.75, 1.00, 1.40, 1.75 and 2.75% drift levels in Fig. 20b. This plot shows the small reduction in strength with increase in the in-plane drift levels up to 1% drift level. However, for a very high drift level of 2.75% it was observed that the out-of-plane capacity was negligible due to separation of infill from the bounding frame due to which the arching action was not developed. The interaction curve was obtained by plotting the maximum pressure sustained by the wall for each in-plane drift level.

The interaction curve for the unstrengthened and strengthened specimens are shown in Fig. 21. In unstrengthened specimen (CS) with no in-plane damage, the out-of-plane strength of the infill wall was 16 kPa. The wall retained its strength with increasing in-plane damage till drift level of 0.5%. However, with further increase in

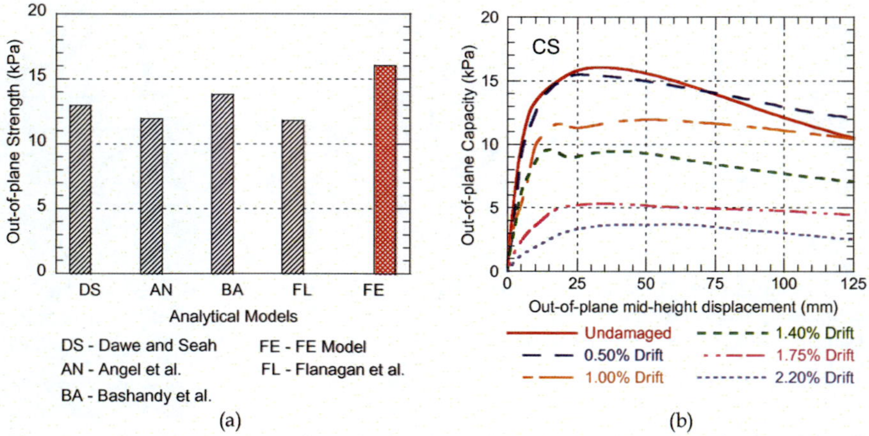

Fig. 20 Out-of-plane strength of infill wall specimen CS **a** comparison with analytical predictions; **b** reduction in capacity with increasing in-plane damage

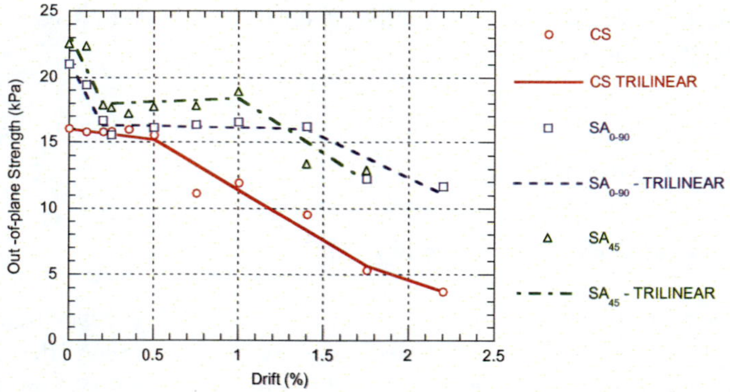

Fig. 21 Interaction curve for **a** unstrengthened masonry infilled RC frame (CS); **b** FRCM strengthened masonry infilled RC frame with anchors (SA_{0-90}); **c** obliquely oriented FRCM strengthened masonry infilled RC frame with anchors (SA_{45})

the in-plane damage, the strength degradation started and reduced to less than 50% of the initial strength of undamaged state.

5.3.2 FRCM Strengthened Infill with Anchors ($SA_{0\text{-}90}$/SA_{45})

The out-of-plane analyses of the walls with anchored FRCM reinforcement subjected to prior damage exhibited the beneficial effect of anchors. The anchored specimen enabled the fabric to develop its peak strength and ductility for OP forces. The FRCM reinforcement with anchors exhibited OP strength of 21 kPa in the undamaged state, which is 78% higher than that of unreinforced infill. The OP capacity reduced gradually till in-plane drift level of 0.25% but continued to retain the capacity of 15 kPa till damage level corresponding to 1.40% drift. At high damage state of 2.2%, a residual capacity of 11.69 kPa was available which corresponds to 3 times the capacity of unstrengthened infill at the same level. This provides adequate factor of safety to avoid falling debris collapse in a seismic event.

The out-of-plane performance of the 45° oriented FRCM reinforced infilled frame was found to be no different than the specimen with orthogonal orientation (0–90°). After 2.20% drift level damage state, the infill failed under out-of-plane force by failure of fabric reinforcement. Due to the enhanced contribution of fabric in the diagonal tension load resistance at high drift levels, the strands ruptured and did not contribute to the OP resistance at ultimate damage levels. Because of the rupture of fabric, the specimen was assumed to have failed and the strength is not included in the interaction curve. Hence, 0–90° orientation of fabric was found to be more suitable for better out-of-plane resistance of the infill with prior in-plane damage.

6 Performance Based Design for FRCM Strengthening of Infills

Fragility analysis was performed for FRCM reinforced infill walls to quantify the improvement of performance in the out-of-plane direction and a performance-based design framework is proposed to include the interaction effect. Identification of the performance of the structure under seismic forces requires structural analysis carried out by either nonlinear static analysis or nonlinear dynamic analysis. The capacity spectrum method was used for the evaluation of the performance points due to the simplified nonlinear analysis and programmable calculation of performance points (Shinozuka et al. 2002; Rota et al. 2010; Frankie et al. 2012). Recommendations for improved capacity spectrum method proposed by FEMA 440 (2005) was used along with the ATC-40 (1996) procedure in this study.

6.1 Capacity and Demand Spectrum

The prototype frame assumed for the experimental model was analysed for vulnerability assessment. SAP 2000 (CSI 2009) program was used to model the masonry infilled RC frame in this study. Beams and columns were modeled using frame elements and masonry infill walls were modeled as pin-joint single-strut elements. Thickness of the strut was taken as the thickness of the wall and the width of the strut was assumed to be one-fourth of the diagonal length of the wall as proposed by Paulay and Priestley (1992). Adopting lumped plasticity model for beams and columns, displacement-based flexural hinges and force-based shear hinges were assigned at half the plastic hinge length away from the ends. Masonry strut carries only axial compressive force, hence the nonlinearity was lumped at the center of the diagonal length of the strut using axial (P-δ or σ-ε) hinge. The plastic zone length of diagonal strut was assumed to be three-fourth of the diagonal length of the strut. The axial load-deformation curve of the infill was obtained by subtracting the bare frame response from the infilled frame load deformation response and converting to axial stress-strain curve. The pushover curve obtained was converted to an equivalent capacity curve.

The seismic demand spectrum can be obtained from mean spectra and mean ± standard deviation spectra generated from a set of different ground motion spectra representing the event to event and site to site variation (Shinozuka et al. 2002). The 5% damped elastic acceleration spectra specified in Indian seismic design code IS 1893 (BIS 2007) for the medium soil site was scaled for different PGA levels ranging from 0.1 to 1 g with increments of 0.1 g and used as mean demand spectrum. The mean + standard deviation spectrum was obtained by multiplying the spectral acceleration ordinate values of mean spectrum by 1.5. Seismic demand from elastic spectrum was reduced to inelastic demand spectrum by considering inherent viscous damping of 5% and hysteretic damping obtained from pushover curve. The intersection of the reduced demand spectrum with the capacity spectrum gives the *performance point* of the SDOF system which can be converted to roof drift of the building.

6.2 Performance Limit States

Capacity spectrum analysis gives the maximum roof drift which can occur for a given earthquake spectrum. Therefore, the performance levels should correspond to a roof drift, so that the fragility curves can be obtained using capacity spectrum method.

6.2.1 In-Plane Performance Levels

FEMA 306 (1998), explaining the behavior modes of infilled frames, recommends the deformation capacities of solid masonry infills based on the experimental observations. As per FEMA 306, the diagonal cracking begins at 0.25% inter-storey drifts when the masonry starts exhibiting nonlinear behavior. The cracking completes from corner to corner at 0.50% and the corner crushing begins. The maximum inter-storey drift limit for the general shear mode is specified as 1.50% from experimental evidence for brick masonry infills. These drift values are used in this study as drift limits corresponding to *Light Damage*, *Moderate Damage* and *Severe Damage* states.

The performance points calculated from capacity spectrum method provide roof drift of the structure. For a given roof drift, the inter-storey drift limits were calculated using fundamental mode shape of the frame. The lowest mode is usually the easiest mode to excite by seismic ground motions and it is also a conservative assumption than assuming higher modes. The second storey was found to be the critical storey for in-plane drift according to the fundamental mode shape for the study frame's stiffness and mass distribution.

6.2.2 Out-of-Plane Performance Level

In this study, the out-of-plane performance level was defined as the drift level at which the out-of-plane strength reduces to 80% of the initial strength at undamaged state. The out-of-plane acceleration demand in a building linearly increases along the height to about 3 times the ground acceleration at the top storey. ASCE 7 (2010) recommends the seismic demand for non-structural components to be calculated using linear increase in the acceleration demand along the height, with twice the base acceleration at mid-height of the building and thrice the base acceleration at the top. This means that for a given earthquake peak ground acceleration, the ground storey infill is apparently three times stronger than the top storey infill. To consider this variation in out-of-plane acceleration at different storey levels, a term *Apparent Capacity* is introduced. This assumption is valid only if the infills in all the storeys are of the same dimensions, strength, and material properties. The apparent out-of-plane capacity was derived with respect to the capacity of the top storey (Q_t) and demand in each storey with respect to the demand in first storey (A_b).

A critical wall can be anywhere in the frame but predominantly it will be at middle level storeys due to the higher level of in-plane damage and higher acceleration demand. To find the critical storey level, the apparent out-of-plane strength of infills was normalized with respect to the strength of the top storey infill for derivation of out-of-plane performance level drift. The inter-storey drift levels for the reduction of apparent strength to 80% of the true strength of infill at undamaged state was derived for infills at different storey levels of the study building frame, as shown in Fig. 22. The storey level which reached the 80% of true undamaged strength at the minimum roof drift was taken as the critical storey to develop the fragility curve. Thus, for the

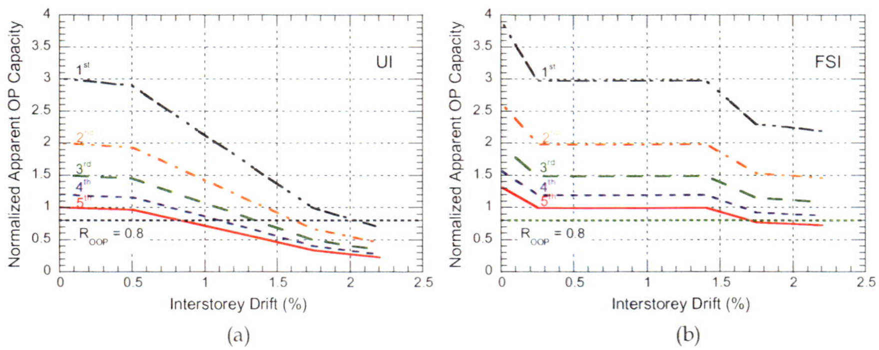

Fig. 22 Derivation of out-of-plane performance level in terms of inter-storey drift limits for each storey of study frame using, **a** unstrengthened infills (UI) and **b** FRCM strengthened infills (FSI)

study frame with unstrengthened infills the critical roof drift was found to be 1.12% and that for the frame with FRCM strengthened infills was 3.50%.

The above definition of out-of-plane performance level includes the effect of amplification of acceleration at higher storeys, effect of in-plane damage on the out-of-plane strength, improvement in strength and less degradation of strength with in-plane damage due to strengthening by FRCM reinforcement. Using this procedure, the critical infills can be identified, which will be vulnerable to out-of-plane collapse during an earthquake. It is also useful in selecting the infills which are required to be reinforced with FRCM.

6.3 *Fragility Curves*

The probability of exceedance of a damage state at a particular level of seismic demand was calculated from the limiting value of spectral roof displacement and distribution parameters of the spectral displacement for the demand. The generated fragility points were fitted with a fragility function which is a log-normal cumulative distribution function as given in FEMA 440. The peak ground acceleration (PGA) was taken as the engineering demand parameter in this study. The fragility functions developed for in-plane performance levels for RC frames is shown in Fig. 23. The fragility functions developed for the out-of-plane strength performance level of the study frame using unstrengthened infill (UI) and FRCM strengthened infill (FSI) is given in Fig. 24.

The fragility curve of the study frame for in-plane performance levels shows the probability of exceedance of the slight damage is about 74% at the Maximum Considered Earthquake (MCE) of 0.36 g as per IS 1893 (BIS 2002a, b). The acceptance criteria for common buildings can be assumed to be 50% probability of exceedance of moderate damage at the design level PGA, which is satisfied for even MCE level PGA in this study frame.

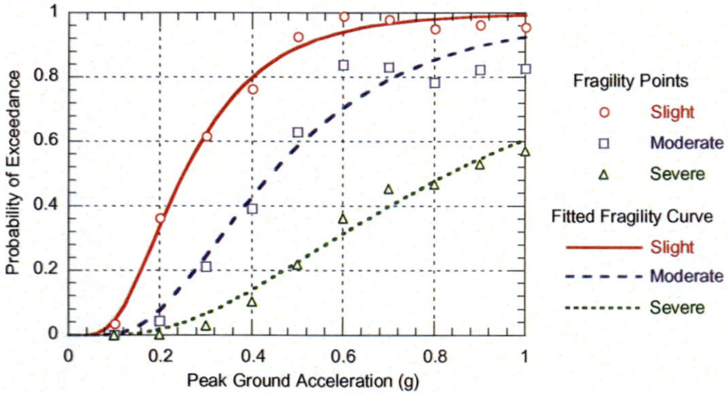

Fig. 23 Fragility curves of existing in-plane performance levels for the study RC frame

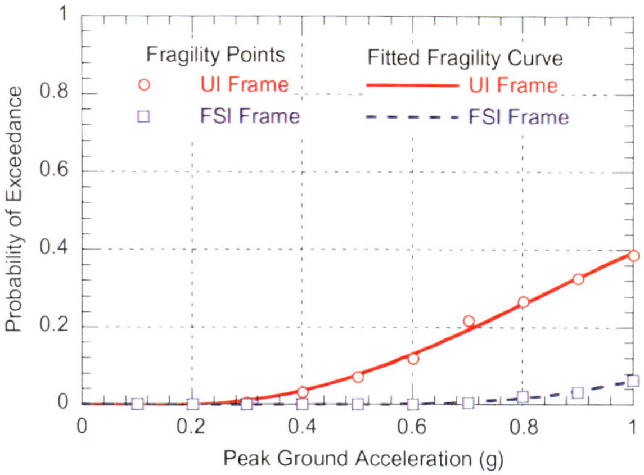

Fig. 24 Fragility function for out-of-plane performance level for the study RC frame with unstrengthened infills (UI) and FRCM strengthened infills (FSI)

The frame with unstrengthened infills reached 10% probability of exceedance of out-of-plane performance level at 0.56 g PGA, whereas the frame with FRCM strengthened infills did not exceed 10% probability even at 1 g PGA. This methodology can be used for actual structures to understand the vulnerability under design level earthquake and for new structures to ensure satisfactory performance.

7 Concluding Remarks

Fabric reinforced cementitious matrix (FRCM) as a composite material has significant potential in improving the seismic response of unreinforced masonry infilled RC frames. FRCM is embedded in mortar plaster directly applied to masonry wall substrate and anchored to the surrounding RC frame. Infilled walls strengthened with FRCM in such a manner were able to control the out-of-plane displacements of cracked masonry under simulated seismic loads. Analytical simulations on models suitably calibrated with experimental observations can be used to develop performance-based evaluation and design procedures for FRCM strengthened infilled walls.

References

AC434 (2013) Acceptance criteria for masonry and concrete strengthening using fiber-reinforced cementitious matrix (FRCM) composite systems. ICC-Evaluation Service, Whittier, CA

Agnihotri P, Singhal V, Rai DC (2013) Effect of in-plane damage on out-of-plane strength of unreinforced masonry walls. Eng Struct 57:1–11

Aldea CM, Mobasher B, Jain N (2007) Cement-based matrix-grid system for masonry rehabilitation. Text Reinf Concr (TRC) 244(9):141–156

American Concrete Institute (ACI) (2005) Acceptance criteria for moment frames based on structural testing and commentary. ACI 374.1-05, Farmington Hills, MI
American Concrete Institute (ACI) (2013) Design and construction guide of externally bonded FRCM systems for concrete and masonry repair and strengthening. ACI 549, Farmington Hills, MI
American Society for Testing and Materials (ASTM) (2010) Standard test methods for diagonal tension (shear) in masonry assemblages. ASTM E519/ E519M, West Conshohocken, PA
American Society for Testing and Materials (ASTM) (2013) Standard test method for tensile strength of concrete surfaces and the bond strength or tensile strength of concrete repair and overlay materials by direct tension (pull-off method). ASTM C 1583-13, West Conshohocken, PA
Angel R, Abrams DP, Shapiro D, Uzarski J, Webster M (1994) Behavior of reinforced concrete frames with masonry infills. Technical Report Civil Engineering Studies SRS-589, 1994-03, University of Illinois Engineering Experiment Station, College of Engineering, University of Illinois at Urbana-Champaign, IL, USA
ASCE/SEI 7-05 (2010) Minimum design loads for buildings and other structures. Structural Engineering Institute, American Society of Civil Engineers ASCE, Reston, VA, USA
ATC (1996) Seismic evaluation and retrofit of concrete buildings. ATC-40 Report, vols 1 and 2, Applied Technology Council, Redwood City, CA, USA
Babaeidarabad S, De Caso F, Nanni A (2014a) URM walls strengthened with fabric-reinforced cementitious matrix composite subjected to diagonal compression. J Compos Constr 18(4):04013057
Babaeidarabad S, De Caso F, Nanni A (2014b) Out-of-plane behavior of URM walls strengthened with fabric-reinforced cementitious matrix (FRCM) composite. J Compos Constr 18(2):04013045
Bashandy T, Rubiano NR, Klingner RE (1995) Evaluation and analytical verification of infilled frame test data. Phil M. Ferguson Structural Engineering Laboratory, University of Texas at Austin, Austin, TX, USA
Bhushan RS (2015) Analytical out-of-plane response of Fabric Reinforced Cementitious Matrix (FRCM) strengthened masonry infilled reinforced concrete frames with prior in-plane damage. M.Tech Dissertation, Dept. of Civil Engineering, Indian Institute of Technology Kanpur, Kanpur, India
Bischof P, Suter R, Chatzi E, Lestuzzi P (2014) On the use of CFRP sheets for the seismic retrofitting of masonry walls and the influence of mechanical anchorage. Polymers 6(7):1972–1998
Braga F, Manfredi V, Masi A, Salvatori A, Vona M (2011) Performance of non-structural elements in RC buildings during the L'Aquila, 2009 earthquake. Bull Earthq Eng 9(1):307–324
British Standard Institution (BSI) (1999) Methods of test for masonry—Part 2: Determination of flexural strength. BS EN 1052-2, London
Bureau of Indian Standards (BIS) (1987) Indian standard code of practice for structural use of unreinforced masonry. IS 1905, New Delhi, India
Bureau of Indian Standards (BIS) (2000) Indian standard plain and reinforced concrete—code of practice. IS 456, New Delhi, India
Bureau of Indian Standards (BIS) (2002a) Indian standard criteria for earthquake resistant design of structure, Part 1: General provisions and buildings. IS 1893, 5th Rev, New Delhi, India
Bureau of Indian Standards (BIS) (2002b) Ductile detailing of reinforced concrete structures subjected to seismic forces—code of practice. IS 13920, Edition 1.2, New Delhi, India
Calvi GM, Bolognini D (2001) Seismic response of reinforced concrete frames infilled with weakly reinforced masonry panels. J Earthq Eng 5(02):153–185
Carozzi FG, Milani G, Poggi C (2014) Mechanical properties and numerical modeling of Fabric Reinforced Cementitious Matrix (FRCM) systems for strengthening of masonry structures. Compos Struct 107:711–725
Crisafulli FJ, Carr AJ, Park R (2000) Analytical modelling of infilled frame structures—a general review. Bull N Z Soc Earthq Eng 33(1):30–47
CSI (2009) Integrated software for structural analysis and design, SAP 2000. Computers and Structures Inc., Walnut Creek, CA, USA

Dawe JL, Seah CK (1989) Out-of-plane resistance of concrete masonry infilled panels. Can J Civ Eng 16(6):854–864

El-Dakhakhni WW, Hamid AA, Hakam ZHR, Elgaaly M (2006) Hazard mitigation and strengthening of unreinforced masonry walls using composites. Compos Struct 73(4):458–477

FEMA 306 (1998) Evaluation of earthquake damaged concrete and masonry wall buildings: basic procedures manual. FEMA 306, Applied Technology Council (ATC-43 Project), Washington, DC, USA

FEMA 440 (2005) Improvement of nonlinear static seismic analysis procedures. FEMA 440, prepared by Applied Technology Council (ATC-55 Project), Washington, DC

Flanagan R, Bennett R (1999) Arching of masonry infilled frames: comparison of analytical methods. Pract Period Struct Des Constr 4(3):105–110

Frankie TM, Gencturk B, Elnashai AS (2012) Simulation-based fragility relationships for unreinforced masonry buildings. J Struct Eng 139(3):400–410

Gopalaratnam VS, Shah SP (1985) Softening response of plain concrete in direct tension. ACI J 82(3):310–323

Hamoush SA, McGinley MW, Mlakar P, Scott D, Murray K (2001) Out of-plane strengthening of masonry walls with reinforced composites. J Compos Constr 5(3):139–145

Harajli M, ElKhatib H, San-Jose JT (2010) Static and cyclic out-of-plane response of masonry walls strengthened using textile-mortar system. J Mater Civ Eng 22(11):1171–1180

Italian National Research Council (2004) Guide for the design and construction of externally bonded FRP systems for strengthening existing structures, materials, RC and PC structures, masonry structures. CNR-DT200, Rome, Italy

Kalali A, Kabir MZ (2012) Cyclic behavior of perforated masonry walls strengthened with glass fiber reinforced polymers. Sci Iran 19(2):151–165

Kaushik HB, Rai DC, Jain SK (2007a) Stress-strain characteristics of clay brick masonry under uniaxial compression. J Mater Civ Eng 19(9):728–739

Kaushik HB, Rai DC, Jain SK (2007b) Uniaxial compressive stress-strain model for clay brick masonry. Curr Sci 92(4):497–501

Kent DC, Park R (1971) Flexural members with confined concrete. J Struct Div 97(7):1969–1990

Komaraneni S, Rai DC, Singhal V (2011) Seismic behavior of framed masonry panels with prior damage when subjected to out-of-plane loading. Earthq Spectra 27(4):1077–1103

Kunnath SK, Heo Y, Mohle JF (2009) Nonlinear uniaxial material model for reinforcing steel bars. J Struct Eng 135(4):335–343

Li B, Wang Z, Mosalam KM, Xie H (2008) Wenchuan earthquake field reconnaissance on reinforced concrete framed buildings with and without masonry infill walls. In: The 14th world conference on earthquake engineering. Beijing, China

Lin Y, Biggs D, Wotherspoon L, Ingham J (2014) In-plane strengthening of unreinforced concrete masonry wallettes using ECC shotcrete. J Struct Eng. https://doi.org/10.1061/(ASCE)ST.1943-541X.0001004

Komaraneni S, Rai DC, Singhal V (2011b) Seismic behaviour of framed masonry panels with prior damage when subjected to out-of-plane loading. Earthq Spectra 27(4):1077–1103

Koutas L, Bousias SN, Triantafillou TC (2014a) Seismic strengthening of masonry-infilled RC frames with TRM: experimental study. J Compos Constr 19(2):04014048

Koutas L, Pitytzogia A, Triantafillou TC, Bousias SN (2014b) Strengthening of infilled reinforced concrete frames with TRM: study on the development and testing of textile-based anchors. J Compos Constr 18(3):A4013015

Milani G, Rotunno T, Sacco E, Tralli A (2006) Failure load of FRP strengthened masonry walls: experimental results and numerical models. Struct Durab Health Monit 2(1):29–50

Mosallam AS (2007) Out-of-plane flexural behavior of unreinforced red brick walls strengthened with FRP composites. Compos Part B: Eng 38(5):559–574

Özkan C, Ökten MS, Gençoğlu M, Güler K (2012) Experimental investigation of infill walls strengthened by cementitious matrix-fabric composites. In: Proceedings of the 15th world conference on earthquake engineering, International Association for Earthquake Engineering, Portugal

Papanicolaou CG, Triantafillou TC, Karlos K, Papathanasiou M (2007) Textile reinforced mortar (TRM) versus FRP as strengthening material of URM walls: in-plane cyclic loading. Mater Struct 40(10):1081–1097

Papanicolaou CG, Triantafillou TC, Papathanasiou M, Karlos K (2008) Textile reinforced mortar (TRM) versus FRP as strengthening material of URM walls: out-of-plane cyclic loading. Mater Struct 41(1):143–157

Parisi F, Iovinella A, Balsamo A, Augenti N, Prota A (2013) In-plane behavior of tuff masonry strengthened with inorganic matrix–grid composites. Compos B Eng 45(1):1657–1666

Paulay T, Priestley MJN (1992) Seismic design of reinforced concrete and masonry buildings. Wiley-Interscience, NY, USA

Prota A, Marcari G, Fabbrocino G, Manfredi G, Aldea C (2006) Experimental in-plane behavior of tuff masonry strengthened with cementitious matrix–grid composites. J Compos Constr 10(3):223–233

Rai DC, Mondal G, Singhal V, Parool N, Pradhan T, Mitra K (2012) Reconnaissance report of the M6. 9 Sikkim (India–Nepal border) earthquake of 18 September 2011. Geomat Nat Hazards Risk 3(2):99–111

Rai DC, Singhal V, Paikara S, Mukherjee D (2014) Sub-paneling of masonry walls using precast reinforced concrete elements for earthquake resistance. Earthq Spectra 30(2):913–937

Rai DC, Singhal V, Raj SB, Sagar SL (2015) Reconnaissance of the effects of the M7.8 Gorkha (Nepal) earthquake of April 25, 2015. Geomat Nat Hazards Risk 1–15

Rota M, Penna A, Magenes G (2010) A methodology for deriving analytical fragility curves for masonry buildings based on stochastic nonlinear analyses. Eng Struct 32(5):1312–1323

Sagar SL, Singhal V, Rai DC, Gudur P (2017) Diagonal shear and out-of-plane flexural strength of fabric-reinforced cementitious matrix-strengthened masonry walletes. J Compos Constr 21(4):04017016

Sagar SL, Singhal V, Rai DC (2019) In-plane and out-of-plane behavior of masonry-infilled RC frames strengthened with fabric-reinforced cementitious matrix. J Compos Constr 23(1):04018073

Scawthorn C, Johnson GS (2000) Preliminary report: Kocaeli (Izmit) earthquake of 17 August 1999. Eng Struct 22(7):727–745

Shinozuka M, Feng MQ, Kim H, Ueda T (2002) Statistical analysis of fragility curves. Technical Report at Multidisciplinary Center for Earthquake Engineering Research, Buffalo, NY, USA

Simulia (2011) ABAQUS analysis user's manual, version 6.13-3. Dassault Systèmes, Providence, RI, USA

Singhal V, Rai DC (2014a) Suitability of half-scale burnt clay bricks for shake table tests on masonry walls. J Mater Civ Eng 26(4):644–657

Singhal V, Rai DC (2014b) Role of toothing on in-plane and out-of-plane behavior of confined masonry walls. J Struct Eng 140(9):04014053

Stratford T, Pascale G, Manfroni O, Bonfiglioli B (2004) Shear strengthening masonry panels with sheet glass-fiber reinforced polymer. J Compos Constr 8(5):434–443

Tan KH, Patoary MKH (2004) Strengthening of masonry walls against out-of plane loads using fiber-reinforced polymer reinforcement. J Compos Constr 8(1):79–87

Tomaževič M, Gams M, Berset T (2011) Seismic strengthening of brick masonry walls with composites: an experimental study. In: Proceedings, 4th structural engineering world congress, International Association for Shell and Spatial Structures, Madrid, Spain, p 307

Triantafillou TC (1998) Strengthening of masonry structures using externally bonded FRP laminates. J Compos Constr 2(2):96–104

Tumialan JG, Nanni A (2001) In-plane and out-of-plane behavior of masonry walls strengthened with FRP systems. Report CIES, 01-24, Center for Infrastructure Engineering Studies, University of Missouri—Rolla, Rolla, MO, USA

Tumialan JG, Galati N, Nanni A (2003) Fiber-reinforced polymer strengthening of unreinforced masonry walls subject to out of-plane loads. ACI Struct J 100(3):312–329

Markov Chain Modelling of Evolution of Deflection in Ferrocement Flexural Members

K. Balaji Rao

Abstract Service life design and health monitoring of structures are among the current research topics in structural engineering. Both these require prediction of state of the structure under different load levels. There is a need for developing methodologies that take into account different randomness in different basic variables in the prediction of the states of the structure. This requires a suitable stochastic model to be used for predicting the evolution of linear/nonlinear response (that determines the state) of the structure when subjected to different load levels. This Chapter presents results of some of the investigations related to application of Non Homogeneous Markov Chain model for modelling evolution of central deflection of ferrocement flexural members. In the Appendix simple Chebyshev bounds on deflection proposed by Steliga and Szynal (Int J Pure Appl Math 58:137–152, 2010) are computed and compared with the respective experimental values.

1 Introduction

Ferrocement is a thin composite in which the brittle cement mortar matrix is reinforced with ductile mesh wires. Ferrocement has found its applications in structural and architectural engineering (Paul and Pama 1978; ACI 549.1R 2018). The current practice in the design of ferrocement (FC) flexural members is to design against relevant ultimate limit states (ULS) and check for serviceability limit states (SLS), viz. cracking and deflection. Large body of literature exists on the reliability analysis and reliability-based design of RC flexural members against ULS (viz. Ditlevsen and Madsen 1996; Nowak and Collins 2000). However, similar studies dealing with SLS are comparatively less. The serviceability issues such as cracking (due to mechanical loading and environmental effects) and deflection are gaining importance due to the

K. Balaji Rao (✉)
Structural Engineering Research Centre, CSIR Campus, Taramani, Chennai 600113, India
e-mail: balaji@serc.res.in; balajiserc@gmail.com

S. B. Singh et al. (eds.), *Emerging Trends of Advanced Composite Materials in Structural Applications*, Composites Science and Technology,
https://doi.org/10.1007/978-981-16-1688-4_3

emphasis on service life design. Hence, there is a need to develop models that integrate the phenomenological aspects and the uncertainties associated with cracking and deflection.

Service life design requires application of suitable stochastic models. It is known that Markov Chain (MC) models are the simplest stochastic models that are extensively applied in engineering (Ang and Tang 1984). This Chapter highlights the possibility of application of Non Homogeneous Markov Chain (NHMC) for modelling the evolution of central deflection of ferrocement flexural members subjected to monotonically applied loading. An attempt has been made to compare the results of NHMC with relevant experimental results. In this chapter, the structural system or simply system and ferrocement flexural member are used synonymously. The main aim of this chapter to demonstrate how NHMC can be used to model the stochastic evolution of deflection of ferrocement flexural members. In the Appendix simple Chebyshev bounds on deflection proposed by Steliga and Szynal (2010) are computed and compared with the respective experimental values. The bounds are found to be wide to be practically useful in design decision making and thus brings out the importance of NHMC for better prediction of state of the flexural member.

1.1 Markov Chain—Preliminaries

Markov Chain (MC) is a stochastic process model in which both the state space and the index space are discrete (viz. Ang and Tang 1984). The stochastic evolution of the system, modelled by homogeneous MC can be completely described by the Transition Probability Matrix (TPM), P. For a system with N states, the transition probability matrix P is given by,

$$P = \begin{bmatrix} p_{11} & p_{12} & \cdots & p_{1N} \\ p_{21} & p_{22} & \cdots & p_{2N} \\ \vdots & \vdots & & \vdots \\ p_{N1} & p_{N2} & \cdots & p_{NN} \end{bmatrix} \quad (1)$$

Typically, the element p_{ij} represents the probability that the state of the system will be j in the next step given that the state at present is i. That is,

$$p_{ij} = P(X_{k+1} = j|X_k = i) \quad (2)$$

P is a one-step transition probability matrix the elements of which can be evaluated analytically once the deterministic model of the evolutionary phenomenon is available. The n-step TPM is given by

$$P^n = P \times P \times \cdots \times P \ \ (n\text{-times}) \quad (3)$$

where $P = \{p_{ij}\}_{i\in 1,\ldots,N;\, j\in 1,\ldots,N}$; $P^n = \{p_{ij}(n)\}_{i\in 1,\ldots,N;\, j\in 1,\ldots,N}$; $p_{ij}(n)$ is the Probability {system state will by 'j' after n-transitions, starting from state 'i'}. If X_n denotes the state of the system after n-steps, then

$$p_{ij}(n) = P(X_{k+n} = j|X_k = i) \tag{4}$$

As the states of the system are mutually exclusive and collectively exhaustive after each transition, the probabilities in each row add up to 1.0. For a system whose evolution is defined by a homogeneous MC (HMC), the unconditional state vector of the system at any future step/time can be determined using the one-step TPM, once the initial state vector is known. The long-run behaviour of the system described by TPM, P, is characterised by a vector $\pi = (\pi_1, \pi_2, \ldots, \pi_N)$. This vector also characterises the steady-state behaviour of the system. The elements of the vector are obtained by solving the following set of simultaneous equations:

$$\left.\begin{aligned} \pi P &= \pi \\ \sum_{i=1}^{N} \pi_i &= 1 \end{aligned}\right\} \tag{5}$$

In general, the following are the steps involved in modelling the stochastic behaviour of the system using HMC:

- Identification and definition of states of the system (i.e. identification of state space)
- Identification and definition of index set of the system (i.e. identification of index space)
- Determination of elements of TPM (may be obtained either by using the observational record or by using analytical model developed or using both)
- Determination of the unconditional probability vector at any specified value of index set by multiplying the TPM with initial state vector
- Estimation of the statistical quantities of interest of the response of the system using the unconditional probability vector at any given value of index set
- Interpretation of the results obtained.

If the phenomenon considered is nonlinear/nonhomogeneous, a nonhomogeneous MC (NHMC) needs to be used for modelling the stochastic evolution of the response of the system. In such a case, the n-step TPM is given by

$$P^n = P_1 \times P_2 \times P_3 \times \cdots \times P_{n-1} \tag{6}$$

where P_m ($m = 1, \ldots, n-1$) represents the 1-step TPM estimated at the mth stage of loading. The unconditional probability vector at any stage is obtained by multiplying the P_m with the initial state vector.

Some of the books dealing with fundamentals and applications of MC are Benjamin and Cornell (1970), Ang and Tang (1984), Bogdanoff and Kozin (1985), Ditlevsen and Madsen (1996), and Nowak and Collins (2000). The author has been involved in the application of HMC/NHMC to model the evolution of strains in reinforced concrete flexural members (Anoop et al. 2012), cracking in reinforced concrete flexural members under static loading (Desayi and Balaji Rao 1989) and under fatigue loading (Balaji Rao and Appa Rao 2004), developing a stochastic model for service life design of reinforced concrete girders against limit state of shear (Yogalakshmi and Balaji Rao 2020), development of models for assessment of existing bridge girders (Balaji Rao and Appa Rao 1999; Balaji Rao et al. 2004), and settlement of footings on cohesion less soils (Geetha Manjari et al. 2014).

2 NHMC Modelling of Deflection in Ferrocement Flexural Members

Deflection is a basic measurable quantity that is often used in condition assessment of structures/structural members. For example, the deflection of a ferrocement flexural member at the centre of the span (which is a serviceability criterion) is obtained using the equations presented in the next section. In this study, the stochastic modelling of evolution central deflection of ferrocement flexural member under monotonic loading is considered. The index space of this stochastic process is the loading stages, which can be considered as discrete, $\{N_1, N_2, \ldots, N_n\}$. The state space of the stochastic process represents the deflection at the centre of the span at different stages of loading. By dividing the state space into finite number of discrete states (each state represents a finite range of central span deflection of the member), the central deflection can be treated as a stochastic process with discrete state space and discrete index space. The discreteness of the state space is introduced by aggregating ranges of deflection as different states.

In this study, the stochastic evolution of central deflection is obtained by making the following *assumptions*:

i. The stochastic process can be described as a one-step memory process. This implies that the process is Markov and the present state of the system can be completely determined by the immediate past state. This assumption is justified since the deflection at ith stage (i.e. at loading stage N_i) more or less depends on deflection at (i − 1)th stage (i.e. at loading stage N_{i-1}) as the nonlinear response can be approximated by multi-linear response.
ii. The stochastic evolution of the deflection can be described using bi-level NHMC. This is because: (i) the deflection response of ferrocement flexural member can be approximated by a bi-linear curve (Sect. 3.2), and (ii) the deflection at any stage of loading is a function of the applied load.
iii. The classification of the index space is easy compared to that of state space since monotonic loading case is considered, the index set is $\{N_1, N_2, \ldots, N_n\}$; where

N represents the magnitude of the applied load (considered to be deterministic since good control is exercised during experiments to maintain the load level) and the subscript n represents the loading stage. For instance the index set for the member MI1 is {2 kN, 8 kN, 10 kN, 15 kN}; for MI2 it is {1 kN, 8 kN, 10.41 kN, 13 kN}; for MI3 it is {2 kN, 10 kN, 15 kN, 17.62 kN}; for MI4 it is {1 kN, 8 kN, 9.9 kN, 10 kN}; for MI5 it is {3 kN, 15 kN, 16 kN, 17.79 kN}; for MI6 it is {2 kN, 9 kN, 9.6 kN, 14 kN}; In all these sets the first two elements belong to linear elastic behaviour of the flexural member while the other two elements belong to the inelastic behaviour of the member.

iv. Classification of state space takes into account the bi-linear approximation of the load deflection behaviour of the ferrocement member (Sect. 3.4). For each flexural member the state space has been classified into two ranges namely elastic range and inelastic range. In both the ranges the classification has been done in such a way that the range of deflections are representative of the elastic and inelastic behaviour of the member. Also, the classification of states has been done in such way that they are mutually exclusive and collectively exhaustive. In order to avoid the bias and to satisfy the basic property of state classification, the initial value of the deflection has been set to zero. The state classification for the members MI1 to MI6 are presented in Tables 6, 7, 8, 9, 10 and 11.

v. At a given stage of loading (i.e. as the load on the beam increases from N_i to N_{i+1}), starting from a given state, the system can remain in the same state or can make a transition to all higher states since we are modelling the stiffness degrading system. This assumption is used to determine the elements of TPM at different stages of loading; in both linear and nonlinear ranges.

2.1 Determination of Elements of TPM

A typical element of TPM of Markov Chain considered can be written as,

$$p_{ij}(N_k, N_{k+1}) = \frac{P\{\Delta(N_{k+1}) = j \cap \Delta(N_k) = i\}}{P\{\varepsilon(N_k) = i\}} \tag{7}$$

which gives the probability of deflection of the system being in state ‘j’ at loading stage N_{k+1} given that the system was in deflection state ‘i’ at the loading stage N_k. To compute these probabilities, the information regarding joint probability density function (jpdf) of deflection states at any two successive loading stages, (N_k, N_{k+1}), and also at a loading stage, N_k, should be known. Since it is difficult to generate this information from the test data, in the present investigation, it is assumed that deflection states at successive loading stages follow bivariate normal distributions and at any loading stage, the deflection state follows a normal distribution. It is also noted that when means, variances and correlation coefficient values between the deflection states at two successive loading stages ($\rho_{N,N+1}$) are known, the maximum entropy distribution would be a bivariate normal distribution (Kapur 1993). Knowing the jpdf and pdf, and using the above equation, the elements of TPM can be computed. A typical element of the conditional 1-step TPM is given by

$$p_{ij}(N_k, N_{k+1}) = \frac{\int_{\delta_{i-1}}^{\delta_i} \int_{\delta_{j-1}}^{\delta_j} f_{k,k+1}(\delta_k, \delta_{k+1}) d\delta_k d\delta_{k+1}}{\int_{\delta_{i-1}}^{\delta_i} f_k(\delta_k) d\delta_k} \tag{8}$$

where $f_{k,k+1}(\delta_k, \delta_{k+1})$ is the bivariate normal distribution given by

$$\begin{aligned} & f_{k,k+1}(\delta_k, \delta_{k+1}) \\ & = \frac{1}{2\pi\sigma_k\sigma_{k+1}\sqrt{1-\rho_{k,k+1}^2}} \exp\left\{-\frac{1}{2(1-\rho_{k,k+1}^2)}\left[\left(\frac{\delta_k-\mu_k}{\sigma_k}\right)^2\right.\right. \\ & \left.\left. -2\rho_{k,k+1}\left(\frac{\delta_k-\mu_k}{\sigma_k}\right)\left(\frac{\delta_{k+1}-\mu_{k+1}}{\sigma_{k+1}}\right)+\left(\frac{\delta_{k+1}-\mu_{k+1}}{\sigma_{k+1}}\right)^2\right]\right\} \\ & -\infty \le \delta_k \le \infty; -\infty \le \delta_{k+1} \le \infty; \quad -1 \le \rho_{k,k+1} \le 1; \end{aligned} \tag{9}$$

and $f_k(\delta_k)$ is the univariate normal distribution, given by

$$f_k(\delta_k) = \frac{1}{\sqrt{2\pi}\sigma_k} \exp\left\{-\frac{1}{2}\left(\frac{\delta_k-\mu_k}{\sigma_k}\right)^2\right\}; \quad -\infty \le \delta_k \le \infty \tag{10}$$

2.2 *Determination of Statistical Properties*

Using the formulations presented above, it is possible to compute the statistical properties of the state of deflection of the system at any loading stage using the following procedure proposed by Balaji Rao and Appa Rao (2004). Though these equations are not used in the present chapter, they are presented for the sake of completeness.

i. Divide the state space into mutually exclusive and collectively exhaustive event sets and compute the central value of each event set, namely, $\beta\delta_1, \beta\delta_2, \ldots, \beta\delta_{m-1}, \beta\delta_m$
ii. Compute unconditional probability vector of the states of the system at any loading stage, that is, $(p_1, p_2, p_3, \ldots, p_m)$
iii. Compute the mean and standard deviation of strain in the beam at the level considered using,

$$\text{Mean} = \langle \Delta_{1,k} \rangle = \sum_{i=1}^{m} (\beta\delta_i) p_i(N_1, N_k) \tag{11}$$

$$\text{Standard deviation} = SD_{1,k} = \sqrt{\left(\sum_{i=1}^{m} (\beta\delta_i)^2 p_i(N_1, N_k)\right) - \langle \Delta_{1,k} \rangle^2} \tag{12}$$

Since the mth state of the system may be unbounded, it can be made bounded by approximating it to be the limiting value of deflection. This approximation can be made only to compute the mean and standard deviation of deflection at any stage of loading.

3 Application

3.1 MC Modelling of Deflection of Ferrocement Flexural Members

Ferrocement (FC) elements can be used as roofing/floor elements to cover larger spans due to their thinness. The slenderness of these elements may adversely affect their performance under working loads. Hence, there is need to study the serviceability behaviour of the FC elements, especially the load (P)-deflection (δ) behaviour.

When the experimental P-δ behaviour of FC flexural members are carefully analysed, it is found that, as soon as the member cracks, the load applied on the member reduces, for the same value of deflection. This behaviour is attributed to the sudden loss in stiffness of the member due to cracking. The load corresponding to this stage is called cracking load. While the P-δ behaviour of the FC before cracking is linear (i.e., elastic behaviour), after the member cracks, the P-δ behaviour would be essentially non-linear (i.e., in-elastic behaviour). This observation needs to be taken into account in stochastic modelling of evolution of deflection of FC elements. In this study, a MC model is proposed to predict the P-δ behaviour of FC flexural members. The performance of this model is studied by comparing the predictions with results of experimental investigations conducted at the Indian Institute of Science, Bangalore (Balaji Rao 1990), on ferrocement monolithic I-joists.

3.2 Prediction of P-δ Behaviour of FC Flexural Members

In developing the method for prediction of deflection, the test data of 6 roofing elements of trapezoidal cross-section, 9 floor elements of channel cross-section, and ferrocement joists comprising of 16 built-up I-sections and 6 monolithic I-sections were used. All the 37 specimens were tested in flexure under symmetrical two-point loading. Based on the detailed investigations carried out at Indian Institute of Science, Bangalore, Desayi and Balaji Rao (1988) proposed a bi-linear equation for predicting the behaviour of FC flexural elements. While the details of investigations are available in Ref. Desayi and Balaji Rao (1988), the final equation for short-term deflection of FC elements is as follows:

$$\delta = \mathrm{K\,ML^2/(E_c I_g)}; \quad \mathrm{M < M_{cr}} \tag{13}$$

$$\delta = \mathrm{K\,M_{cr}L^2/(E_c I_g) + K(M - M_{cr})L^2/\alpha E_c I_{ct}}; \quad \text{for } \mathrm{M \geq M_{cr}}$$

where M is the applied moment, L is the effective span, I_g is the moment of inertia of gross-transformed section (for roofing and floor elements) or gross section (for I-joists), I_{ct} is the moment of inertia of cracked transformed section, K is a constant depending on loading and boundary conditions, α is a constant (equal to 0.821) and E_c is the modulus of elasticity of mortar, given by $E_c = 5700\sqrt{f_{cu}}\ N/\mathrm{mm}^2; f_{cu}$ is the compressive strength of mortar cube in MPa and M_{cr} is the cracking moment of the FC element and is obtained using the method proposed by Desayi and Balaji Rao (1988).

3.3 *Experimental Work Related to the Ferrocement Flexural Elements Considered in This Investigation*

The experimental work carried out at Indian Institute of Science, Bangalore (Balaji Rao 1990) consisted of casting and testing six monolithic I-joists, designated MI1 to MI6. The length (span) of the joists was varied from 2300 (2000 mm) to 4500 mm (4200 mm) and two different span-to-depth ratios, viz. 10 and 14 were used. The actual and the equivalent cross-section of the I-joist used in computation are shown Fig. 1. The cross-sectional dimension and the arrangement of reinforcement of actual and equivalent sections are shown in Fig. 1, typically for the joist MI3. The dimensions of the specimens tested are given Table 1. The material used for casting the monolithic I-joist were cement-sand mortar, machine woven galvanized wire mesh and square weld-mesh. The mortar used had cement to sand ratio of 0.5 (by weight) and water to cement ratio of 0.5 (by weight). Dimensions and strengths of wires of wire-mesh and bars of weld-mesh, details of meshes used in individual specimens and the strength of mortar are given in Tables 2, 3 and 4, respectively.

All specimens were tested in 20 t (199.4 kN) capacity frame. The load was applied through a 10 t (99.7 kN) jack, at one-third points. The constant bending moment zone was divided into number of zones of equal length (viz. 203.2 mm). At each stage of loading surface strains, deflections at the centre as well as under the load points, and

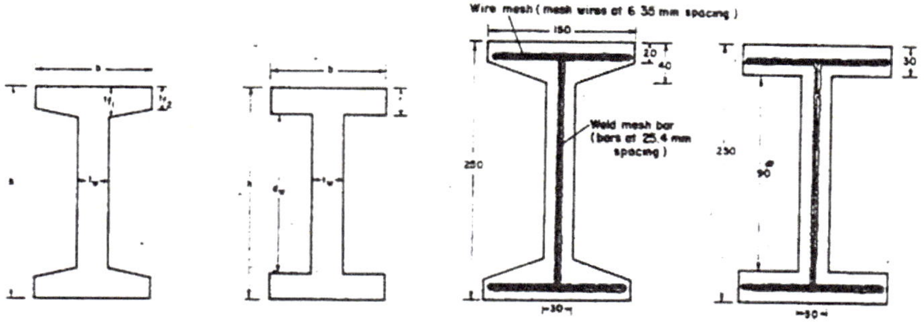

Fig. 1 (1) Actual and equivalent sections of monolithic I-sections and (2) actual and equivalent sections for joist MI3 (Desayi and Balaji Rao 1993; Balaji Rao 1990)

Table 1 Cross-sectional details of I-sections (Balaji Rao 1990)

S. No.	Specimen designation	Total length (mm)	Effective span (L) (mm)	Total depth (D) (mm)	L/D	b (mm)	d_w (mm)
1	MI1	2300	2000	200	10	120	140
2	MI2	3100	2800	200	14	120	140
3	MI3	2800	2500	250	10	150	190
4	MI4	3800	3500	250	14	150	190
5	MI5	3300	3000	300	10	180	240
6	MI6	4500	4200	300	14	180	240

Note For all the specimens $t_{f1} = 40$ mm, $t_{f2} = 20$ mm (hence $t = 30$ mm) and $t_w = 30$ mm

Table 2 Strength details of wire mesh and weld-mesh bars (Balaji Rao 1990)

S. No.	Wiremesh/weldmesh	Proof stress (MPa)	Ultimate stress (MPa)	Modulus of elasticity (MPa)
1	4/20 wire-mesh (20 gauge or 0.76 mm diameter wires at 6.35 mm c/c spacing	498.4	575.5	8240
2	4/22 wire-mesh (22 gauge or 0.61 mm diameter wires at 6.35 mm c/c spacing	392.4	466.0	12,949
3	25.4 mm × 25.4 mm weldmesh (2.54 mm diameter bars at 25.4 mm spacing)	600.9	610.0	130,081

Table 3 Details of mesh wires and weld-mesh bars in the Joist section (Balaji Rao 1990)

Specimen designation	Type of wire mesh used	Number of wire mesh in longitudinal direction		Number of wire mesh bars in longitudinal direction	
		Each flange	Web	Each flange	Web
MI1	4/20	30	46	6	6
MI2	4/20	30	46	6	6
MI3	4/22	38	56	7	7
MI4	4/22	38	56	7	7
MI5	4/20	54	76	8	9
MI6	4/22	54	76	8	9

Table 4 Strength details of mortar and I-joists (Balaji Rao 1990)

Specimen designation	Cube strength (MPa)	Modulus of rupture (MPa)	Dist. of load point from support (mm)	Cracking moment (kN m)	Ultimate moment (kN m)
MI1	31.4	2.80	670	23925.12	77821.26
MI2	23.2	3.39	920	15505.97	76426.08
MI3	27.7	2.29	870	40714.15	118806.85
MI4	21.6	2.01	1170	21838.90	94722.61
MI5	29.8	2.29	1000	43503.43	139150.31
MI6	32.3	2.51	1400	55146.95	158214.31

crackwidths (after the appearance of cracks) were measured at different levels and on both the faces of the specimen.

All the specimens failed in flexure. For specimen MI3, failure occurred under one of the load points and for the others in between the two load points. From the comparison of deflections predicted with the experimental values, at a load, approximately equal to 66% of the experimental ultimate load (which corresponds to the working load, considering a load factor of 1.5), it was concluded that Eq. 13 predicts satisfactorily the deflections of FC specimens. Hence, this equation was used in Monte Carlo simulation to determine the statistical properties of deflection at different stages of loading (Balaji Rao 1990). In simulation, the cross-sectional dimensions, strengths of mortar, wire mesh and square weld-mesh are treated as random variables. The experimental details of applied loads, central deflection under these applied loads, for specimens MI1 to MI6, are presented in Table 5. The results of Monte Carlo simulation of deflection at different stages of loading, for these six specimens, are also presented in this table.

Table 5 Applied loads and corresponding deflections for specimens MI1 to MI6

Specimen designation	Applied load (kN)	Exp. deflection (mm)	Results of Monte Carlo simulation of load-deflection behaviour (Balaji Rao 1990)	
			Mean, μ (mm)	Standard deviation, σ (mm)
MI1	2	0.038	0.015	0.004
	8	0.216	0.317	0.157
	10	0.407	0.555	0.184
	15	0.9	1.154	0.246
MI2	1	0.035	0.024	0.005
	8	1.273	1.542	0.378
	10.41	1.965	2.336	0.460
	13	2.809	3.190	0.549
MI3	2	0.021	0.015	0.003
	10	0.261	0.422	0.204
	15	0.728	1.047	0.270
	17.62	1.02	1.375	0.304
MI4	1	0.036	0.020	0.005
	8	1.4	1.083	0.363
	9.9	1.45	1.645	0.419
	10	1.63	1.673	0.423
MI5	3	0.04	0.020	0.005
	15	0.645	0.699	0.286
	16	0.701	0.820	0.299
	17.79	0.825	1.035	0.321
MI6	2	0.077	0.035	0.008
	9	0.73	0.731	0.478
	9.6	0.932	0.915	0.516
	14	2.8	2.356	0.678

3.4 Stochastic Modelling of P-δ Behaviour

A NHMC model, presented in Sect. 2, is proposed to predict the P-δ behaviour. The state space considered is such that the states are mutually exclusive and collectively exhaustive (here each state represents a range of central deflection). The discretisation of the state space for the specimens MI1-MI6 are given in Tables 6, 7, 8, 9, 10 and 11. The loading stages have been considered, for MI1–MI6, in such a way that, first two stages correspond to elastic behaviour while the latter two stages corresponds to the in-elastic behaviour. The initial state vector, required to compute the unconditional probabilities, are different for elastic and in-elastic stages. The

Table 6 Discretisation of the state space for the specimen MI1

Elastic stage		In-elastic stage	
State number	Range of central deflection (mm)	State number	Range of central deflection (mm)
1	0.000–0.02	1	0.000–0.30
2	0.021–0.04	2	0.301–0.60
3	0.041–0.06	3	0.601–0.90
4	0.061–0.08	4	0.901–1.20
5	0.081–0.10	5	1.201–1.50
6	>0.10	6	>1.50

Table 7 Discretisation of the state space for the specimen MI2

Elastic stage		In-elastic stage	
State number	Range of central deflection (mm)	State number	Range of central deflection (mm)
1	0.000–0.02	1	0.000–0.40
2	0.021–0.04	2	0.401–0.80
3	0.041–0.06	3	0.801–1.20
4	0.061–0.08	4	1.200–1.60
5	0.081–0.10	5	1.601–2.00
6	>0.10	6	>2.00

Table 8 Discretisation of the state space for the specimen MI3

Elastic stage		In-elastic stage	
State number	Range of central deflection (mm)	State number	Range of central deflection (mm)
1	0.000–0.02	1	0.000–0.20
2	0.021–0.04	2	0.201–0.40
3	0.041–0.06	3	0.401–0.60
4	0.061–0.08	4	0.601–0.80
5	0.081–0.10	5	0.801–1.00
6	>0.10	6	>1.00

statistical properties (namely, mean and coefficient of variation) required to compute TPM are obtained from Monte Carlo simulation of central deflection of specimens MI1–MI6, at different stages of loading (Balaji Rao 1990). Four values of $\rho_{k,k+1}$, namely 0.0, 0.5, 0.80 and 0.95, are considered to study the effect of the same on the unconditional probabilities.

Table 9 Discretisation of the state space for the specimen MI4

Elastic stage		In-elastic stage	
State number	Range of central deflection[a] (mm)	State number	Range of central deflection[a] (mm)
1	0.000–0.02	1	0.000–0.35
2	0.021–0.04	2	0.351–0.70
3	0.041–0.06	3	0.701–1.05
4	0.061–0.08	4	1.051–1.40
5	0.081–0.10	5	1.401–1.75
6	>0.10	6	>1.75

[a]For this specimen, the central deflection was not measured properly and hence in the present study deflections measured under the load point was considered

Table 10 Discretisation of the state space for the specimen MI5

Elastic stage		In-elastic stage	
State number	Range of central deflection (mm)	State number	Range of central deflection (mm)
1	0.000–0.02	1	0.000–0.20
2	0.021–0.04	2	0.201–0.40
3	0.041–0.06	3	0.401–0.60
4	0.061–0.08	4	0.601–0.80
5	0.081–0.10	5	0.801–1.00
6	>0.10	6	>1.00

Table 11 Discretisation of the state space for the specimen MI6

Elastic stage		In-elastic stage	
State number	Range of central deflection (mm)	State number	Range of central deflection (mm)
1	0.000–0.04	1	0.000–0.50
2	0.041–0.08	2	0.501–1.00
3	0.081–0.12	3	1.001–1.50
4	0.121–0.16	4	1.501–2.00
5	0.161–0.20	5	2.001–2.50
6	>0.20	6	>2.50

4 Results and Discussion

The unconditional probabilities computed for the specimens MI1 to MI6 (for different values of $\rho_{k,k+1}$) at different stages of loading are presented in Tables 12, 13, 14, 15,

Table 12 Unconditional state probability vectors of specimen MI1 for four different values of correlation coefficient (loading stage 1—before cracking)

State	Rho			
	0.00	0.50	0.80	0.95
1	0	0	0	0
2	**0.416**	**0.23**	**0.045**	**0**
3	0.311	0.426	0.381	0.191
4	0.122	0.268	0.465	0.684
5	0.052	0.062	0.103	0.119
6	0.098	0.014	0.006	0.006

Table 13 Unconditional state probability vectors of specimen MI1 for four different values of correlation coefficient (loading stage 2—before cracking)

State	Rho			
	0.00	0.50	0.80	0.95
1	0	0	0	0
2	0.004	0.002	0	0
3	0.01	0.004	0	0
4	0.015	0.006	0	0
5	0.019	0.008	0.001	0
6	**0.951**	**0.98**	**0.999**	**1**

Table 14 Unconditional state probability vectors of specimen MI1 for four different values of correlation coefficient (loading stage 3—after cracking (nonlinear range))

State	Rho			
	0.00	0.50	0.80	0.95
1	0	0	0	0
2	**0.516**	**0.435**	**0.354**	**0.283**
3	0.381	0.484	0.592	0.682
4	0.047	0.048	0.032	0.014
5	0.016	0.006	0.001	0
6	0.039	0.027	0.021	0.021

16, 17, 18, 19, 20, 21, 22, 23, 24, 25, 26, 27, 28, 29, 30, 31, 32, 33, 34 and 35. In these tables, the experimental deflection state of specimens MI1–MI6, at different stages of loading are highlighted for the purpose of inference. From the results presented in these Tables, it is noted that the predicted states agree satisfactorily with the experimentally observed states. It is noted that in in-elastic range, the predicted states are slightly conservative (though not excessively conservative). The degree

Table 15 Unconditional state probability vectors of specimen MI1 for four different values of correlation coefficient (loading stage 4—after cracking (nonlinear range))

State	Rho			
	0.00	0.50	0.80	0.95
1	0.006	0.005	0.001	0
2	0.125	0.099	0.062	0.039
3	0.403	0.388	0.356	0.282
4	**0.341**	**0.377**	**0.447**	**0.553**
5	0.126	0.131	0.135	0.126
6	0.006	0.005	0.001	0

Table 16 Unconditional state probability vectors of specimen MI2 for four different values of correlation coefficient (loading stage 1—before cracking)

State	Rho			
	0.00	0.50	0.80	0.95
1	0	0	0	0
2	**0.648**	**0.597**	**0.577**	**0.575**
3	0.328	0.39	0.418	0.423
4	0.01	0.007	0.004	0.002
5	0.007	0.003	0	0
6	0.007	0.003	0	0

Table 17 Unconditional state probability vectors of specimen MI2 for four different values of correlation coefficient (loading stage 2—before cracking)

State	Rho			
	0.00	0.50	0.80	0.95
1	0	0	0	0
2	0	0	0	0
3	0	0	0	0
4	0	0	0	0
5	0	0	0	0
6	**1**	**1**	**1**	**1**

of conservatism depends on discretisation scheme used for a given specimen in the in-elastic range. From the results, it is noted that as the value of $\rho_{k,k+1}$ increases the dispersion of pdf decreases. This implies that with the increase in $\rho_{k,k+1}$, there is a tendency for the process to converge to a preferential state.

Table 18 Unconditional state probability vectors of specimen MI2 for four different values of correlation coefficient (loading stage 3—after cracking (nonlinear range))

State	Rho			
	0.00	0.50	0.80	0.95
1	0	0	0	0
2	0	0	0	0
3	0	0	0	0
4	0.05	0.049	0.02	0.001
5	**0.18**	**0.211**	**0.219**	**0.173**
6	0.77	0.74	0.761	0.826

Table 19 Unconditional state probability vectors of specimen MI2 for four different values of correlation coefficient (loading stage 4—after cracking (nonlinear range))

State	Rho			
	0.00	0.50	0.80	0.95
1	0	0	0	0
2	0	0	0	0
3	0	0	0	0
4	0	0.001	0	0
5	0.003	0.01	0.007	0
6	**0.997**	**0.99**	**0.993**	**1**

Table 20 Unconditional state probability vectors of specimen MI3 for four different values of correlation coefficient (loading stage 1—before cracking)

State	Rho			
	0.00	0.50	0.80	0.95
1	0	0	0	0
2	**0.443**	**0.214**	**0.007**	**0**
3	0.348	0.475	0.049	0.242
4	0.111	0.264	0.172	0.687
5	0.04	0.04	0.307	0.069
6	0.058	0.006	0.464	0.002

5 Summary

In this chapter, an attempt has been to model the stochastic evolution of deflection of ferrocement flexural members subjected to monotonic loading. The states with highest probabilities predicted using NHMC compare satisfactorily, both in elastic and in-elastic ranges, with the experimental observations. Hence, NHMC is proposed

Table 21 Unconditional state probability vectors of specimen MI3 for four different values of correlation coefficient (loading stage 2—before cracking)

State	Rho			
	0.00	0.50	0.80	0.95
1	0	0	0	0
2	0.003	0.001	0	0
3	0.007	0.003	0.002	0
4	0.01	0.004	0.018	0
5	0.012	0.005	0.079	0
6	**0.968**	**0.987**	**0.901**	**1**

Table 22 Unconditional state probability vectors of specimen MI3 for four different values of correlation coefficient (loading stage 3—after cracking (nonlinear range))

State	Rho			
	0.00	0.50	0.80	0.95
1	0	0	0	0
2	0.007	0.007	0.001	0
3	0.04	0.049	0.027	0.002
4	**0.129**	**0.172**	**0.197**	**0.169**
5	0.246	0.307	0.423	0.612
6	0.577	0.464	0.352	0.217

Table 23 Unconditional state probability vectors of specimen MI3 for four different values of correlation coefficient (loading stage 4—after cracking (nonlinear range))

State	Rho			
	0.00	0.50	0.80	0.95
1	0	0	0	0
2	0	0	0	0
3	0	0.002	0.002	0
4	0.004	0.018	0.02	0.005
5	0.037	0.079	0.103	0.082
6	**0.958**	**0.901**	**0.875**	**0.912**

to predict the deflection state at any stage of loading. From the results presented in Tables 12, 13, 14, 15, 16, 17, 18, 19, 20, 21, 22, 23, 24, 25, 26, 27, 28, 29, 30, 31, 32, 33, 34 and 35, it is inferred that $\rho_{k,k+1} = 0.95$ can be used in the modelling. The results presented in the Appendix are useful since the Chebyshev inequality can be used to obtain the first cut information about the interval within which the actual

Table 24 Unconditional state probability vectors of specimen MI4 for four different values of correlation coefficient (loading stage 1—before cracking)

State	Rho			
	0.00	0.50	0.80	0.95
1	0	0	0	0
2	**0.866**	**0.849**	**0.841**	**0.84**
3	0.093	0.14	0.159	0.16
4	0.014	0.004	0	0
5	0.014	0.004	0	0
6	0.014	0.004	0	0

Table 25 Unconditional state probability vectors of specimen MI4 for four different values of correlation coefficient (loading stage 2—before cracking)

State	Rho			
	0.00	0.50	0.80	0.95
1	0	0	0	0
2	0	0	0	0
3	0	0	0	0
4	0.001	0	0	0
5	0.001	0	0	0
6	**0.998**	**1**	**1**	**1**

Table 26 Unconditional state probability vectors of specimen MI4 for four different values of correlation coefficient (loading stage 3—after cracking (nonlinear range))

State	Rho			
	0.00	0.50	0.80	0.95
1	0	0	0	0
2	0	0	0	0
3	0	0	0	0
4	0.227	0.169	0.083	0.009
5	**0.345**	**0.351**	**0.391**	**0.405**
6	0.428	0.48	0.526	0.586

value of deflection would lie. In most cases the lower unsymmetrical bound seem to corroborate well with the experimental value.

Table 27 Unconditional state probability vectors of specimen MI4 for four different values of correlation coefficient (loading stage 4—after cracking (nonlinear range))

State	Rho			
	0.00	0.50	0.80	0.95
1	0	0	0	0
2	0	0	0	0
3	0	0	0	0
4	0.048	0.057	0.04	0.006
5	**0.229**	**0.238**	**0.262**	**0.307**
6	0.722	0.705	0.697	0.687

Table 28 Unconditional state probability vectors of specimen MI5 for four different values of correlation coefficient (loading stage 1—before cracking)

State	Rho			
	0.00	0.50	0.80	0.95
1	**0**	**0**	**0**	**0**
2	0.858	0.912	0.946	0.954
3	0.053	0.054	0.048	0.046
4	0.03	0.012	0.002	0
5	0.03	0.012	0.002	0
6	0.03	0.012	0.002	0

Table 29 Unconditional state probability vectors of specimen MI5 for four different values of correlation coefficient (loading stage 2—before cracking)

State	Rho			
	0.00	0.50	0.80	0.95
1	0	0	0	0
2	0.002	0.001	0	0
3	0.002	0.001	0	0
4	0.003	0.001	0	0
5	0.003	0.002	0.001	0
6	**0.99**	**0.995**	**0.999**	**1**

Table 30 Unconditional state probability vectors of specimen MI5 for four different values of correlation coefficient (loading stage 3—after cracking (nonlinear range))

State	Rho			
	0.00	0.50	0.80	0.95
1	0	0	0	0
2	0	0	0	0
3	0	0	0	0
4	**0.316**	**0.334**	**0.376**	**0.412**
5	0.327	0.349	0.414	0.527
6	0.357	0.317	0.21	0.062

Table 31 Unconditional state probability vectors of specimen MI5 for four different values of correlation coefficient (loading stage 4—after cracking (nonlinear range))

State	Rho			
	0.00	0.50	0.80	0.95
1	0	0	0	0
2	0	0	0	0
3	0	0	0	0
4	0.055	0.069	0.081	0.064
5	**0.191**	**0.209**	**0.26**	**0.333**
6	0.755	0.722	0.659	0.603

Table 32 Unconditional state probability vectors of specimen MI6 for four different values of correlation coefficient (loading stage 1—before cracking)

State	Rho			
	0.00	0.50	0.80	0.95
1	0	0	0	0
2	**0.854**	**0.916**	**0.928**	**0.928**
3	0.052	0.062	0.072	0.072
4	0.031	0.007	0	0
5	0.031	0.007	0	0
6	0.031	0.008	0	0

Table 33 Unconditional state probability vectors of specimen MI6 for four different values of correlation coefficient (loading stage 2—before cracking)

State	Rho			
	0.00	0.50	0.80	0.95
1	0	0	0	0
2	0.013	0.01	0.007	0.002
3	0.015	0.012	0.008	0.004
4	0.017	0.013	0.01	0.005
5	0.02	0.015	0.012	0.008
6	**0.935**	**0.95**	**0.963**	**0.981**

Table 34 Unconditional state probability vectors of specimen MI6 for four different values of correlation coefficient (loading stage 3—after cracking (nonlinear range))

State	Rho			
	0.00	0.50	0.80	0.95
1	0	0	0	0
2	**0.366**	**0.4**	**0.483**	**0.582**
3	0.32	0.344	0.377	0.373
4	0.136	0.118	0.06	0.007
5	0.048	0.037	0.019	0.004
6	0.13	0.101	0.061	0.034

Table 35 Unconditional state probability vectors of specimen MI6 for four different values of correlation coefficient (loading stage 4—after cracking (nonlinear range))

State	Rho			
	0.00	0.50	0.80	0.95
1	0	0	0	0
2	0.007	0.006	0.001	0
3	0.055	0.044	0.025	0.004
4	0.162	0.142	0.146	0.153
5	0.251	0.248	0.306	0.408
6	**0.525**	**0.56**	**0.521**	**0.435**

Acknowledgements The author is grateful to his Ph.D. thesis guide, late Prof. Prakash Desayi, and to former and present Directors of CSIR-SERC, Chennai, for giving the encouragement. He is thankful to his colleague Dr M. B. Anoop. He is thankful to the Director, CSIR-SERC, Prof. Santosh Kapuria and CSIR for the Emeritus scientist award.

Appendix: Application of Chebyshev Inequalities (Balaji Rao et al. 2013)

In this appendix, the Chebyshev inequalities, presented by Steliga and Szynal (2010) are used to predict the bounds on the deflection of ferrocement members at different stages of loading.

Chebyshev Inequality—Some Preliminaries

Let X be a random variable representing an action or response quantity. Example of action quantity can be load (or loading intensity), external bending moment or external traction force. The response quantity can be deflection, rotation, warping, strain, crack width. In most engineering applications we may not be knowing the actual probability density function (pdf) of X; yet, we will be asked to answer questions like $P[g(x) \geq r] = ?$. It may be noted that g(X) is a function of random variable and r is a specified value. Such decision making probabilities are required in limit state design of structural components (viz. Ang and Tang 1984).

In the face of non-availability of pdf of X can we make probabilistic inferences about $P[g(X) \geq r]$. It can be shown that (viz. Gnedenko 1976)

$$P[g(X) \geq r] \leq \frac{E[g(X)]}{r} \tag{14}$$

Let $g(x) = [x - \mu]^2$; $r = z^2\sigma^2$ where μ and σ are mean and standard deviation of X. According to Chebyschev's inequality this probability computed from (Gnedenko 1976)

$$P[(X - \mu)^2 \geq z^2\sigma^2] \leq \frac{E[(X - \mu)^2]}{z^2\sigma^2}$$

$$\text{(or)}\ \ P[(X - \mu)^2 \geq z^2\sigma^2] \leq \frac{1}{z^2} \tag{15}$$

Therefore, Chebyshev's inequality gives weak upper bound on the desired probability. We note from Eq. (15) that z should be greater than or equal to 1 since if $z <$ 1 (though positive) the interpretation as the value of probability will not be proper. Also, since we are not having any information on type of pdf of X the bounds will be weak. Efforts have been made in the literature to sharpen the bounds and to determine two-sided bounds and also, to determine bounds for multivariate case (with- and without-correlation effect).

While Eq. (15) gives the one-sided bound, let us consider two-sided case. When the distribution is symmetrical about the mean, the symmetrical bounds around the mean are given by (Steliga and Szynal 2010),

$$P[k_1 < x < k_2] \geq 1 - \frac{4\sigma^2}{(k_2 - k_1)^2} \quad \forall (k_2 - k_1) > 2\sigma \tag{16}$$

When the pdf of the random variable is not symmetrical about the mean, the bounds are given by,

$$P[k_1 < x < k_2] \geq \frac{4[(\mu - k_1)(k_2 - \mu) - \sigma^2]}{(k_2 - k_1)^2} \tag{17}$$

where $k_1 < \mu < k_2; \quad (\mu - k_1)(k_2 - \mu) > \sigma^2$.

Application

In this section two example problems demonstrating the use of Chebyshev inequalities in determining the weak upper bound probabilities, those required for engineering decision making, are presented. One of the highlights of these examples is, to infuse confidence in engineering applications, to compare the results with the respective experimental values.

Example In this example an attempt has been made to estimate the weak upper bound probabilities on random central deflection of ferrocement I-beams used for roofing in low-cost housing. This example is considered since the test data on central deflection, at different stages of loading, was available for six specimens. These specimens were tested at the structural engineering laboratory of Indian Institute of Science, Bangalore, in 1980s. The details of tests and the test results are presented in Sect. 3.3. Also, an effort was made to determine statistical properties of deflections using Monte Carlo simulation technique. More details about basic random variables considered and details of simulation are presented in Balaji Rao (1990). The final results of simulation (viz. mean and standard deviations of deflection) for six specimens considered here, at different stages of loading, are presented in Table 5. Also presented in this table are experimental values of central deflections. The weak upper bound probabilities associated with bounds of lengths 2.25σ, 2.5σ, 2.75σ and 3σ are computed using Chebyshev inequalities. These probabilities are computed for two conditions: (a) assuming that the pdf of deflection, at different stages of loading, are symmetrical about the mean, and, (b) assuming that the pdf of deflection is unknown or unsymmetrical about the mean. The values of the bounds and their corresponding probabilities are presented in Table 36 typically for first two interval lengths. Since a bound of length 3σ is very often used in engineering decision making, the same are compared for the cases (a) and (b) in Figs. 2, 3, 4, 5, 6 and 7 for the specimens considered. Also shown in these figures are the experimentally observed deflections. From these figures it is observed that at almost all stages of loading, the estimated bounds contain the observed deflection suggesting that the estimated weak upper bound probabilities are acceptable and can be used in engineering decision making. If it is

Table 36 Bounds of different interval lengths and their comparison with experimental results

Specimen designation	Applied load (kN)	Exp. deflection (mm)	Bounds—symmetrical (length of interval = 2.25σ)[1]		Bounds—unsymmetrical (length of interval = 2.25σ)[2]	
			Lower bound ($\mu - 1.125\sigma$) (mm)	Upper bound ($\mu + 1.125\sigma$) (mm)	Lower bound ($\mu - \sigma$) (mm)	Upper bound ($\mu + 1.25\sigma$) (mm)
MI1	2	0.038	0.011	0.019	0.012	0.019
	4	0.069	0.010	0.056	0.012	0.058
	8	0.216	0.140	0.493	0.160	0.513
	10	0.407	0.348	0.762	0.371	0.785
	15	0.9	0.878	1.431	0.908	1.462
MI2	1	0.035	0.018	0.030	0.018	0.035
	1.5	0.075	0.026	0.046	0.027	0.054
	8	1.273	1.117	1.967	1.164	2.297
	10.41	1.965	1.818	2.854	1.876	3.257
	13	2.809	2.370	4.013	2.641	4.287
MI3	2	0.021	0.011	0.019	0.011	0.020
	4.5	0.065	0.015	0.055	0.018	0.058
	10	0.261	0.192	0.651	0.217	0.677
	15	0.728	0.743	1.350	0.777	1.384
	17.62	1.02	1.033	1.717	1.071	1.755
MI4	1	0.036	0.015	0.025	0.015	0.026
	1.5	0.05	0.023	0.038	0.023	0.039
	8	1.4	0.675	1.491	0.720	1.536
	9.9	1.45	1.173	2.116	1.225	2.169
	10	1.63	1.196	2.148	1.249	2.201
MI5	3	0.04	0.015	0.025	0.016	0.026
	4	0.062	0.012	0.034	0.020	0.035
	15	0.645	0.378	1.021	0.413	1.057
	16	0.701	0.483	1.156	0.520	1.193
	17.79	0.825	0.674	1.396	0.714	1.436
MI6	2	0.077	0.026	0.045	0.027	0.045
	3	0.12	0.038	0.068	0.040	0.070
	9	0.73	0.193	1.268	0.253	1.328
	9.6	0.932	0.334	1.496	0.399	1.560
	14	2.8	1.593	3.119	1.677	3.204

Note 1, 2—associated weak upper bound probabilities are 0.21 and 0.20, respectively

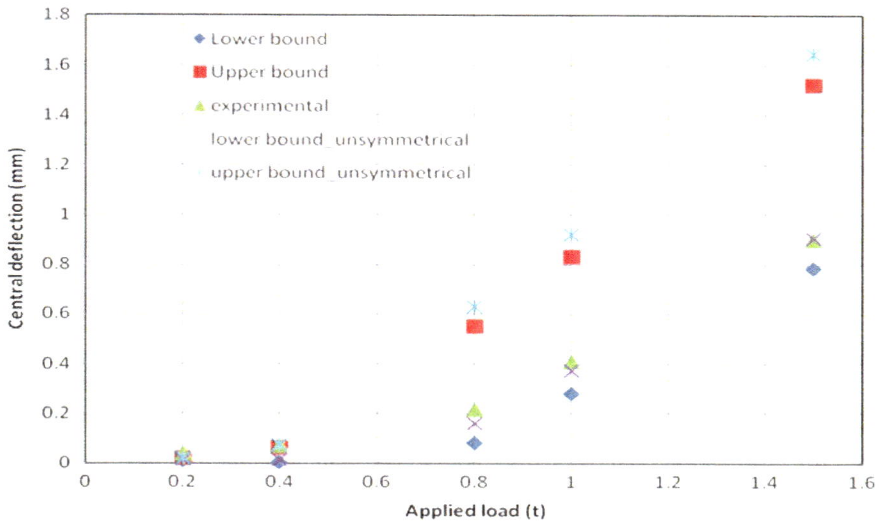

Fig. 2 Load versus deflection plot for specimen MI1 with ($\mu - 1.5\sigma$, $\mu - 1.5\sigma$) symmetrical bounds (associated minimum probability 0.56) and ($\mu - \sigma$, $\mu + 2.0\sigma$) for unsymmetrical bounds (associated minimum probability 0.44) with experimental values

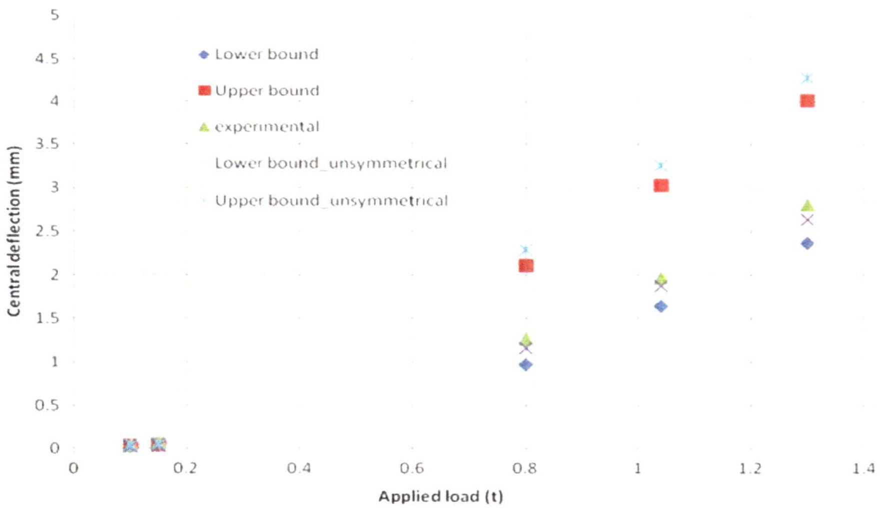

Fig. 3 Load versus deflection plot for specimen MI2 with ($\mu - 1.5\sigma$, $\mu - 1.5\sigma$) symmetrical bounds (associated minimum probability 0.56) and ($\mu - \sigma$, $\mu + 2.0\sigma$) for unsymmetrical bounds (associated minimum probability 0.44) with experimental values

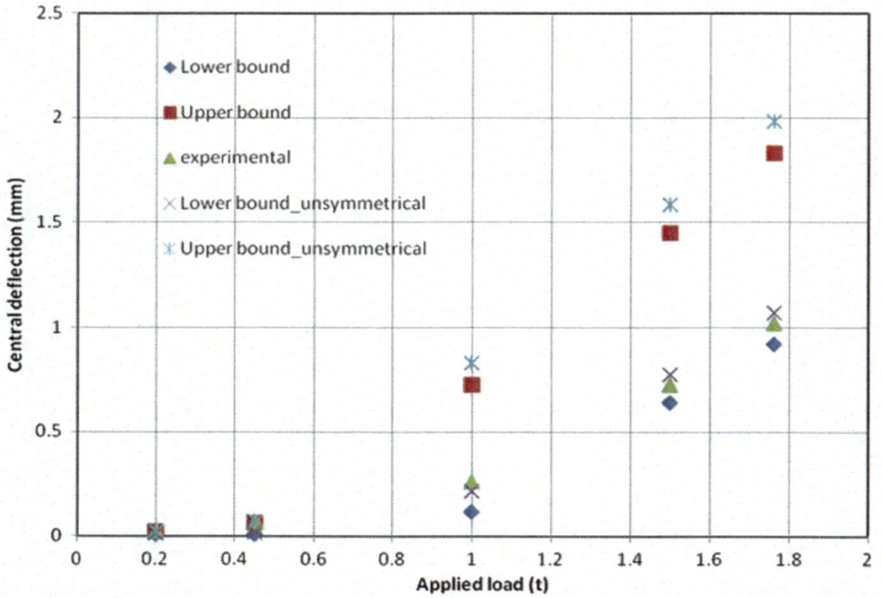

Fig. 4 Load versus deflection plot for specimen MI3 with ($\mu - 1.5\sigma$, $\mu - 1.5\sigma$) symmetrical bounds (associated minimum probability 0.56) and ($\mu - \sigma$, $\mu + 2.0\sigma$) for unsymmetrical bounds (associated minimum probability 0.44) with experimental values

felt that the length of interval of 3σ is high, from Table 36 it is noted that, at higher stages of loading, even though the lengths of interval are small, the bounds enclose the experimentally observed deflections. It may be noted that the weak upper bound probabilities for the two intervals presented in Table 36 vary between 20 and 21% (which is small though). *These observations suggest that the Chebyshev's inequalities can be used for getting an idea about the interval within which the actual value of deflection would lie. The bounds are wide since the coefficient of variation of deflection at different stages of loading is high (Table 5). In most cases the lower unsymmetrical bound seem to corroborate well with the experimental value. Hence, the NHMC model proposed in this chapter, with $\rho_{k,k+1} = 0.95$ is suggested for better prediction of the deflection state of the beam.*

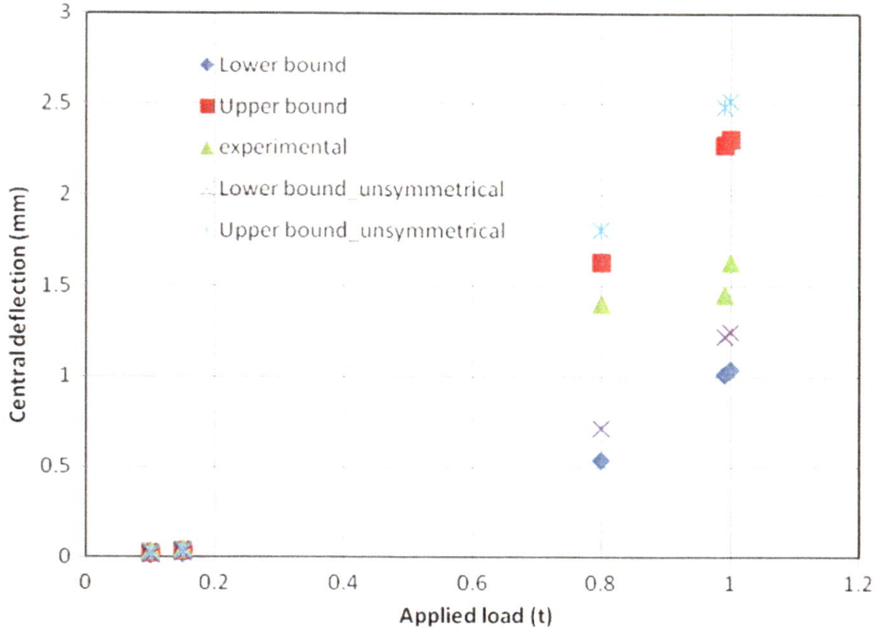

Fig. 5 Load versus deflection plot for specimen MI4 with ($\mu - 1.5\sigma$, $\mu - 1.5\sigma$) symmetrical bounds (associated minimum probability 0.56) and ($\mu - \sigma$, $\mu + 2.0\sigma$) for unsymmetrical bounds (associated minimum probability 0.44) with experimental values

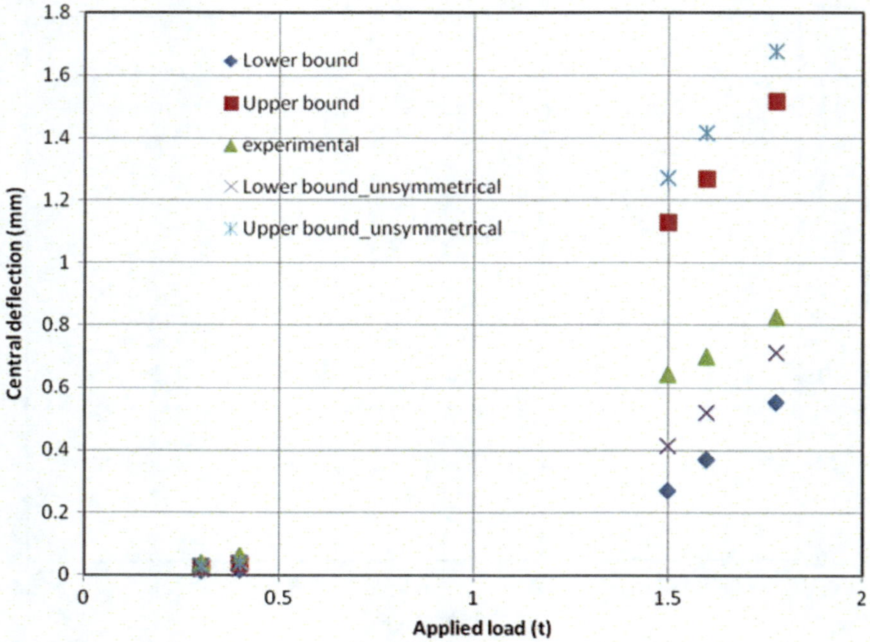

Fig. 6 Load versus deflection plot for specimen MI5 with ($\mu - 1.5\sigma$, $\mu - 1.5\sigma$) symmetrical bounds (associated minimum probability 0.56) and ($\mu - \sigma$, $\mu + 2.0\sigma$) for unsymmetrical bounds (associated minimum probability 0.44) with experimental values

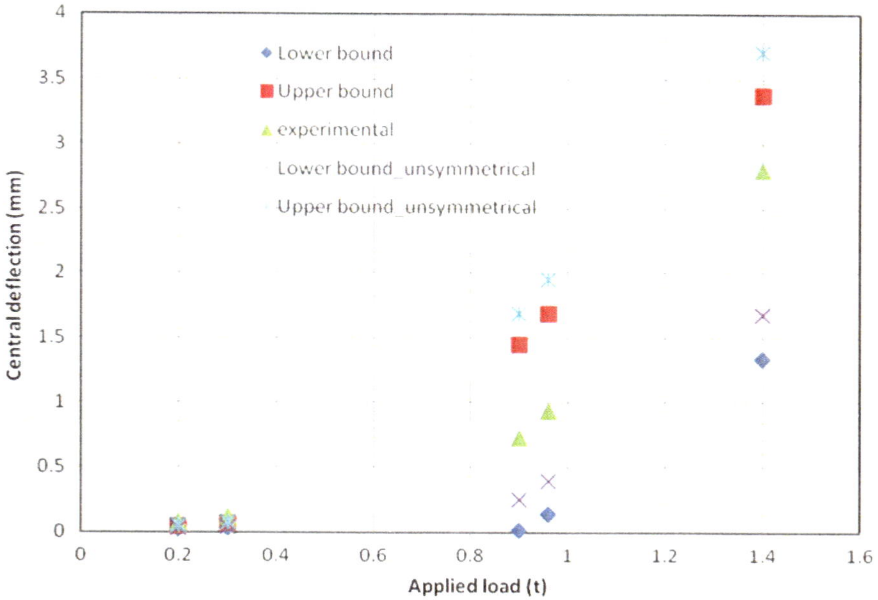

Fig. 7 Load versus deflection plot for specimen MI6 with ($\mu - 1.5\sigma$, $\mu + 1.5\sigma$) symmetrical bounds (associated minimum probability 0.56) and ($\mu - \sigma$, $\mu + 2.0\sigma$) for unsymmetrical bounds (associated minimum probability 0.44) with experimental values

References

ACI 549.1R (2018) Design guide for ferrocement, 32 pp

Ang AH-S, Tang WH (1984) Probability concepts in engineering planning and design, vol II. Wiley, New York

Anoop MB, Balaji Rao K, Lakshmanan N, Raghuprasad BK (2012) Markov chain modelling for evolution of strains in reinforced concrete flexural members. Mater Constr 62(307):443–453

Balaji Rao K, Appa Rao TVSR (1999) A methodology for condition assessment of RC bridge girders with limited inspection data. Bridge Struct Eng Ing-IABSE 29(4):13–26

Balaji Rao K, Appa Rao TVSR (2004) Stochastic modelling of crackwidth in reinforced concrete beams subjected to fatigue loading. Eng Struct 26(5):665–673

Balaji Rao K, Anoop MB, Lakshmanan L, Gopalakrishnan S, Appa Rao TVSR (2004) Risk-based remaining life assessment of corrosion affected reinforced concrete structural members. J Struct Eng 31(1):51–64

Balaji Rao K (1990) Studies on reliability of reinforced concrete beams in cracking and ferrocement elements in tension and flexure. Ph.D. thesis, Department of Civil Engineering, Indian Institute of Science, Bangalore, 872 pp

Balaji Rao K, Anoop MB, Iyer, NR (2013) Application of Chebyshev- and Markov- type inequalities in structural engineering Rel The App 8(1):1–11

Benjamin JR, Cornell AC (1970) Probability, statistics and decision for civil engineers, 1st edn. McGraw-Hill, New York

Bogdanoff JL, Kozin F (1985) Probabilistic models of cumulative damage

Desayi P, Balaji Rao K (1988) Prediction of cracking and ultimate moments, and, load-deflection behaviour of ferrocement elements. In: Third international symposium on ferrocement, New Delhi, proceedings, pp 90–98

Desayi P, Balaji Rao K (1989) Markov chain model for cracking of RC beams. J Struct Eng ASCE 115(9):2129–2144

Desayi P, Balaji Rao K (1993) Estimation of spacing and width of cracks in ferrocement monolithic I-joists. J Ferrocement 23(1):1–13

Ditlevsen O, Madsen HO (1996) Structural reliability methods. Wiley, New York

Geetha Manjari K, Balaji Rao K, Sivakumar Babu GL (2014) Stochastic model for settlement: footings on cohesive soils. GeoRisk 8(4):269–283

Gnedenko BV (1976) The theory of probability. English translation, Mir Publishers, Moscow

Kapur JN (1993) Maximum entropy models in science and engineering. Wiley Eastern Limited, New Delhi

Nowak AS, Collins KR (2000) Reliability of structures. McGraw-Hill, New York

Paul BK, Pama R (1978) Ferrocement. International Ferrocement Information Center, Asian Institute of Technology, 149 pp

Steliga K, Szynal D (2010) On Markov-type inequalities. Int J Pure Appl Math 58(2):137–152

Yogalakshmi NJ, Balaji Rao K (2020) Stochastic model for service life design of reinforced concrete girders against limit state of shear (in preparation)

Pseudo-Ductility through Progressive Rupture of Basalt Fiber-Reinforced Polymer Tendons in Partially Prestressed Functionally-Graded Concrete Beam

Ali Alraie, Nikhil Garg, and Vasant Matsagar

Abstract Corrosion of steel tendons/reinforcements has been a major concern in the prestressed concrete (PSC) and reinforced concrete (RC) beams used in bridge construction. In order to overcome this drawback, basalt fiber-reinforced polymer (BFRP) tendons have been introduced to replace the steel tendons/reinforcements. However, the major limitation with such BFRP composites is their brittle failure without showing any yielding, i.e. lack of ductility. Most of the codes/standards recommend to design the beams prestressed/reinforced with the fiber-reinforced polymer (FRP) as an over-reinforced section to utilize the ductility of concrete in its nonlinear range giving adequate warning prior to the complete failure through extensive cracking and deformations. However, such over-reinforced section need not necessarily provide sufficient ductility to the structure. In contrast, the present study is focused on the FRP beam designed as an under-reinforced section with some techniques to avoid the sudden failure and introduce a type of pseudo-ductility based on the nonlinear behavior of the PSC beam. The beams are designed in such a way that progressive ruptures of the BFRP tendons occur, i.e. they rupture one after the other in succession, resulting in a progressive decrease in the stiffness of the beam, and thereby introducing additional nonlinearity. The ductility is defined in terms of the energy ratio, which is enhanced owing to the progressive rupture of the BFRP tendons. The present analytical investigation shows that the BFRP-prestressed concrete beam with multiple layers of the BFRP tendons, vertically distributed across the depth of the section, and/or partial prestressing applied result in a significant nonlinearity contributing to the additional inelastic energy, which improves the ductility significantly. Additionally, the functionally-graded concrete section proved effective in

A. Alraie
Department of Civil Engineering, Indian Institute of Technology (IIT) Delhi, Hauz Khas, New Delhi 110 016, India

N. Garg
Department of Civil Engineering, Graphic Era University, Dehradun, Uttarakhand 248 002, India

V. Matsagar (✉)
Dogra Chair Professor, Department of Civil Engineering, Indian Institute of Technology (IIT) Delhi, Hauz Khas, New Delhi 110 016, India
e-mail: matsagar@civil.iitd.ac.in

S. B. Singh et al. (eds.), *Emerging Trends of Advanced Composite Materials in Structural Applications*, Composites Science and Technology,
https://doi.org/10.1007/978-981-16-1688-4_4

improving the ductility while maximizing the load-carrying capacity, by providing more depth to the cross-section to compensate the capacity lowered due to the partial prestressing and/or the tendons layered towards the neutral axis.

1 Introduction

Nowadays, prestressed concrete systems are being widely used for the construction of structures such as high-rise buildings, stadiums, bridges, and similar major infrastructure projects. Detailed investigation on the features and characters of the prestressed concrete (PSC) members is crucial in assessing their properties to withstand working and extreme loading conditions and in the determination of safe load-carrying capacities. However, the corrosion-induced deterioration of the steel-prestressed and/or reinforced concrete structures has become one of the major issues which can reduce the life-span of structures, particularly that of the bridges, which are lifeline structures. One possible solution for this problem is to use the non-corrosive fiber-reinforced polymer (FRP) tendons/reinforcements in place of the conventional steel reinforcement. The salient features of the FRP reinforcements are the non-corrosiveness along with their high strength, light weight, and ease of installation, which made them appealing alternative to the conventional steel reinforcement, especially for structures exposed to severe/harsh environments. Considerable research has been conducted on the application of carbon fiber-reinforced polymer (CFRP) composite materials in the infrastructure projects (Grace et al. 2008). Particularly, PSC bridges with the CFRP tendons have been extensively investigated for their structural performance in the laboratory conditions by Grace et al. (2010a, b, 2013a, b).

An alternate non-corrosive reinforcing material in reinforced concrete (RC) and PSC structures is basalt fiber-reinforced polymer (BFRP), which has been the newest FRP material recently introduced to the construction industry owing to its superior properties and lower cost as compared to the CFRP composites. The non-conductivity of the BFRP composites in magnetic and electrical fields makes them an attractive alternative to the conventional steel reinforcement. Also, in the applications sensitive to electromagnetic fields, such as magnetic resonance imaging units, they prove to be effective. There are various forms of basalt fibers that could be utilized in civil engineering constructions, ranging from chopped fibers, fabrics, laminates, rods, tendons, and strands made from the basalt fibers. Evaluating mechanical properties of the products made from the basalt fibers helps in designing the structural members with such FRP composites (Jalasutram et al. 2017).

It is evident from the comparison of the stress–strain curve of steel bars and FRP bars, as shown in Fig. 1 (Zhishen et al. 2012), that though the BFRP bars have high strength as compared to steel bars, they are not ductile enough and do not deform plastically under the applied tensile stress. Hence, they do not have

Fig. 1 Comparison of steel with different FRP composites

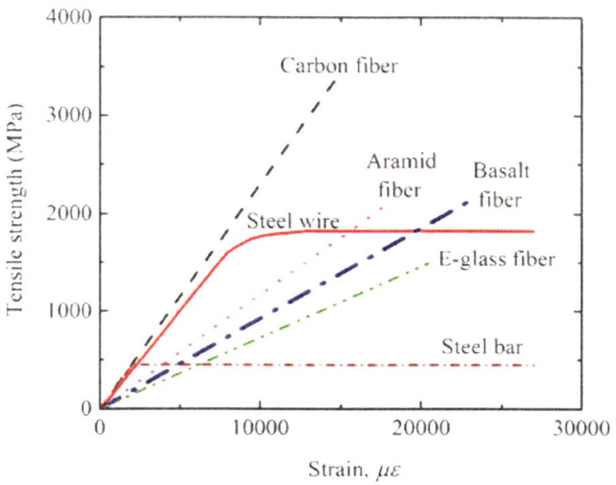

inherent ability to dissipate internal strain energy, like steel bars, which is considered as a major limitation in structural applications. The large structural deformations and significant load-carrying capacity prior to ultimate failure is very critical in civil engineering structures, otherwise a sudden failure causes catastrophe with lack of warning. Hence, the designers of the BFRP composite structures must look for alternative means to dissipate the internal strain energy, and design the structure to undergo large deformations along with carrying the anticipated design loads. This can be attained by rational design of the BFRP composite member, making the composite structure to fail in a controlled progressive manner, and hence resulting in gradual dissipation of the internal strain energy. The over-reinforced design, recommended by ACI 440.4R (2004) and ACI 440.5 (2008) provisions for the concrete beams reinforced with the FRP reinforcement, may not necessarily provide sufficient ductility to the structure even though concrete is more ductile than FRP. If the design philosophy for the RC members and the PSC members with the conventional steel is compared with the FRP-reinforced/prestressed concrete members, the notable fundamental difference relates to the under-reinforced and over-reinforced flexural members recommended, respectively. In the steel-prestressed/reinforced concrete members, the brittle behavior exhibited by concrete during failure is avoided by designing an under-reinforced section, whereby yielding of steel reinforcements is relied upon to provide required ductility. Hence, in case of the PSC members with the FRP as well, the brittle failure of the concrete can still be avoided without compromising with the desired ductility to be achieved by employing inventive techniques suitably.

In this present work, the proposed design is focusing on the partial prestressing and layering of the tendons across the depth of an under-reinforced section, enabling a type of progressive failure of the PSC beam with FRP reinforcements, rupturing sequentially. Both the techniques do help improve the failure behavior by introducing pseudo-ductility, albeit with reduced load-carrying capacity of the beam, which can be compensated using the functionally-graded concrete, by using a high-grade concrete for the compression zone and low-grade concrete for the tension zone. The use of low-grade concrete for the tension zone enables increasing the depth of the cross-section while being economically equivalent to the plain cross-section fully made from high-grade concrete. This extra depth of the cross-section increases the lever arm, and hence, compensates the reduction in the load-carrying capacity due to the partial prestressing and/or layering of tendons. The functionally-graded concrete with higher depth of cross-section increases the shear capacity as well, albeit partial lower grade of concrete possesses low shear strength.

2 BFRP in Prestressed Concrete

For prestressing purpose, BFRP rods and strands are available in different diameters and forms. The characteristics of the BFRP reinforcements and prestressed concrete are discussed here.

2.1 Characteristics of BFRP

The mechanical properties of the BFRP composites have been improved over the recent years with advancements in the production technology. The ultimate tensile strength and elastic modulus of the BFRP sheets have reached about 2100 MPa and 90 GPa, respectively. For the BFRP tendons with vinyl ester, the tensile strength and elastic modulus have reached 1350 MPa and 55 GPa, respectively; whereas, the same have reached 1500 MPa and 50 GPa for the tendons with epoxy resin. Out of the different types of the FRPs, the glass fiber-reinforced polymer (GFRP) has the lowest cost, and the carbon fiber-reinforced polymer (CFRP) has the highest cost with highest strength and elastic modulus though, which depend upon the type of fibers and matrix used in the composite. However, the BFRP composites have better structural properties than the GFRP and lesser cost and strength than the CFRP (Zhishen et al. 2012). The density of the BFRP is about 2.75 g/cm^3, and the friction coefficient may be between 0.42 and 0.50.

2.2 Prestressed Concrete (PSC)

Prestressed concrete is the most common type of concrete used today for the bridge construction purposes and similar mega-infrastructure development. It is a method of overcoming concrete's inherent weakness in tension using external addition of compression force typically. Prestressed concrete technology is used to produce beams, columns, slabs, and alike structural members of large spans as it is lighter in weight as compared to the normal RC structural elements. It consists of mainly prestressing tendons composed of high tensile strength steel strands/cables or rods, which apply the prestressing force. This produces an initial compressive stress in the member section that balances the tensile stress experienced by the structural element under service loads. Prestressed concrete beams are particularly durable in nature as the cracking in concrete is reduced due to the prevailing compressive stresses and are also adaptable and fire resistant. Since reinforced concrete is less durable and heavy in weight, they can only be constructed for shorter spans.

Prestressing can be accomplished in three different systems, which are: (a) pre-tensioned, (b) bonded post-tensioned (grouted), and (c) unbonded post-tensioned (ungrouted) concrete systems, either internally or externally. Prestressed members can be fully or partially prestressed, where reinforcing elements can be judiciously decided for the level of prestressing force to be applied. The fully prestressed beams possess higher capacity and exhibit lesser deflection than the partially prestressed beams. This is because the maximum capacity involves the full contribution of all the prestressing tendons simultaneously in the fully prestressed beams, whereas in the partially prestressed beams, the prestressing tendons reach their ultimate tensile strength and rupture before the non-prestressing bars reach their ultimate tensile strength.

2.3 *BFRP Over Steel Reinforcements*

Steel reinforcements are relatively cost-efficient, which is also attributed their extensive use; however, they are vulnerable to oxidation when exposed to chlorides. Marine areas and snow-prone roads, where salt is used for deicing, are examples of the most susceptible areas to reinforcement corrosion. When the reinforcements are properly protected against the attack of chlorides, they last for decades without any signs of failure; however, such protection to reinforcement is not ideally possible and not completely reliable too. The lack of proper cover to concrete, improper concrete mix, poor workmanship, and presence of reactive agents can result in deterioration of steel reinforcements, and thereby resulting in spalling and cracking of concrete, which leads to unsafe and unserviceable structure. Several parts of the world are facing serious problem of ageing infrastructure primarily attributed to the more deterioration experienced in the steel reinforcement than other construction materials used conventionally.

The BFRP reinforcing bars have a potential to overcome the corrosion-related deficiencies; however, they suffer from the major drawback of non-ductility. This raises the concern on the degree of safety of the member reinforced with the BFRP composites in structural engineering applications. So far, the concrete members reinforced/prestressed with the BFRP bars/tendons have been designed as over-reinforced structure due to the lack of ductility of the BFRP bars, and the ductility of concrete has been used for giving warning during failure through extensive cracking, spalling of concrete, and visible deformations. However, hardly any studies have been found on the under-reinforced section designed using the BFRP reinforcement particularly, and with the design philosophy proposed herewith employing (a) partial and varying prestressing, (b) layering of the reinforcements across the cross-section, (c) varying the reinforcing bar diameters, and (d) functional grading of concrete. This study has been initiated to find some suitable solution to achieve pseudo-ductility comparable to the conventional RC structural members. Based on this premise, the PSC flexural members with the BFRP composites are required to be analyzed for assessing their ductility with the aforementioned variabilities.

The major objective of this book chapter is to analytically investigate the ductility of the BFRP-prestressed concrete beams and propose achieving a pseudo-ductility for the beams by: (a) partially prestressing the beam and studying the effect of prestress ratio on the ductility, (b) layering of tendons and studying the effect of number of layers on the ductility, and (c) functionally-grading the concrete and investigating the improvement in the ductility and load-carrying capacity as compared to the conventional plain RC cross-section.

3 Ductility and Deformability

Initial studies on the ductility and energy dissipation of the prestressed concrete beam were carried out by Janney et al. (1956), who conducted a comparative study on the strength and flexural behaviors with different straight tendons, particularly on the ultimate load-carrying capacity of beam, deflection recovery, and load–deflection relationship. The high-strength beam gave a high deflection recovered under the permissible load–deflection scenario. A study on the ductility was conducted by Park (1989), who discussed various ductility factors with different scenarios. Matthys and Taerwe (2006) studied the ductility requirements in the existing design guidelines for the flexural concrete members strengthened with the FRPs. The ductility of the flexural concrete members strengthened with the FRP composites decreased with increasing amount of FRP strengthening. Moreover, a limitation in the strengthening material was specified in view of the considered guidelines to restrict the overuse of the FRP composites. Iskhakov and Ribakov (2007) carried out theoretical investigations on two-layer prestressed beams with high-strength concrete in the compression zone and normal strength concrete mixed with steel fibers in the tension zone. The results of the investigations stated that the mix design used for the tension zone resulted in 20% additional plastic energy dissipation as compared to the one-layer section of high strength concrete. Additional studies were carried out by Eurviriyanukul and Askes (2010) on optimizing layout of tendons using configuration forces on determinate and indeterminate beams, which is most relevant to the current study of achieving layer-based ductility in the BFRP-prestressed concrete beam as one of the measures.

Wang and Belarbi (2011) studied the fiber-reinforced concrete (FRC) with FRP-hybrid reinforcements and FRP/plain concrete beams, and evaluated the flexural performance and ductility. The results showed that the crack-width of the plain concrete was greater than that of the FRC beams, whereas the ultimate concrete strain in the plain concrete was smaller than that of the FRC. The ductility was increased by the addition of fibers in the FRP system. Hussein et al. (2012) conducted experimental investigations on the flexural behavior of normal strength concrete (NSC) and high strength concrete (HSC) beams, prestressed with bonded and unbonded tendons, and compared the results with the theoretical analysis in terms of load–deflection relationship. This study concluded that the bonded beams had higher ductility and stiffness than the unbonded beams, and the theoretical estimations of force equilibrium and strain compatibility gave comparable results with the experimental observations. Salem et al. (2013) compared the fully prestressed and partially prestressed concrete beams of HSC to evaluate the effect of the prestressing force. The ductility of the fully prestressed concrete beam was found lesser than that of the partially prestressed one, which is expected. The inelastic portion of the deformation when available is an essential component of the ductility. The conventional ductility index evaluation, which depends on the response at the yielding and ultimate conditions, is not applicable in the case of the FRP-reinforced beams; hence, clear distinction should be made between the ductility and deformability of structures (Burgoyne 2001). The ductility is determined by the plastic behavior that a structure may exhibit before failure, without significant loss in strength, which is expressed by the energy absorbed before collapse. The concept of deformability is related to the amount of deformation that a structure can exhibit prior to failure, without the involvement of the plastic work (Abdelrahman and Rizkalla 1999). Prior to that, Grace et al. (1998) had defined the ductility in beams as the ability to sustain inelastic deformation without loss in the load-carrying capacity, before the failure takes place. With this definition, the ductility may be expressed in terms of energy absorption or deformation.

For steel-reinforced beam, the ductility is calculated as the ratio of the ultimate deformation to the deformation at yield. The two main approaches for evaluating the ductility index of RC beams, given by Park (1989), are as follows.

1. Curvature ductility: the ductility index is given as:

$$\mu_{\Psi} = \frac{\Psi_u}{\Psi_y} \tag{1}$$

   where μ_{Ψ} is the curvature-based ductility index; Ψ_u is the curvature at the ultimate moment; and Ψ_y is the curvature at the yielding moment.
2. Displacement ductility: the ductility index is given as:

$$\mu_d = \frac{\Delta_u}{\Delta_y} \tag{2}$$

   where μ_d is the displacement-based ductility index; Δ_u is the displacement at the ultimate load; and Δ_y is the displacement at the yielding load.

In case of FRP-reinforced/prestressed beams, these approaches are inapplicable as FRPs do not possess a yield point; therefore, researchers have given different approaches in order to quantify the ductility. The most prominent approach is based on the energy dissipation.

3.1 Energy-Based Method

The ductility is defined as the ratio of the inelastic energy to the total energy, expressed by the areas under the entire load–deflection curve. Here, the net elastic energy is separated from the total energy. Grace et al. (1998) proposed a way for quantifying the ductility based on the load–deflection curve, as shown in Fig. 2, in which the slope of the line separating the elastic energy from the inelastic energy can be obtained as

$$S = \alpha\beta\gamma\frac{E_f}{E_s} \times \frac{f_y}{f_{ds}} \times \frac{P_1S_1 + (P_2 - P_1)S_2 + (P_3 - P_2)S_3}{P_3} \quad (3)$$

where P_1, P_2, and P_3 are loads shown in Fig. 2; S_1, S_2, and S_3 are the corresponding slopes; E_f is the elastic modulus of the FRP; E_s is the elastic modulus of steel; f_y is the yield strength of steel; f_{ds} is the design strength of the FRP; α is the stirrup factor; β is the failure mode factor; and γ is the factor of the type of reinforcement. The ductility factors, α, β, and γ, are given in Table 1.

Another method for separating the elastic energy from the inelastic energy can be obtained from the model proposed by Abdelrahman and Rizkalla (1999) for calculating the residual deflection under repeated loads, as shown in Fig. 3. The unloading path can be obtained based on moment of inertia for repeated loading (I_{rep}) which can be calculated as

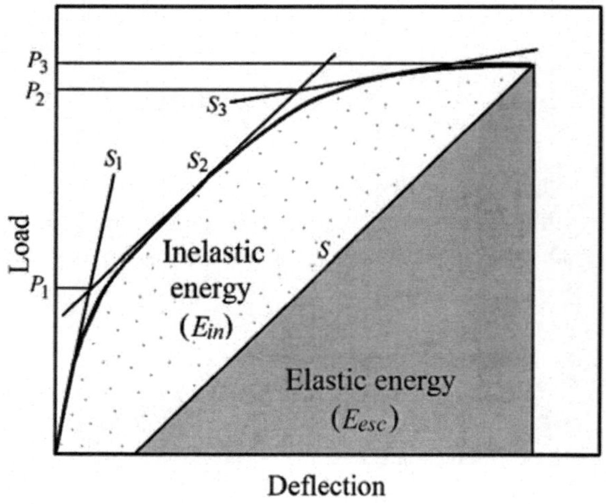

Fig. 2 Elastic and inelastic energies (Grace et al. 1998)

Table 1 Ductility factors (Grace et al. 1998)

Stirrup material	α	Failure mode	β	Reinforcement material	γ
Steel	1.0	Compressive flexure	1.0	Steel	1.0
GFRP	0.95	Flexural shear	0.95	GFRP	4.0
CFRP	0.98	Shear	0.98	CFRP	2.1

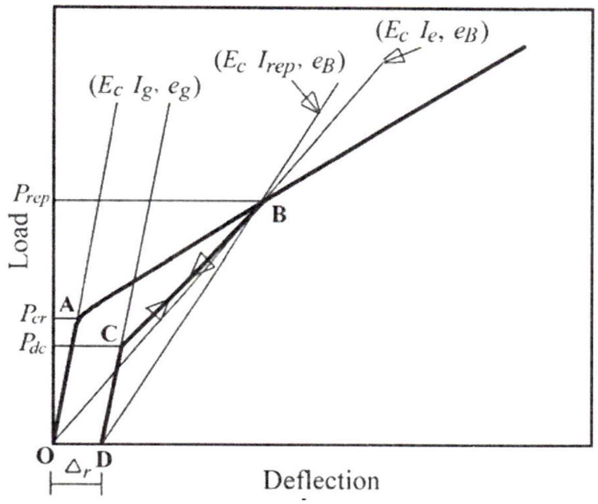

Fig. 3 Proposed model by Abdelrahman and Rizkalla (1999) for load–deflection relationship under repeated loads

$$I_{rep} = \Psi^{2.5} I_g + (1 - \Psi^{2.5}) I_{cr} \le I_g \tag{4}$$

where I_g is the gross moment of inertia; I_{cr} is the cracked moment of inertia; and Ψ is an interpolation factor which can be obtained as

$$\Psi = \left(\frac{M_{cr} - M_{dc}}{M_s - M_{dc}} \right) \tag{5}$$

where M_{cr} is the cracking moment; M_s is the service moment; and M_{dc} is the decompression moment corresponding to zero stress at the extreme bottom fiber. The calculation of both M_{cr} and M_{dc} are discussed later in this study.

The energy-based method proposed by Grace et al. (1998) is used to quantify the ductility of the BFRP-prestressed concrete beams, in which the ductility index, in terms of energy ratio, is determined by the ratio of the absorbed inelastic energy to the total energy as

$$\text{Energy ratio} = \frac{E_{inelastic}}{E_{total}} = \frac{E_{in} + E_{in(addl)}}{E_{in} + E_{esc} + E_{in(addl)}} \tag{6}$$

where $E_{inelastic}$ is the total inelastic energy; E_{in} is the initial inelastic energy absorbed before reaching the ultimate load-carrying capacity; $E_{in(addl)}$ is the additional inelastic energy absorbed after reaching the ultimate load-carrying capacity; E_{esc} is the elastic energy; and E_{total} is the total energy.

The beam is considered to exhibit: (i) a ductile failure if the energy ratio is greater than 75%, (ii) semi-ductile failure for the energy ratio ranging in 70–74%, and (iii) brittle failure for the energy ratio lesser than 69%. In this study, the energy-based approach is adopted for evaluating the ductility of the BFRP-prestressed concrete beams.

3.2 Proposed Solution for Ductility Improvement

The brittle behavior of the BFRP reinforcements is undesirable, which can be overcome by achieving a progressive type of failure in the PSC beam. An experimental work was carried out in the Multi-Hazard Protective Structures (MHPS) Laboratory at Indian Institute of Technology (IIT) Delhi on partially prestressed concrete beams with two layers of BFRP tendons. The beams were designed as an under-reinforced section to evaluate the effectiveness of partial prestressing and layering of tendons on the ductility. The average load–deflection response of the tested PSC beams with the BFRP composite reinforcement is shown in Fig. 4.

It can be seen from Fig. 4 that the partial prestressing and layering of tendons resulted in a significant additional inelastic energy, $E_{in(addl)}$, owing to the successive ruptures of the tendons, and thereby a high energy ratio of 80.1% could be achieved. Hence, the PSC beams with BFRP tendons have been able to exhibit a ductile behavior. However, no studies have been carried out on the techniques to impart ductile behavior into under-reinforced prestressed concrete beams with the

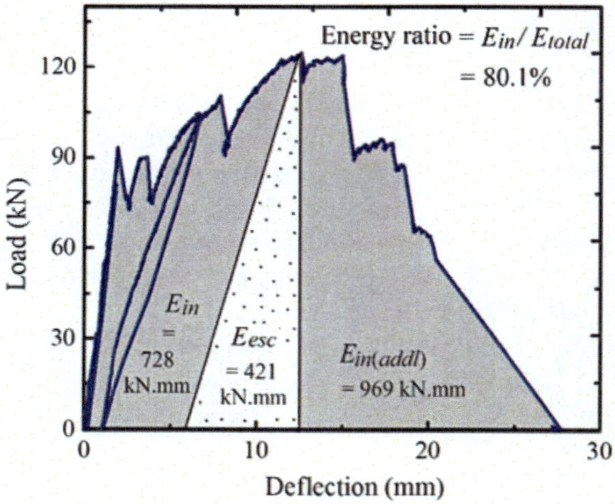

Fig. 4 Load–deflection response of tested BFRP-prestressed concrete beam

BFRP tendons, and to economize the section design simultaneously. Therefore, this present study is undertaken to address this research gap and for proposing a suitable solution to enhance the ductility by conducting a parametric analysis for evaluating the effects of several parameters on the enhancement in energy ratio of the PSC beams.

4 Analytical Investigations

The analysis of the BFRP-prestressed concrete beam is discussed here, incorporating some new measures to increase ductility. The analytical procedure includes the calculation of the cracking moment, ultimate moment capacity of the under-reinforced section, and the deflection under the applied loads. The following procedure is applicable for partially and/or fully prestressed concrete beams with BFRP tendons in single and/or multiple layers, for uniform/plain and/or functionally-graded concrete sections.

4.1 Cracking Moment

The cracking moment of the BFRP-prestressed concrete beams can be obtained as

$$M_{cr} = (f_r + \sum \sigma_{bp})S_b \tag{7}$$

where $f_r = 7.5\sqrt{f'_{c2}}$ (ACI 318-08 2008) is the modulus of rupture; f'_{c2} is the specified compressive strength of concrete in the tension zone expressed in psi unit; and S_b is the section modulus corresponding to the extreme bottom fiber. Moreover, $\sum \sigma_{bp} = \frac{F_{pe}}{A} + \frac{F_{pe} \times e \times y_b}{I}$, where F_{pe} is the total effective prestressing force, after accounting for the prestress losses; A is the cross-sectional area of the structural member; e is the eccentricity of the prestressing force across the cross-section; y_b is the distance from the centroid of the cross-section to the extreme bottom fiber; and I is the moment of inertia of the beam section.

4.2 Flexural Capacity

The flexural capacity of the BFRP-prestressed concrete beam is computed as follows.

4.2.1 Balanced Ratio

The balanced ratio (ρ_b) is determined based on the strain compatibility in the cross-section, and it represents the reinforcement ratio at which the compression failure of concrete and rupture of reinforcement in tension occur simultaneously. The balanced ratio is based on several assumptions, which are: (a) the ultimate compression strain in the extreme concrete fiber is 0.003, (b) the nonlinear behavior of concrete is considered as equivalent rectangular Whitney's stress block, and (c) the ultimate tensile strain in the tendons is ε_{fu}. The balanced ratio is expressed as (ACI 440.4R 2004)

$$\rho_b = 0.85\beta_1 \frac{f'_{c1}}{f_{fu}} \frac{\varepsilon_{cu1}}{\varepsilon_{cu1} + \varepsilon_{fu} - \varepsilon_{pmi}} \tag{8}$$

where β_1 is the ratio of the depth of the equivalent rectangular stress block to the neutral axis depth which is given in Eq. (9); f'_{c1} is the specified compressive strength of the concrete in the compression zone; f_{fu} is the tensile strength of the BFRP prestressing and non-prestressing tendons; ε_{cu1} is the ultimate compression strain in the extreme concrete fiber of the compression zone; ε_{fu} is the ultimate tensile strain of the tendons; and ε_{pmi} is the initial prestressing strain in the lowest prestressing tendons, after accounting for the prestress losses. Here, ACI 440.4R (2004) recommends using

$$\beta_1 = \left(0.85 - \frac{f'_{c1} - 27.5}{6.9} \times 0.05\right) \geq 0.65. \tag{9}$$

4.2.2 Reinforcement Ratio

The reinforcement ratio (ρ) can be obtained as

$$\rho = \frac{\sum_{j=1}^{p} A_{fj}\alpha_j}{bd_m} \tag{10}$$

where A_{fj} is the cross-sectional area of the tendon in a layer j, which is considered as positive in the tension zone and negative in the compression zone; $\alpha_j = f_{bj}/f_{fu}$; f_{bj} is the stress in the tendon in a layer j at the balanced condition; b is the width of the beam; d_m is the distance from the lowest prestressing tendons to the extreme compression fiber; and p is the total number of reinforcing materials.

When $\rho > \rho_b$, the beam section is considered as over-reinforced; however, when $\rho < \rho_b$, the beam section is considered as under-reinforced, which is the focus of this present study.

4.2.3 Depth of Neutral Axis

The depth of the neutral axis can be obtained using the unified design approach or the strain compatibility method, referring to Fig. 5, which illustrates the cross-section, strain, and stress diagrams of the studied PSC beam.

Unified design approach

For an under-reinforced beam section, the depth of the neutral axis (c) can be obtained, as per the unified design approach proposed by Grace and Singh (2003), as $c = k_u d_m$, using the factor k_u defined by

$$k_u = \frac{-B \pm \sqrt{B^2 - 4AC}}{2A} \tag{11}$$

where

$$A = 0.85 f'_{c1} b \beta_1 d_m^2;$$

$$B = -\left[A + F_{pe} d_m + \varepsilon_f d_m \left(\sum_{j=1}^{q} A_{fj} E_{fj}\right)\right];$$

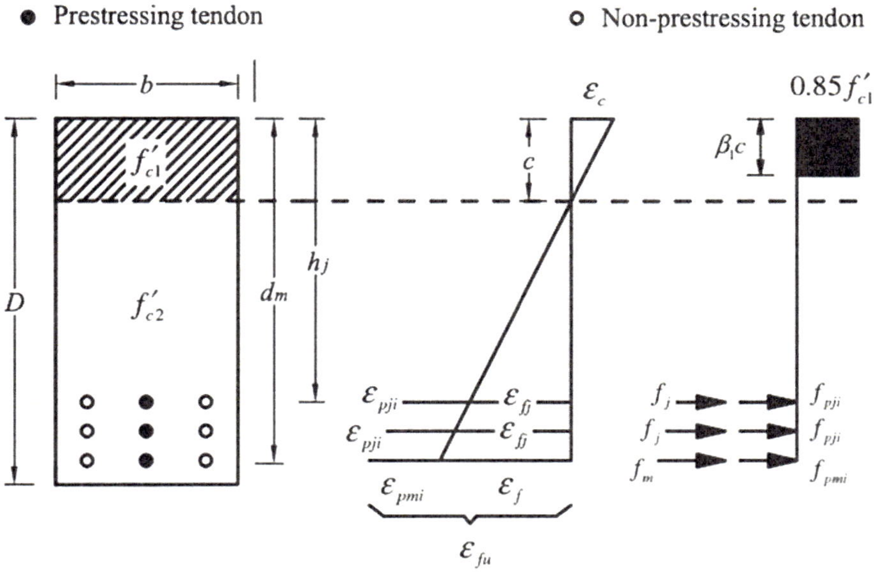

Fig. 5 Cross-section, strain, and stress diagrams of the proposed BFRP-prestressed concrete beam

$$C = \left[F_{pe}d_m - \varepsilon_f\left(\sum_{j=1}^{q} A_{fj}E_{fj}h_j\right)\right];$$

ε_f is the difference between the ultimate rupture strain and initial prestressing strain of the lowest prestressing tendons; E_{fj} is the elastic modulus of the tendons in an individual layer j; F_{pe} is the total effective prestressing force; h_j is the distance of the reinforcement of each individual layer j from the extreme compression fiber; q is number of the tendons layers; and the remaining terms are defined earlier and shown in Fig. 5.

After obtaining the depth of the neutral axis, the strains in the concrete and tendons are obtained as follows. The strain in the prestressing tendons of a layer,

$$\varepsilon_{pj} = (\varepsilon_{fu} - \varepsilon_{pmi})\frac{(h_j - c)}{(d_m - c)} + \varepsilon_{pji}.$$

The strain in the non-prestressing tendons of a layer,

$$\varepsilon_{nj} = (\varepsilon_{fu} - \varepsilon_{pmi})\frac{(h_j - c)}{(d_m - c)}.$$

The strain in the extreme compression fiber of concrete,

$$\varepsilon_c = (\varepsilon_{fu} - \varepsilon_{pmi})\frac{c}{(d_m - c)}$$

where c is the depth of the neutral axis. All the other terms of the above expressions are defined earlier and shown in Fig. 5.

The stresses and forces in the concrete and tendons are obtained as follows. The stresses and forces in the prestressing tendons of a layer,

$f_{pj} = E_f \times \varepsilon_{pj} \le f_{fu}$ and $F_{pj} = f_{pj} \times A_{fj}$, respectively.

The stresses and forces in the non-prestressing tendons of a layer,

$f_{nj} = E_f \times \varepsilon_{nj} \le f_{fu}$ and $F_{nj} = f_{nj} \times A_{fj}$, respectively.

The compression force resultant in the concrete can be obtained as:

$$C_c = 0.85 f'_{c1}\beta_1 bc. \tag{12}$$

Strain compatibility method

The depth of the neutral axis can be obtained using the strain compatibility method by assuming the depth of the neutral axis and calculating the strains, stresses, and forces in the concrete and tendons, as mentioned previously, then subsequently checking for the force equilibrium. Several trials are required till the equation $F_{pj} + F_{nj} = C_c$ is satisfied.

4.2.4 Ultimate Moment-Carrying Capacity

Let F_R be the total tensile force of all the tendons, d be the centroid of the total tensile force, and $\bar{d}$ be the centroid of the concrete compression force, from the extreme concrete top fiber. Then, d can be obtained as

$$d = \frac{\sum_{j=1}^{p} F_j h_j}{\sum_{j=1}^{p} F_j} \tag{13}$$

where F_j are the tensile forces in the prestressing and non-prestressing tendons of a layer. The centroid of the concrete compression force, from the top fiber, is obtained as $\bar{d} = \beta_1 c/2$.

The ultimate moment-carrying capacity (M_u) of the BFRP-prestressed concrete beam can subsequently be obtained as

$$M_u = F_R(d - \bar{d}). \tag{14}$$

4.3 *Deflection*

The beam deflection prior to cracking load can be computed using the following expression in the linear stage. Deflection due to dead load,

$$\delta_d = \frac{5}{384} \frac{W_d L^4}{E_c I_c} \tag{15}$$

where W_d is the uniformly distributed load due to self-weight; L is the span-length of the beam; I_c is the section moment of inertia; and E_c is the elastic modulus of concrete, which should be obtained for the functionally-graded concrete based on the elastic moduli of the high-grade and low-grade concretes as

$$E_c = \frac{E_{c1} c + E_{c2}(D - c)}{D} \tag{16}$$

where E_{c1} is the elastic modulus of the high-grade concrete in the compression zone; E_{c2} is the elastic modulus of the low-grade concrete in the tension zone; c is the neutral axis depth; and D is the overall depth of the beam section.

Deflection due to prestressing force applied at a constant eccentricity,

$$\delta_p = \frac{F_{pe} e L^2}{8 E_c I_c} \tag{17}$$

where all the terms are defined earlier.

Deflection due to live load applied as a concentrated force, P, at let's say the mid-span,

$$\delta_a = \frac{PL^3}{48E_cI_c}. \tag{18}$$

The total deflection of the beam, $\delta = \delta_a + \delta_d - \delta_p$. After cracking, the cracked section analysis is carried out as follows.

4.3.1 Cracked Section Analysis

The deflection of the beam can be obtained, after cracking, based on the effective moment of inertia (I_{eff}) which can be calculated as (Abdelrahman and Rizkalla 1999)

$$I_{eff} = \Psi^3 I_g + (1 - \Psi^3) I_{cr} \leq I_g \tag{19}$$

where I_g is the gross moment of inertia; I_{cr} is the cracked moment of inertia; and Ψ is the interpolation factor which is obtained from Eq. (5), wherein the decompression moment (M_{dc}) can be obtained as

$$M_{dc} = \left(\frac{F_{pe}}{A} + \frac{F_{pe} \times e \times y_b}{I}\right) S_b \tag{20}$$

where all the terms are defined earlier.

Bischoff (2005) proposed an alternative expression for evaluating the effective moment of inertia, for both steel and FRP-reinforced concrete beams, based on a rational concept of providing a weighted average of flexibility instead of stiffness as

$$I_{eff} = \frac{I_{cr}}{1 - \left(1 - \frac{I_{cr}}{I_g}\right)\left(\frac{M_{cr}}{M_s}\right)^2} \leq I_g \tag{21}$$

where all the terms are defined earlier.

The cracked moment of inertia (I_{cr}) is calculated based on the transformed cross-section, as shown in Fig. 6 (ACI 435R-95 2003), using the following equations

$$B = \frac{b}{mA_f} \tag{22a}$$

$$a = \frac{\sqrt{2dB + 1} - 1}{B} \tag{22b}$$

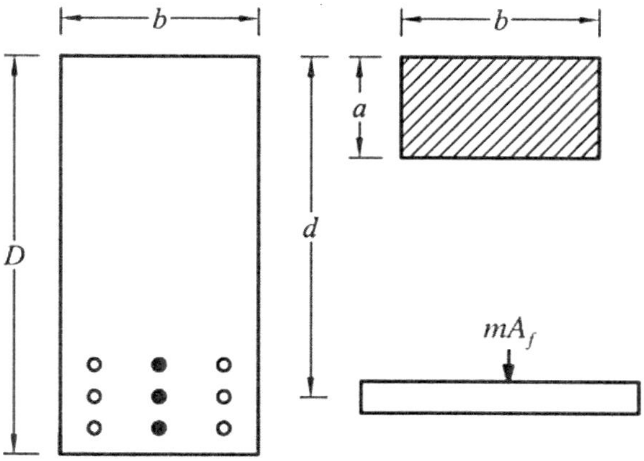

Fig. 6 Transformed cross-section for cracked section analysis

$$I_{cr} = \frac{ba^3}{3} + mA_f(d - a)^2 \quad (22c)$$

where A_f is the area of reinforcement and m is the modular ratio, $m = E_f/E_c$.

The eccentricity of the prestressing force changes after cracking, and is calculated with respect to the centroid of the transformed cross-section. The effective eccentricity of the prestressing force (y_e) is obtained after cracking as

$$y_e = \Psi^2 y_g + (1 - \Psi^2)y_{cr} \leq y_g \quad (23)$$

where $y_g = D/2$ is the uncracked centroidal distance, and y_{cr} is the cracked centroidal distance calculated as

$$y_{cr} = \frac{a \times b \times a/2 + m \times A_f \times d}{a \times b + m \times A_f} \quad (24)$$

where d is the distance from the extreme concrete fiber to the centroid of the reinforcements.

The residual deformation after unloading (Δ_r) can be obtained based on the difference between the calculated deflections using I_{eff}, obtained from Eq. (19) or Eq. (21), and I_{rep}, obtained from Eq. (4). The residual deformation after unloading is expressed as

$$\Delta_r = \left(\frac{5}{384}\frac{W_d L^4}{E_c} - \frac{F_{pe}(d - y_e)L^2}{8E_c} + \frac{PL^3}{48E_c}\right)\left(\frac{1}{I_{eff}} - \frac{1}{I_{rep}}\right) \quad (25)$$

where all the terms are defined earlier.

The load–deflection curve is obtained based on the previous equations, and the ductility index is obtained based on the elastic and inelastic energies, with the help of Eq. (3) for separating the energies under the curve.

5 Validation of the Proposed Model

The analytical approach introduced here is translated in a MATLAB code based on the previously discussed equations and available in the secured data shared along with this book chapter. The results obtained from the code were required to be validated before being adopted for further studies. The validation was conducted with the results of the experimental work carried out at MHPS Laboratory, IIT Delhi, as well as with the results reported in the literature by Johnson (2011) and Guomundsson (2012). Equation (21) was used for calculating the effective moment of inertia for the determination of deflection. All the data used for validation purpose was for the BFRP-reinforced and prestressed concrete beams with under-reinforced section. The validation of the results obtained analytically with all the details of the experimental models is shown in Fig. 7a for results reported by Johnson (2011), Fig. 7b for results reported by Guomundsson (2012), and Fig. 7c for results reported within the current experimental work. It can be seen from Fig. 7 that the results obtained from the analytical approach are in close agreement with the experimental work and literature. The analytically obtained load–deflection curves are coinciding on the cracking and ultimate points with the experimental and literature curves, with marginal difference within 10%. Hence, the proposed approach is deemed valid and can be adopted for further analysis on the ductility evaluation for varying the parameters mentioned in the research objectives.

6 Numerical Study on Ductility

The ductility analysis was conducted using the analytical approach validated earlier. The ductility was evaluated here based on separating the elastic and inelastic energies (Grace et al. 1998), which were obtained from the load–deflection curves and evaluated based on Eq. (3) with reference to Fig. 2 and Table 1. The ductility index, which is the energy ratio, is given in Eq. (6). In order to improve the ductility, two techniques are introduced, to partially prestress the concrete beams and/or to vertically distribute the reinforcements and tendons in multiple layers across the cross-section. In both the techniques, the functionally-graded concrete is introduced to compensate the reduction in the load-carrying capacity.

The effect of the prestress ratio on the ductility was investigated by evaluating the ductility under four different prestress ratios; 35%, 45%, 55%, and 65% of the ultimate tensile strength of the corresponding BFRP tendons. The beam here is fully prestressed, i.e. all the tendons are prestressed to the same ratio. The slope S, which separates the elastic energy from the inelastic energy, is obtained from Eq. (3). Because the material factors in Table 1 are given for steel, CFRP, and GFRP only, and there are no factors given for BFRP; thereupon, it was decided for this study to consider the factors of the GFRP, which has similar properties to the BFRP, as $\alpha = 0.95$ (GFRP), $\beta = 1$, and $\gamma = 4$ (GFRP).

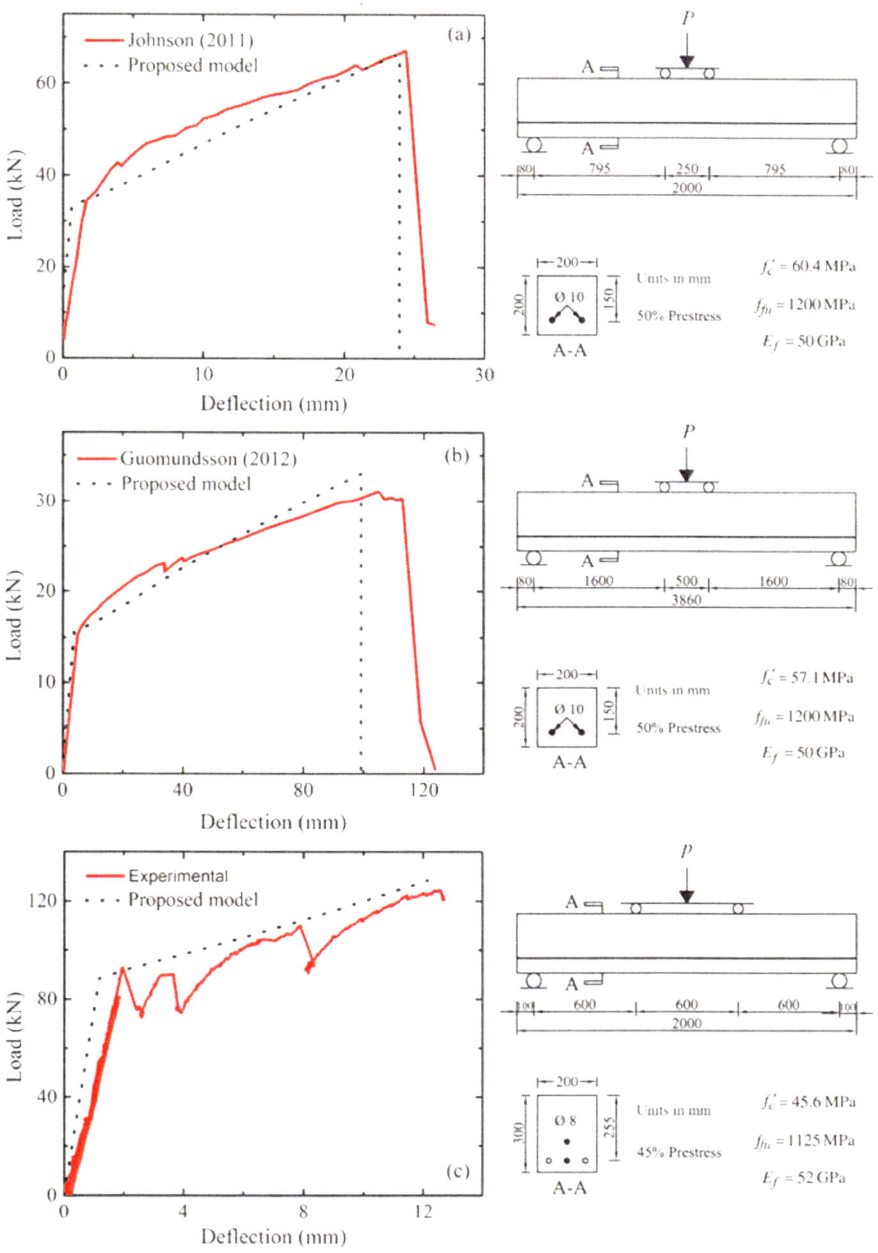

Fig. 7 Validation of the proposed model with: **a** Johnson (2011), **b** Guomundsson (2012), and **c** current experimental work

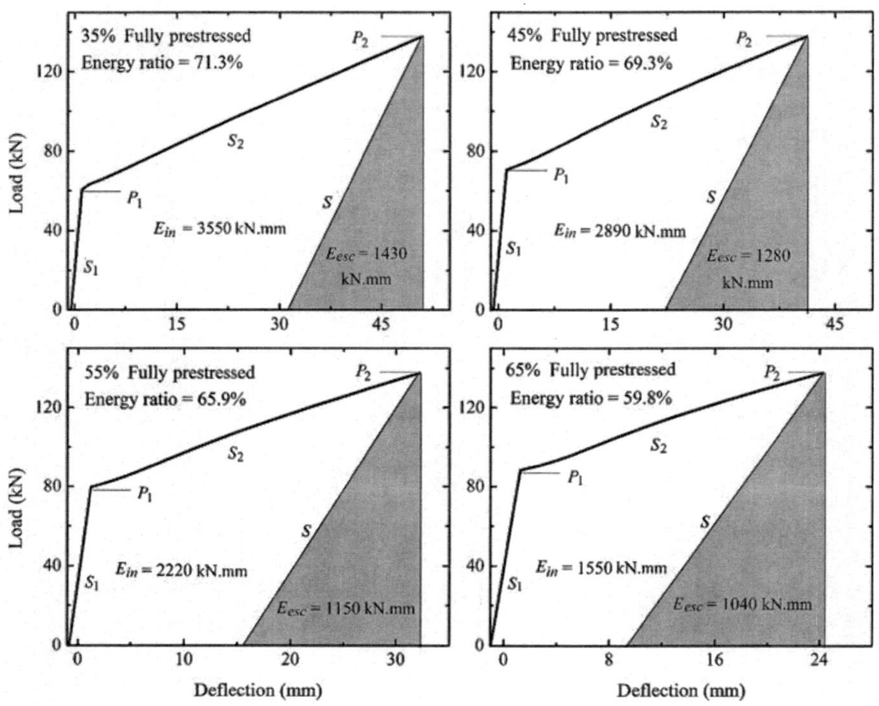

Fig. 8 Effect of prestress ratio on ductility in fully prestressed beams

The variation in ductility with the prestress ratio is shown in Fig. 8. It can be seen from Fig. 8 that the energy ratio has decreased with increasing the prestress ratio. The beam prestressed at 35% exhibited the maximum ductility with energy ratio of 71.3%, which is classified as semi-ductile according to Grace et al. (1998), whereas the remaining beams with higher prestress ratio exhibit lower ductility with energy ratio of below 69%, which is classified as brittle behavior.

6.1 *Partial Prestress*

Figure 9 shows the ductility of the BFRP-reinforced/prestressed concrete beams, as well as the dimensions and properties of the analyzed models. The ductility evaluation is shown in Fig. 9a for non-prestressed concrete beams and in Fig. 9b for fully prestressed concrete beam. In a fully prestressed concrete beam, with all the tendons placed in a single layer, the tendons rupture simultaneously, when it is designed as an under-reinforced section; wherein, the beam reaches a higher load-carrying capacity as compared to the partially prestressed beam. The load-carrying capacity of the non-prestressed and fully prestressed beams is the same, because in both the cases, all the tendons fully contribute to the beam flexural strength. The advantage

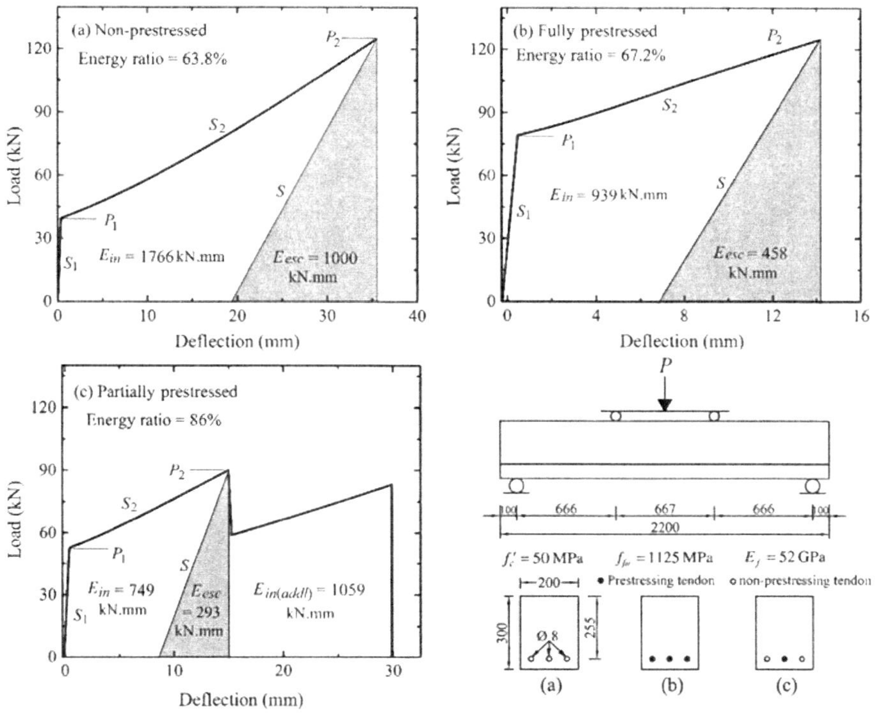

Fig. 9 Ductility evaluation of **a** non-prestressed, **b** fully prestressed, and **c** partially prestressed concrete beams with BFRP tendons

of the prestressed beam over the non-prestressed beam is in limiting the deflection, as seen in Fig. 9. However, the ductility index for both was found below 69%, which indicates lack of ductility of the BFRP-reinforced/prestressed concrete beams.

When the beam is partially prestressed, the prestressing tendons rupture before the non-prestressing tendons within the same layer, and hence, the beam reaches its load-carrying capacity before the non-prestressing tendons reach their ultimate tensile strength. Therefore, the non-prestressing tendons partially contribute to the flexural strength of the beam, and their remaining strength is further employed in the post-peak stage, which explains the reason behind the reduction in the flexural strength of the partially prestressed beam. However, this post-peak behavior adds a significant amount of so-termed inelastic energy, and thereby improves the ductility evaluation, as shown in Fig. 9c, wherein the partially prestressed beam exhibited a significant energy ratio, as high as 86%, attributed to high $E_{in(addl)}$.

The variation in ductility with different prestress ratios was investigated. It was observed that with increasing the number of prestress ratios, i.e. having multiple prestress magnitudes for the different tendons within the beam section, different post-peak behaviors were exhibited, and the ductility and load-carrying capacity changed, as shown in Fig. 10. Four different prestress ratios were investigated, wherein the

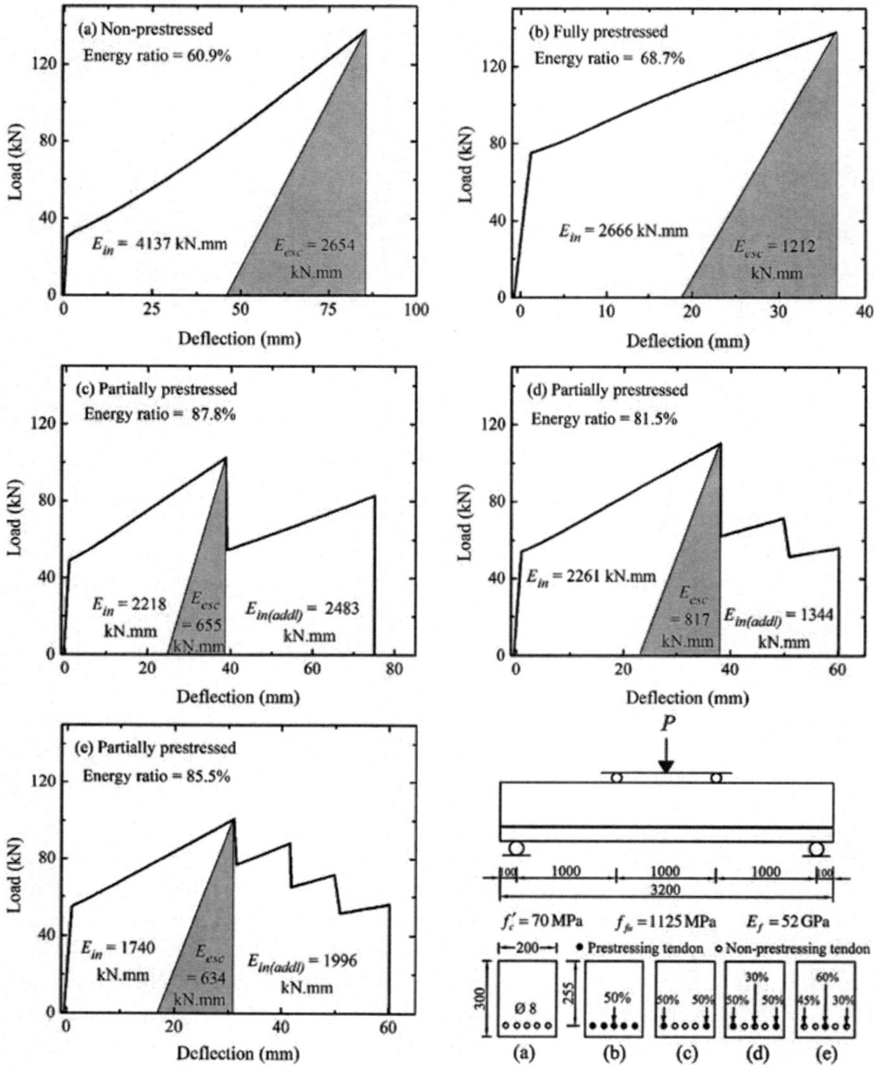

Fig. 10 Variation in ductility with different prestress ratios

number one indicates the fully prestressed and the other numbers indicate the partially prestressed beams. The variation in the ductility and load-carrying capacity with the different prestress ratios is shown in Fig. 11. It is notable from Fig. 11 that the PSC beams cross-sections are showing how many different prestressing forces have been applied to the tendons just over each case of prestress ratio: 1, 2, 3, and 4 marked on x-axis. Thus, with respect to the number of prestress ratios, 1, 2, 3, and 4 on the abscissa corresponding prestressing applied varies in the range of 60%, 50%, 45%, 30%, and 0% as shown in the beam cross-sections for each number of prestress ratio.

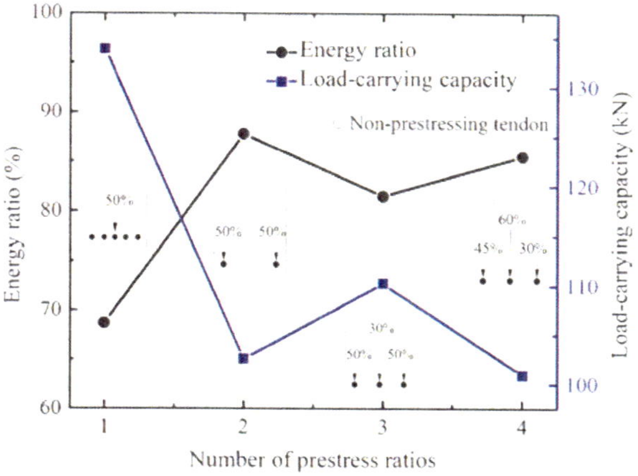

Fig. 11 Variation in ductility (energy ratio) and load-carrying capacity of the PSC beam with number of prestress ratios (prestress magnitudes)

It can be seen from Figs. 10 and 11 that the fully prestressed beam exhibits the lowest ductility of 68.7%, however with the highest load-carrying capacity of 133.87 kN, and the partially prestressed beam exhibits increased ductility, however with the reduced load-carrying capacity of the beam. Importantly, for the partially prestressed beam with the three different prestress ratios, the ductility has always been greater than 80% and the change in the load-carrying capacity is marginal. The change in the ductility and load-carrying capacity is dependent on the number of tendons for each prestress ratio, which is probability-governed; however, it can be seen from Fig. 11 that in case of three prestress ratios, acceptable ductility and load-carrying capacity could yet be achieved. For the sake of simplicity and practicality, it can be recommended to opt for two different prestress ratios for the PSC beams with the dimensions and detailing similar to that taken here as an example. Nonetheless, the design of the PSC beams for a specific purpose thus can be made such that both, load-carrying capacity as well as energy ratio can be maximized fulfilling the requirements thereof.

The ductility of the partially prestressed beam is dependent also on the number of prestressing tendons out of the total reinforcement provided. For the PSC beam shown in Fig. 10, two tendons were prestressed, out of five, and that significantly helped in increasing the ductility up to 87.8% because more than half of the reinforcement did not rupture at the first peak when the failure was experienced and contributed to the post-peak behavior considerably. For the PSC beam shown in Fig. 12, three tendons, out of five, were prestressed, because of which the ductility has decreased from what was exhibited in Fig. 10; however, the ductility has still been comparable albeit with higher load-carrying capacity. The variation in ductility and load-carrying capacity of the beam with prestress ratio is shown in Fig. 13, wherein the energy ratio has increased, and the flexural strength has decreased with increasing the prestress ratio in the partially prestressed beam. However, the ductility response seen here is different from that shown in Fig. 8, wherein the energy ratio decreased with increasing the prestress ratio in case of the fully prestressed beams. In the partially prestressed

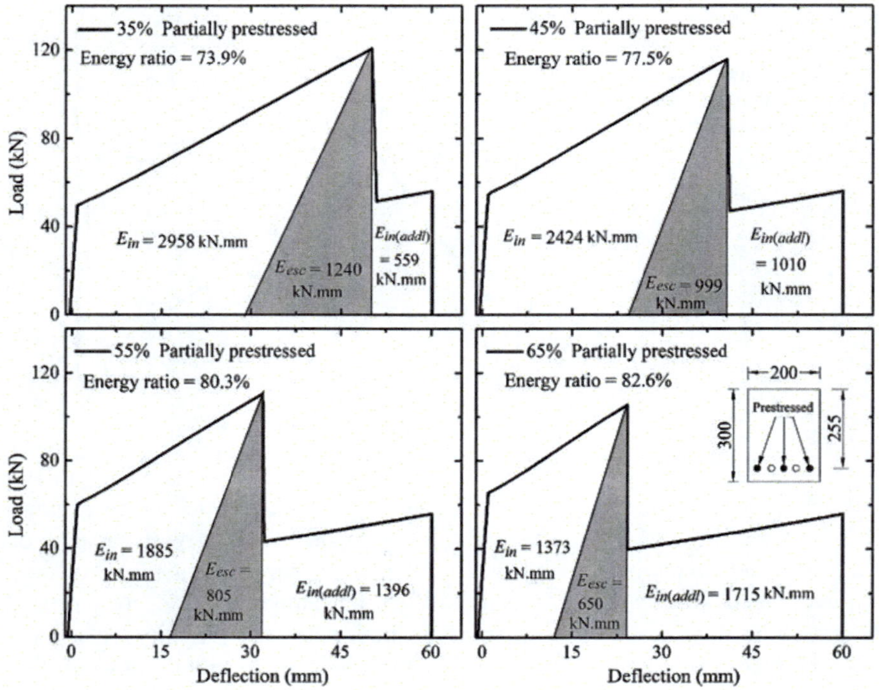

Fig. 12 Variation in ductility with the prestress ratio in partially prestressed beams

Fig. 13 Variation in ductility (energy ratio) and load-carrying capacity of the PSC beam with varying prestress ratio in partially prestressed beams

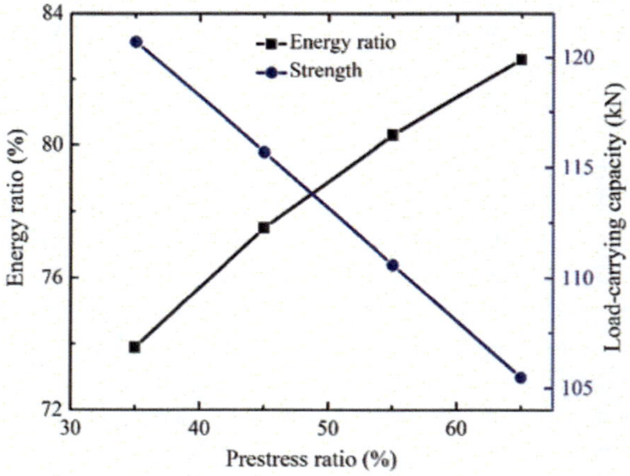

beams, increasing the prestress ratio means increasing the difference in the remaining tensile strength between the prestressing and non-prestressing tendons, and hence, increasing the remaining strength in the non-prestressing tendons, which contribute to the post-peak behavior, and that explains the increase in the ductility with increasing the prestress ratio in the partially prestressed beams.

6.2 Layering of Tendons

The other technique to improve the pseudo-ductility of the BFRP-reinforced/prestressed concrete beams is the layering of the tendons, by vertically distributing the tendons in multiple layers from the neutral axis so that they rupture consecutively contributing to the additional inelastic energy, $E_{in(addl)}$. The concept here is similar to the partial prestressing in controlling the ruptures of the tendons and avoiding the sudden/catastrophic failure. While in the partially prestressed concrete beams, the non-prestressing tendons provide the safety against such catastrophic failure, the reinforcements in the upper layers here control the failure owing to the delay in reaching their ultimate tensile strength as compared to those in the bottommost layer. The ductility evaluation is shown in Fig. 14 for one, two, and three reinforcement layers.

The ductility response here is dependent on the number of tendons in each layer and the spacing between layers, which govern the peak load achieved and the maximum deflection experienced corresponding to each reinforcement layer. The variation in ductility and load-carrying capacity of the PSC beam with the number of layers and prestress ratio is shown in Fig. 15. It can be seen from Fig. 15 that

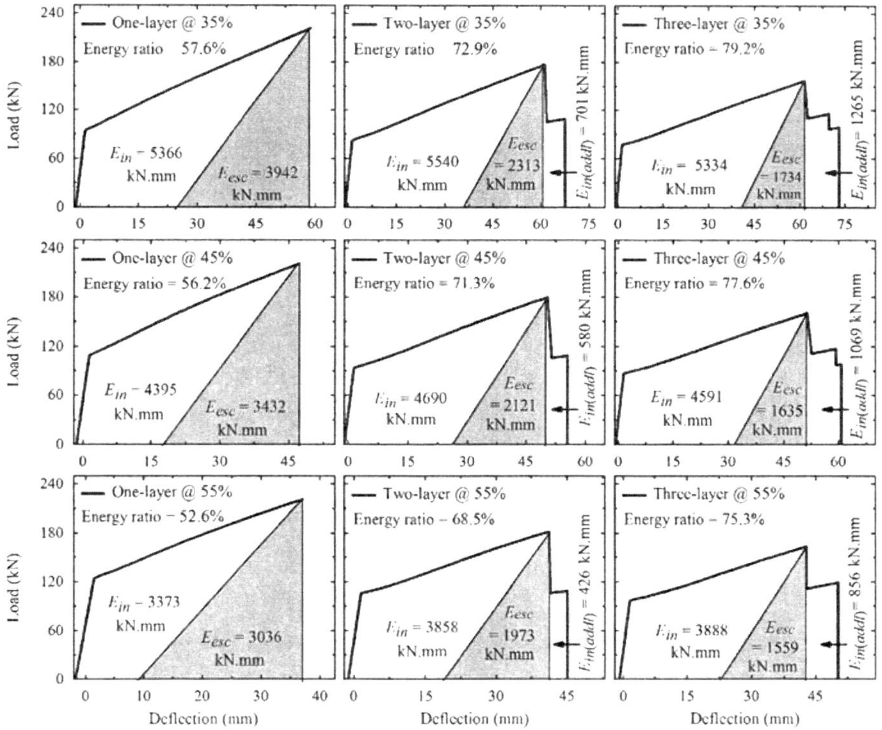

Fig. 14 Variation in ductility with prestress ratio and number of reinforcement layers

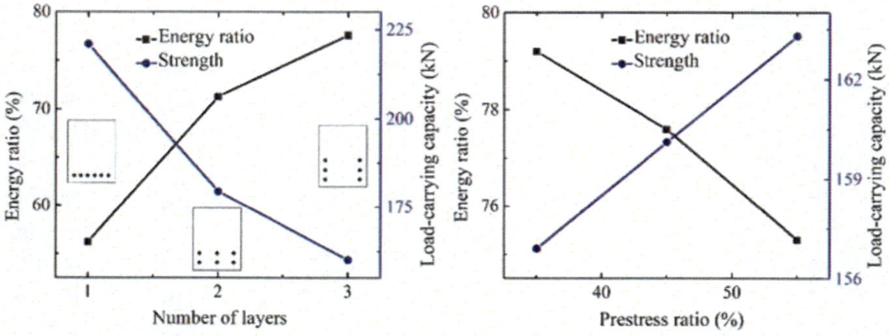

Fig. 15 Variation in ductility (energy ratio) and load-carrying capacity of the PSC beam with number of layers and prestress ratio

with increasing the number of reinforcement layers, the ductility and load-carrying capacity exhibit different trends, while the ductility increases, the load-carrying capacity decreases with the number of layers. Moreover, the ductility decreases and the load-carrying capacity of the beam increases with increasing the prestress ratio, which is contrary to that shown in Fig. 13 for the case of partial prestressing.

6.3 Functionally-Graded Concrete

Functionally-graded concrete is not a unique technique for improving the ductility, however it can be used along with either the partial prestressing or layering of the tendons, or even with both of them, to compensate the decrease in the load-carrying capacity. As discussed earlier, improving the ductility as per the presented approaches has been associated with reduction in the flexural strength of the PSC beam; hence, there is a need to compensate the flexural strength so influenced, while maintaining the ductility to an acceptable level. Functionally-graded concrete is achieved by using different grades of concrete in the cross-section of the PSC beam, usually placing higher grade concrete for the compression zone and lower grade concrete for the tension zone. Of course, within the compression zone of the beam as well further variation in grading of concrete is possible such that superior grade is placed farther from the neutral axis, contributing considerably to the flexural strength. Doing so, it is possible to increase the depth of the cross-section without violating the economy in the design, because the saving in using the lower grade concrete for the tension zone enables enlarging the depth within the same cost and probably with more saving. The enlarged depth increases the lever arm as well, and hence, improves the load-carrying capacity along with the shear strength enhancement of the section (compensating low shear strength of reduced grade concrete).

The performance of the functionally-graded concrete is discussed here briefly. The ductility and load-carrying capacity evaluation of the functionally-graded concrete

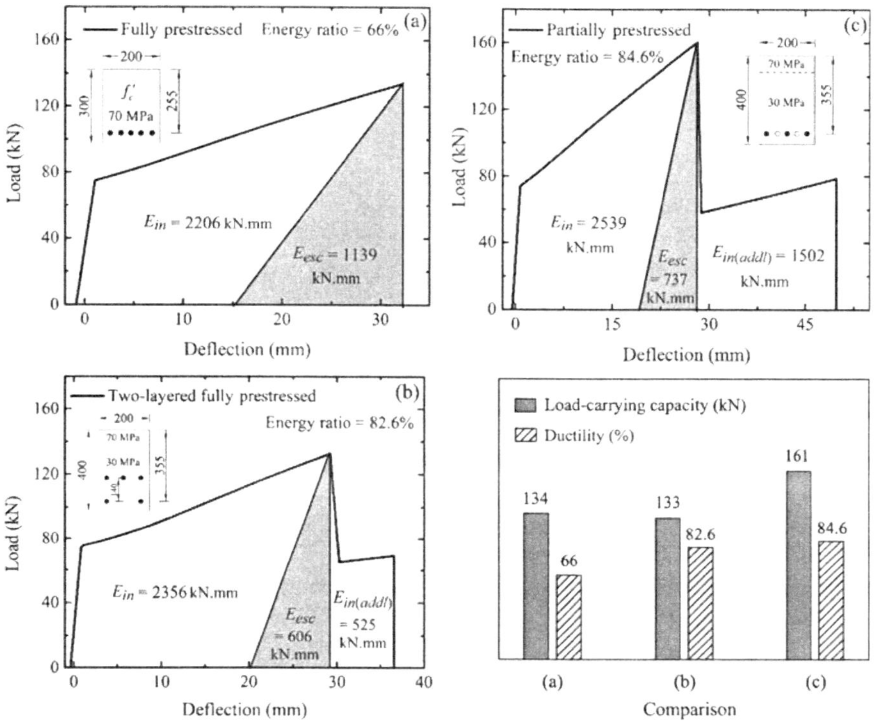

Fig. 16 Comparison between fully prestressed, partially prestressed, and layered system with functionally-graded concrete

beams with either partial prestressing or layering of the tendons is compared to that with the fully prestressed beam with uniform/plain cross-section, as shown in Fig. 16. It can be seen from Fig. 16 that the functional grading of concrete helped in achieving superior performance in terms of ductility and load-carrying capacity of the PSC beam, upon applying the partial prestressing and/or the layering of the tendons. However, it can be observed that with the partial prestressing and functionally-graded concrete, the most desirable performance could be achieved with the highest ductility and load-carrying capacity of the structural member. Table 2 summarizes the comparison between all the techniques in terms of the ductility and load-carrying capacity of the PSC beam achieved. In Table 2, the PSC beam in all the cases has the same number of tendons, five BFRP tendons, and same cross-section and concrete grade, $b \times h = 200 \times 300$ mm, $f_c' = 70$ MPa, with extra depth in the case of functionally-graded concrete, $b \times h = 200 \times 400$ mm, $f_{c1}' = 70$ MPa, and $f_{c2}' = 30$ MPa. It can be seen from Table 2 that the four techniques introduced here in the present investigation have certainly helped in providing adequate ductility, with superiority for the partial prestressing and functionally-graded concrete, which achieved the highest ductility and maintained high load-carrying capacity. Hence, it is more economical

Table 2 Ductility and load-carrying capacity comparison of all the techniques proposed

Technique description	Energy ratio (%)	$P_u/P_{u,max}$[a]	Behavior
Fully prestressed one-layered system	66	0.83	Brittle
Partially prestressed (plain section)	79.9	0.7	Ductile
Multi-layered tendons (plain section)	78.7	0.63	Ductile
Partially prestressed (functionally-graded section)	84.6	1.0	Ductile
Multi-layered tendons (functionally-graded section)	82.6	0.82	Ductile

[a]P_u is the load-carrying capacity for each case and $P_{u,max}$ is the maximum load-carrying capacity

and relatively safer to opt for partial prestressing with functionally-graded concrete rather than layering of the tendons alone.

7 Conclusions

An analytical approach is introduced here for the ductility evaluation for the BFRP-prestressed concrete beams. The analytical approach introduced is translated in a MATLAB code, based on the equations obtained from different codes and standard literature, for evaluating flexural capacity and deflection at failure, and studying the load-deflection response. The MATLAB code was validated with the results obtained from the experimental work conducted and reported in the literature. The ductility is obtained here, like in the case of all other FRP materials, from the energy-based approach, as the ratio of the inelastic energy to the total energy under the load-deflection curve. Different techniques were introduced to improve the ductility in the present study. The following conclusions are drawn from the current study.

1. The BFRP-reinforced/prestressed concrete beams exhibit a brittle behavior with energy ratio of below 69% and high load-carrying capacity, however, the ductility is inversely proportional to the prestress ratio in the fully prestressed beams.
2. The ductility is improved in the partially prestressed beams with energy ratio reaching above 75%. The ductility here is directly proportional to the prestress ratio. However, the load-carrying capacity decreases in the partially prestressed beam as compared to the fully prestressed one.
3. The ductility is improved in the multi-layered system of prestressing tendons, and varies with the number of layers provided. However, the load-carrying capacity decreases in the multi-layered system as compared to the single-layer system of fully prestressed or non-prestressed beams.
4. The functionally-graded concrete is recommended as an effective technique to compensate the reduction in flexural strength of the beam when partial prestressing and/or layering of the tendons is introduced. The functionally-graded concrete can be used along with the two techniques, however, the partially

prestressed beam with functionally-graded section achieved the highest ductility and load-carrying capacity amongst all the introduced techniques.

For future study, optimizing parameters used in the techniques proposed here for maximizing the ductility and load-carrying capacity of the BFRP-reinforced/prestressed beams is anticipated. Such optimal problem can be posed with minimizing cost of the PSC beams with the BFRP composite reinforcements.

References

Abdelrahman AA, Rizkalla SH (1999) Deflection control of concrete beams pretensioned by CFRP reinforcements. J Compos Constr (ASCE) 3(2):55–62

ACI 318-08 (2008) Building code requirements for structural concrete. American Concrete Institute (ACI), Farmington Hills, Michigan (MI), USA

ACI 435R-95 (2003) Control of deflection in concrete structures. American Concrete Institute (ACI), Farmington Hills, Michigan (MI), USA

ACI 440.4R (2004) Prestressing concrete structures with FRP tendons. American Concrete Institute (ACI), Farmington Hills, Michigan (MI), USA

ACI 440.5 (2008) Specification for construction with fiber-reinforced polymer reinforcing bars. American Concrete Institute (ACI), Farmington Hills, Michigan (MI), USA

Bischoff PH (2005) Reevaluation of deflection prediction for concrete beams reinforced with steel and fiber-reinforced polymer bars. J Struct Eng (ASCE) 131(5):752–767

Burgoyne CJ (2001) Ductility and deformability in beams prestressed with FRP tendons. In: FRP composites in civil engineering: proceedings of the international conference on FRP composites in civil engineering, vol 1, pp 15–25

Eurviriyanukul S, Askes H (2010) The equilibration of configurational forces in the tendon layout optimization of prestressed concrete beams. Comput Struct 88(23–24):1412–1418

Grace NF, Singh SB (2003) Design approach for carbon fiber-reinforced polymer prestressed concrete bridge beams. ACI Struct J 100(3):365–376

Grace NF, Soliman AK, Abdel-Sayed G, Saleh KR (1998) Behavior and ductility of simple and continuous FRP reinforced beams. J Compos Constr (ASCE) 2(4):186–194

Grace NF, Jensen EA, Matsagar VA, Bebawy M, Soliman E, Hanson J (2008) Transverse post-tensioning of side-by-side box-beam bridges. Testing and Research Section, Construction and Technology Division, Michigan Department of Transportation (MDOT), USA, 2004-0105. https://trid.trb.org/view.aspx?id=875179

Grace NF, Jensen EA, Enomoto T, Matsagar VA, Soliman EM, Hanson JQ (2010a) Transverse diaphragms and unbonded CFRP post-tensioning in box beam bridges. PCI J 55(2):109–122

Grace NF, Jensen EA, Matsagar VA, Soliman E, Hanson J (2010b) Use of unbonded CFRP strands in transverse post-tensioning in box beam bridges. Int J Earth Sci Eng 3(2):128–137

Grace NF, Jensen EA, Matsagar VA, Penjendra P (2013a) Performance of an AASHTO beam bridge prestressed with CFRP tendons. J Bridge Eng (ASCE) 18(2):110–121

Grace NF, Ushijima K, Matsagar VA, Wu C (2013b) Performance of AASHTO-type bridge model prestressed with carbon fiber-reinforced polymer reinforcement. ACI Struct J 110(3):491–502

Guomundsson SH (2012) Capacity of concrete beams prestressed with BFRP tendons. Master of Science Thesis, Reykjavik University, Iceland

Hussein M, Kunieda M, Nakamura H (2012) Strength and ductility of RC beams strengthened with steel-reinforced strain hardening cementitious composites. Cement Concr Compos 34(9):1061–1066

Iskhakov I, Ribakov Y (2007) A design method for two-layer beams consisting of normal and fibered high strength concrete. Mater Des 28(5):1672–1677

Jalasutram S, Sahoo DR, Matsagar V (2017) Experimental investigation of the mechanical properties of basalt fiber-reinforced concrete. Struct Concr 18(2):292–302

Janney JR, Hognestad E, McHenry D (1956) Ultimate flexural strength of prestressed and conventionally reinforced concrete beams. ACI J Proc 52(2):601–620

Johnson BS (2011) Prestressed BFRP tendons in concrete beams. Master of Science Thesis, Reykjavik University, Iceland

Matthys S, Taerwe L (2006) Evaluation of ductility requirements in current design guidelines for FRP strengthening. Cement Concr Compos 28(10):845–856

Park R (1989) Evaluation of ductility of structures and structural assemblages from laboratory testing. Bull New Zealand Soc Earthq Eng 22(3):155–166

Salem SH, Hilal KM, Hassan TK, Essawy AS (2013) Experimental behavior of partially prestressed high strength concrete beams. Open J Civ Eng 3(3):26–32

Wang H, Belarbi A (2011) Ductility characteristics of fiber-reinforced concrete beams reinforced with FRP rebars. Constr Build Mater 25(5):2391–2401

Zhishen W, Xin W, Gang W (2012) Advancement of structural safety and sustainability with basalt fiber reinforced polymers. In: Proceedings of CICE 2012, 6th international conference on FRP composites in civil engineering, Rome, Italy, 13, pp 15–29

Concrete Filled Unplasticized Poly-Vinyl Chloride (UPVC) Tubes as Column

P. K. Gupta and V. K. Verma

Abstract In present study, a stress–strain model for concrete subjected to compression and confined with unplasticised poly vinyl chloride (UPVC) tube is proposed. To develop this model, total eighteen specimens of concrete filled UPVC tubes (CFUT) having different geometrical properties were tested. To obtain the specimens, UPVC pipes of class 3, 4 and 5 (IS: IS: 4985-2000 (Reaffirmed 2005) "Unplasticized PVC pipes for potable water supplies- specification. Bureau of Indian Standard, New Delhi, India-2000 IS: 4985-2000 (Reaffirmed 2005) "Unplasticized PVC pipes for potable water supplies- specification. Bureau of Indian Standard, New Delhi, India) with nominal/working pressure of 0.6, 0.8 and 1.0 MPa (87, 116 and 145 psi) having diameters of 160, 200 and 225 mm (6.3, 7.87 and 8.86 in.) were taken and cut into 800 mm (31.5 in.) length. Internal hydrostatic pressure test were conducted to obtain the confining pressure of UPVC tubes. M30 and M40 grades of concrete were designed as per IS: 10262-2009 (IS: 10262-2009. Concrete mix proportioning-guidelines. Bureau of Indian Standard. New Delhi, India) to fill the tubes. All the specimens were compressed by application of load on concrete core only, to obtain load–displacement variations and associated mode of deformation. All the specimens were failed by development of shear cracks and macrocracks with slight bulging. A finite element model is developed using proposed stress–strain variation of concrete confined with UPVC tubes to simulate the axial compression of CFUT specimens. On the basis of obtained results, effect of variation of diameter to thickness ratio (D/t) on the failure stress of confined concrete is obtained and discussed with an empirical relationship. Load capacities of different specimens obtained using the proposed Finite Element model and some models available in literature are presented, compared and discussed. It may be concluded that the proposed model is capable to estimate the load capacity and mode of deformation of concrete filled UPVC tubes (CFUT) subjected to axial compression.

P. K. Gupta (✉)
Department of Civil Engineering, IIT Roorkee, Roorkee, India
e-mail: pramod.gupta@ce.iitr.ac.in; spramod_3@yahoo.com

V. K. Verma
Department of Civil Engineering, GBPUA&T, Pantnagar, India

S. B. Singh et al. (eds.), *Emerging Trends of Advanced Composite Materials in Structural Applications*, Composites Science and Technology,
https://doi.org/10.1007/978-981-16-1688-4_5

Keywords CFUT · Confinement · Failure mode · Material degradation parameter (k_3) · Finite element analysis · UPVC

1 Introduction

Plastics have exceptional properties, such as low cost, high resistance to severe environmental attack, and high strength to weight ratio. Due to these properties, plastics are used for different purposes which include commercial and engineering. UPVC pipes are generally employed for water supply across the world. These pipes are designed for induced internal pressure of flowing water. On the basis of resistance against the water pressure and depending upon the nominal pressure exerted by flowing water, these pipes are categorized into different classes (IS: 4985-2000 2000). Authors have tried to use safe nominal/working pressure capacity of these pipes to provide confinement to the concrete core by filling the concrete designed using IS: 10262-2009 (2009), inside these pipes.

Richart et al. (1928) studied the failure behavior of concrete under biaxial and triaxial compression. They took cylinder specimens having 4 in.(101.6 mm) diameter and 8 in. (203 mm) length of three different concrete strengths i.e. 2575, 3660 and 1050 psi (17.75, 25.24 and 7.24 MPa) for triaxial compression test. They applied different confining pressure and axial load on the specimens to get different failure modes. They found that at low confining pressure, most of the specimens failed by development of inclined cracks while at large confining pressure, a pronounced bulging was noticed. Sfer et al. (2002) performed experiments on 150 mm × 300 mm (5.01 in. × 11.82 in.) cylindrical specimens to study their behavior under triaxial compression. A confining pressure of 0–60 MPa (0–8700 psi) was used to obtain the brittle-ductile transition behavior of the material. They found that at zero or lower confinement, the failure was lead by distributed microcracking and several macrocracks while at higher confinement, failure occured by development of fine microcracks and a few macrocracks that seperated the specimens into two or three blocks.

Kurt (1978) studied the behavior of concrete filled plastic columns with the help of experiments and theoretical analysis. He observed that when the concrete core of concrete filled plastic column was loaded axially, an effective compressive pressure was generated on the boundary between the core and pipe due to Poisson effect. He conducted some experiments to see the influence of slenderness ratio on the behavior of concrete filled plastic columns. He found that slenderness effect was more important when the slenderness ratio was greater than 20. Marzouck et al. (2002) conducted experimental study on concrete filled PVC tubes having 100 mm internal diameter and 3 mm thickness. They concluded that PVC tube provided considerable lateral confinement to the concrete core and exhibited large lateral deformation before failure. Wang and Yang (2010) presented an experimental study on plastic pipe confined concrete (PPC). The plastic pipes were taken as HDPE pipes. They reported that the thickness and unconfined compressive strength of concrete affects the ultimate strength and post-peak behavior of PPC. They observed that at lower confinement, the mode of failure was shear while at higher confinement it was drum type. Gupta et al. (2012) tested twelve specimens of unplasticised poly vinyl chloride (UPVC) tubes having 900 mm (35.43 in.) length and filled with steel fiber concrete. They studied the effect of diameter to thickness ratio and length to diameter ratio on strength, confinement and ductility. They found that the maximum displacement till complete failure of concrete core was around 4.5 mm (0.177 in.) while concrete filled UPVC tubes were not completely failed till 11 mm (0.433 in.) of compression. Gupta [Gupta, 2013] carried out experimental study to investigate the effectiveness of UPVC tube for confinement of concrete columns. In his study, UPVC tubes of 140, 160 and 200 mm (5.51, 6.3, 7.87 in.) diameter and filled with concrete of compressive strength of 20, 25 and 40 MPa (2900, 3625, 5800 psi) were tested. He reported that confinement of concrete columns with UPVC tubes enhances their load carrying capacity and ductility. He observed shear type failure of all the tested specimens.

Hu et al. (2003) carried out a nonlinear finite element analysis (FEA) of concrete filled steel tubular (CFT) columns with circular section, square section and square section stiffened by reinforcing ties to study the behavior of CFT columns. They simulated the post peak behavior of concrete confined with steel tube by a linear variation and gave a parameter k_3 to obtain a stress–strain model. They proposed empirical relationships between confining pressure and k_3 as a function of diameter/wall thickness (D/t).

Ragab et al. (2001) tested rigid PVC pipes having 60 mm (2.36 in.) diameter and 5.3 mm (2.09 in.) thickness with 1 MPa (145 psi) working pressure in biaxial stress system covering the four quadrants of the plane-stress space. The loading was considered as tension, compression, torque and internal and external pressurization. They tested specimens at different strain rates and temperature. They plotted the yield loci at various strain rates and temperatures. They observed that yielding of rigid PVC pipes followed the von Mises criterion.

Fam and Rizkalla (2001) introduced an analytical model to study the behavior of axially loaded circular concrete columns confined by fiber reinforced polymer (FRP) tubes. They modified the confinement model for concrete confined by steel reinforcement given by Mander et al. (1988).

Gupta and Verma (2014) developed a finite element model to study the behavior of concrete filled plastic tube (CFPT) using ANSYS software. They verified results against the experimental data available in literature (Wang and Yang 2010). They proposed empirical relationships between confining pressure (f_1) & D/t and material degradation parameter (k_3) and D/t. They reported that confining pressure (f_1) and material degradation parameter (k_3) increases as the outer diameter to thickness (D/t) ratio decreases.

2 Research Significance

From the literature review it can be seen that only a few researchers have studied the behavior of concrete filled plastic tubes subjected to axial compression. Investigation of such columns with the help of computer modeling is still not reported. In this paper a systematic experimental and computational study to understand and model the behavior of concrete filled UPVC tubes (CFUT) subjected to axial compression is attempted.

3 Experimental Procedure

3.1 Materials

UPVC pipes of class 3, 4 and 5 (as per IS: 4985-2000 2000) used for water supply with nominal pressure of 0.6, 0.8 and 1.0 MPa (87, 116, 145 psi) and the diameters of 160, 200 and 225 mm (6.3, 7.87, 8.86 in.) were procured from the market. The pipes were cut and leveled to obtain the specimen length of 800 mm (931.5 in.). The tensile test as per ASTM-D638-10 and internal hydrostatic pressure test as per IS: 12235-part 8-1986 (1986) of UPVC pipes were conducted in the laboratory to get stress–strain curve as well as ultimate tensile strength and confining pressure respectively (Figs. 1 and 2). The ultimate tensile strength of UPVC tube material (f_u) is obtained as 22.5 MPa (3262.5 psi). The confining pressure is given in Table 1. The Ordinary Portland Cement as per IS: 8112-1989 (1989) was used to get the desired grade of concrete. Two different mix designs (M30 & M40) were prepared as per IS: 10262-2009 (2009). The unconfined compressive strengths of concrete for M30 and M40 grades are obtained as 35.08 MPa (5086.6 psi) and 39.05 MPa (5662.25 psi) respectively.

3.2 Specimens

To prepare CFUT specimens, UPVC pipes were properly cut and finished on the lathe machine in such a way that both ends were horizontal and parallel to each other. These pipes are fitted in circular mould before the casting of specimens. Freshly prepared concrete was placed in three layers. Proper vibration and compaction was done in each layer. Specimens were tightly covered with thick polyethylene sheet to avoid the evaporation of water. Typical nomenclatures are used to designate the specimens. For example specimen 160P3PC30: 160-stands for the outer diameter of tube, P3 stands for pipe of class 3 (nominal pressure of 0.6 MPa), PC30 stands for plain concrete of grade M30. Therefore, 200P5PC40 indicates the specimen made with tube of class 5 having nominal diameter 200 mm (7.87 in.), filled with plain concrete of grade

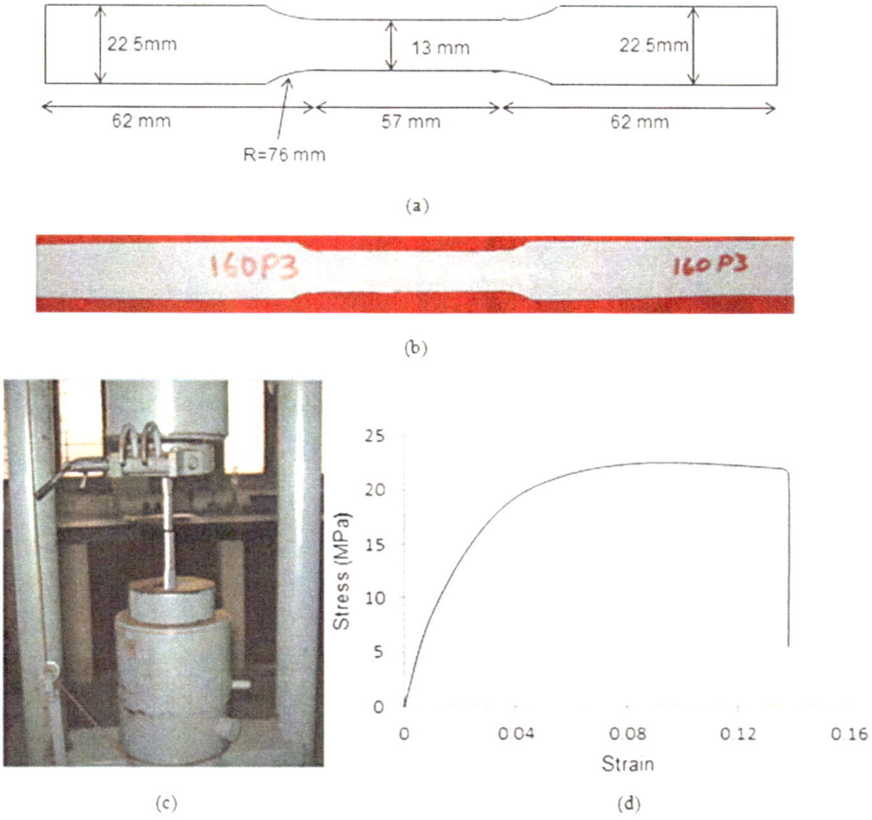

Fig. 1 Tension test of UPVC **a** Line diagram of test specimen, **b** photograph of specimen, **c** testing of specimen, **d** Stress–strain curve of UPVC. *Note* 1 in. = 25.4 mm; 1 MPa = 145 psi

M40. Concrete filled UPVC tube (CFUT) specimens were tested by applying the axial load only on concrete core.

3.3 Test Instruments

CFUT specimens were tested on Universal testing machine of INSTRON make having capacity of 250 tonnes (551,155 lb). During the testing, specimens were kept on bottom platen of the machine. The gauge length of the specimen was the entire length of the specimen. The test was conducted with the strain control method and ramp rate of 0.4 mm/min (0.016 in./min). The plumb-bob and Tri square were used to place the specimen at right angle to the bottom plate. The concentric loading was the main concern for the test to be exact and reliable. Hence, proper leveling of the specimen surface for placing steel plate and placing specimen at right angle

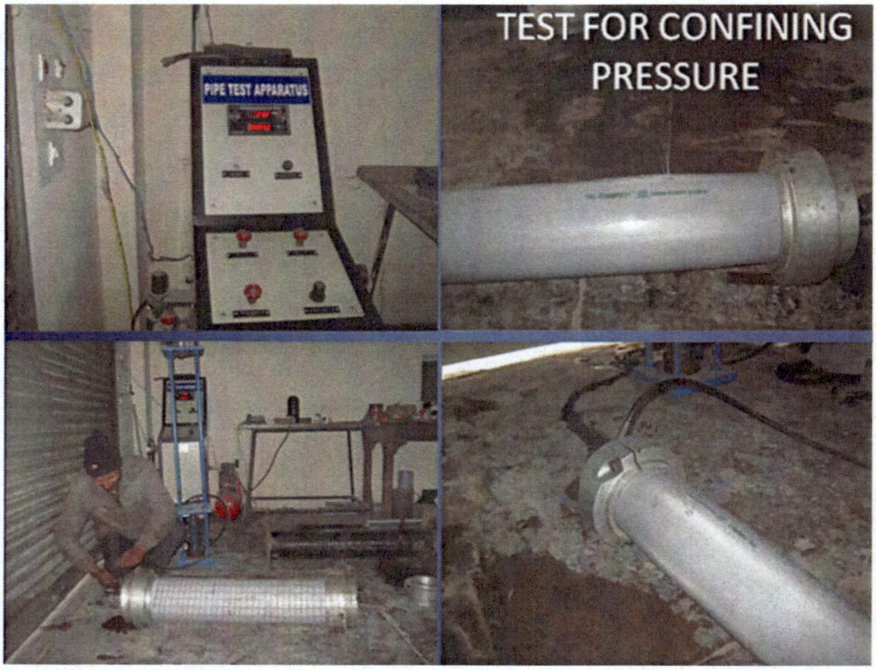

Fig. 2 Typical arrangement of internal hydrostatic pressure test to find confining pressure

Table 1 Confining pressure of UPVC pipes

Class of pipe (nominal pressure)	Dimensions (mm)			Confining pressure (MPa)	
	Mean outer diameter (D)	Internal diameter (d)	Thickness (t)	Experimental	Calculated $f_l = fu(2t/d)$
Three (0.6 MPa)	160.15	152.17	3.99	1.15	1.18
	199.85	187.97	5.94	1.35	1.42
	224.78	207.48	8.65	1.72	1.87
Four (0.8 MPa)	159.88	149.44	5.22	1.52	1.57
	199.56	183.34	8.11	1.99	1.99
	224.58	205.42	9.58	2.00	2.09
Five (1.0 MPa)	160.19	146.15	7.02	2.15	2.16
	199.22	181.44	8.89	2.25	2.20
	224.53	204.59	9.97	2.17	2.19

Note 1 in. = 25.4 mm; 1 MPa = 145 psi

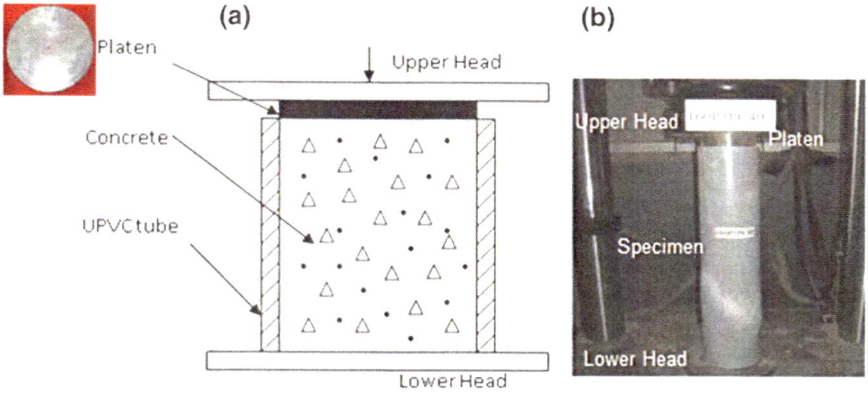

Fig. 3 **a** Loading and **b** testing arrangements

for the concentric loading was given the utmost importance. The load–displacement curves were recorded by the machine automatically. The compression process of CFUT column specimen was continued till the specimen failed. Since the tensile strength of UPVC tube material (f_u) is smaller and approximately 22.5 MPa (3262.5 psi), tubes were used only for confinement purpose. All the specimens were tested by applying the axial load on concrete core only. Total eighteen specimens were tested. Repeatability of the testing was assured by testing three samples of each type. The typical arrangement of loading is depicted in Fig. 3a and testing of specimen is shown in Fig. 3b.

4 Modelling and Simulation

4.1 Finite Element Model

A three dimensional Finite element model was developed using ANSYS software to simulate the concrete filled UPVC tube (CUFT) subjected to axial compression. To model the concrete core, a three dimensional eight node solid element SOLID 65 was used. To model the tube, eight node solid element SOLID 45 was used. The steel platen for the loading of concrete core of specimen was also modeled by SOLID 45 element. The yield stress of platen was taken 1000 MPa (145,000 psi) and stress–strain variation was modeled using elastic-perfectively plastic material option. The platen was behaved as rigid. Mesh size was chosen from 5 to 10 mm (0.197–0.394 in.) for tube, concrete core and steel platen. Two rigid plates were modeled to simulate rigid cross heads of machine. Load was applied to the column through the top loading plate. In the compression test, direct contact exist between the bottom end plate and bottom surface of column while at top surface, contact was exist between steel platen and top surface of concrete core of the column; therefore a contact available in ANSYS was used to simulate the interaction between rigid plate, platen and column surface. The contact was defined as a surface to surface contact.

To activate the confinement of concrete core in finite element model, a contact surface pair comprised of the inner surface of the tube and the outer surface of concrete core was adopted. Flexible behavior in the normal direction was assumed with no penetration allowed between the surfaces. Different friction factors (0.1–0.4) were taken for the contact surface and it was found that friction factor of 0.2 was suitable to achieve a quick convergence. In finite element model, the lower rigid plate contacting the bottom of column was fixed in all six directions by reference node. The upper rigid plate at the top of the column was modeled fixed in five directions and only allowed movement in column axis at reference node. The load was applied as static uniform displacement at upper rigid plate through the reference node at the center of rigid plate. Figure 4 shows a typical finite element model adopted for modeling of CFUT column.

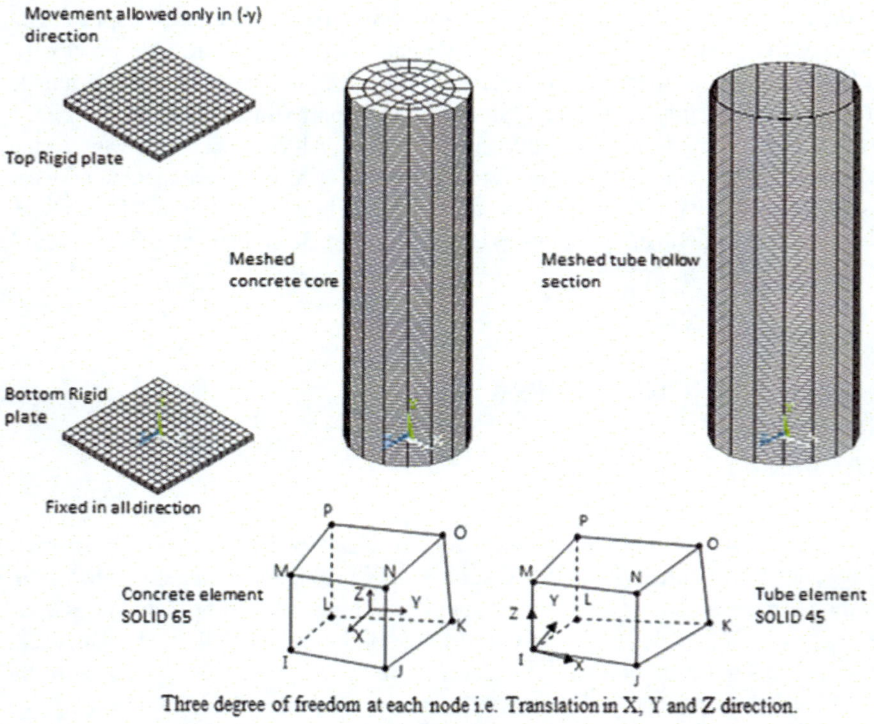

Fig. 4 Typical details of finite element model for concrete filled tube

4.2 Material Model of UPVC

Multilinear elastic–plastic material behavior of the UPVC tube is taken (ANSYS user guide) to model it. Elastic modulus of equal to 880 MPa (127,600 psi) of UPVC was obtained from stress–strain curve (Fig. 1d) and Poisson's ratio is taken as 0.35. When tube is subjected to several stresses, a von Mises criterion F is used to define the elastic limit, which is written as

$$F = \sqrt{3J_2} = \frac{1}{\sqrt{2}}\sqrt{(\sigma_1 - \sigma_2)^2 + (\sigma_2 - \sigma_3)^2 + (\sigma_3 - \sigma_1)^2} = \sigma_y$$

where

J_2	second stress invariant of the stress deviator tensor
σ_1, σ_2, and σ_3	are the Principal stresses
σ_y	yield stress of material under uniaxial tension.

Figure 1d shows the stress–strain variation for UPVC material.

4.3 Material Model of Confined Concrete

Figure 5 shows the uniaxial stress–strain curve for unconfined and confined concrete. The maximum stress of concrete f_{cc} confined by tube and corresponding strain ε_{cc} are calculated by the following equations (Richart et al. 1928).

$$f_{cc} = f_c + k_1 f_l \tag{1}$$

$$\varepsilon_{cc} = \varepsilon_c\left(1 + k_2\frac{fl}{f_c}\right) \tag{2}$$

where

f_c	unconfined compressive cylinder strength of concrete;
k_1, k_2	are constant, generally taken as 4.1 and 20.5

The value of ε_c is usually between 0.002 and 0.003. For confined concrete it may be taken as 0.003 as per ACI-318 (2008).

f_l = lateral confining stress induced due to confinement provided by tube and depends upon D/t & tensile strength of UPVC tube (f_u).

The response of concrete is modeled by an elastic–plastic theory with associated flow and kinematic hardening rule. During plastic deformation, the expansion of yield surface can be guided by the uniaxial stress–strain curve. Confined concrete is subjected to triaxial compressive stresses; the effective stress and effective strain values may be represented by the uniaxial stress and strain values obtained from the stress–strain equation proposed by Saenz (1964). It is given as:

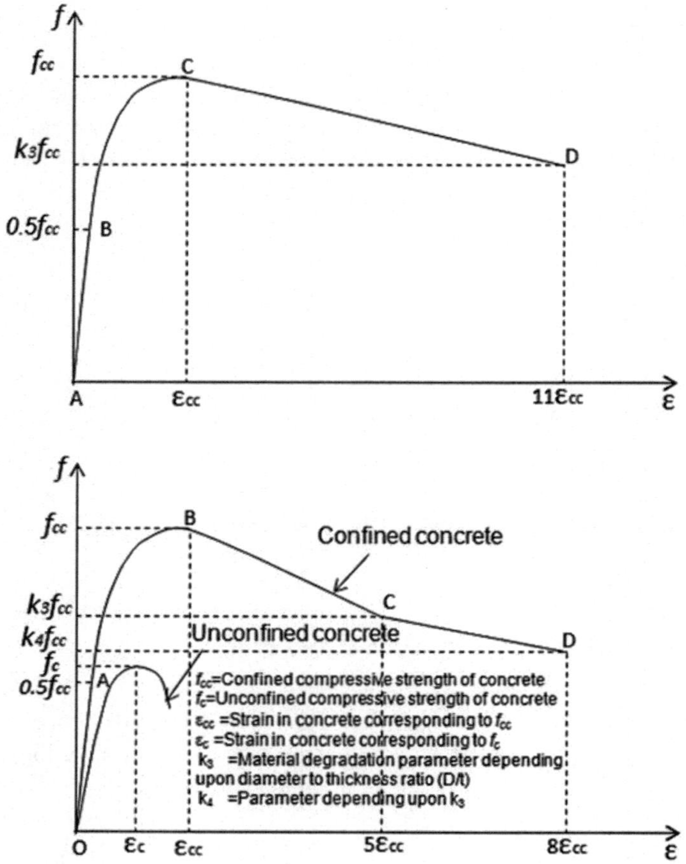

Fig. 5 **a** Stress–strain curves for concrete confined by steel tube, **b** Stress–strain curves for unconfined concrete and concrete confined by UPVC tube

$$f = \frac{E_{cc}\varepsilon}{1 + (R + R_E - 2)\left(\frac{\varepsilon}{\varepsilon_{cc}}\right) - (2R - 1)\left(\frac{\varepsilon}{\varepsilon_c}\right)^2 + R\left(\frac{\varepsilon}{\varepsilon_c}\right)^3} \tag{3}$$

where

$$R_E = \frac{E_{cc}\varepsilon_{cc}}{f_{cc}},\ R = \frac{R_E(R_\sigma - 1)}{(R_\varepsilon - 1)^2} - \frac{1}{R_\varepsilon} \tag{4}$$

values of R_σ and $R_\varepsilon = 4$ (given by Elwi and Murray 1979)

E_{cc} is the initial modulus of elasticity of concrete and calculated from the equation $E_{cc} = 4700\sqrt{f_{cc}}$ as per ACI -318 (2008).

The Stress–strain curve for concrete confined with UPVC tube is obtained by the modification of stress–strain curve for concrete confined with steel tube Hu et al.

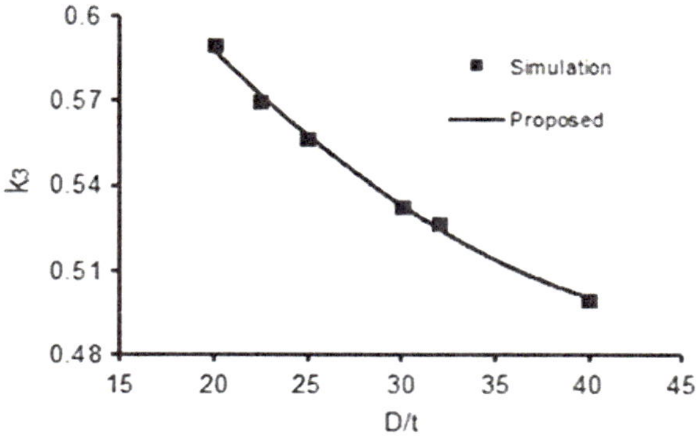

Fig. 6 Variation of k_3 with diameter/wall thickness (D/t)

(2003). The stress–strain curve of concrete confined by steel tube was proposed by Hu et al. (2003) contains three parts as shown in Fig. 5a. In this curve first two parts are independent of the material used for confinement and can be obtained using Eqs. (1), (2), (3) and (4). The third part starts from point (f_{cc}, ε_{cc}) and ends at the point ($k_3 f_{cc}$, $11\varepsilon_{cc}$) is used for post peak part of the curve. This third part is developed for confinement provided by steel tube. Keeping the stress–strain variation of concrete confined by steel tube in consideration, a stress–strain variation of concrete confined by UPVC tube is developed. The stress–strain variation is developed with having four parts. The first two parts 'OA' and 'AB' are represented by Eqs. (1), (2), (3) and (4). In these equations k1 and k2 are taken as 4.1 and 20.5 as suggested by Richart et al. (1928) The first part of stress–strain variation defines the linear property of confined concrete and the proportional limit stress can be assumed $0.5f_{cc}$. The corresponding strain can be obtained using Young's Modulus ACI-318 (2008), $E_{cc} = 4700\sqrt{fcc}$ MPa. The value of f_l has been taken from the experiments by performing water pressure test on the pipes. Generally percentage elongation in steel varies between 22 and 24. On the other hand percentage elongation in UPVC varies between 13 and 16. The last fracture point in stress–strain curve for steel was at $11\varepsilon_{cc}$ strain. This point for UPVC confined concrete may be obtained as $(11\varepsilon_{cc} \times 16)/24 = 7.33\varepsilon_{cc} \approx 8.0\varepsilon_{cc}$. The parts of the curve 'BC' and 'CD' are dependent on values of k_3 and ε_{cc} were obtained using numerical and experimental curve matching. Figure 6 shows the variation of k_3 with D/t ratio for UPVC tube confined concrete. The 'CD' part of the curve was obtained by taking $k_4 = 0.8k_3$. As a result the coordinates of points 'C' and 'D' are ($k_3 f_{cc}$, $5\varepsilon_{cc}$) and ($k_4 f_{cc}$, $8\varepsilon_{cc}$). Figure 5b shows the stress–strain variation of concrete confined with UPVC tube and unconfined concrete.

5 Results and Discussion

The confinement is the main governing factor to study the behavior of concrete filled tubes. Modes of failure, load carrying capacity, stress–strain curve of confined concrete and ductility, all are dependent on confining pressure applied by the tube on concrete core. In the present study, some experiments were carried out to obtain the behavior of concrete confined with UPVC tube. In this process a total 18 specimens are cast and tested by applying axial loading on concrete core only. The internal hydrostatic pressure tests were conducted to get the confining pressure (Fig. 2). The confining pressures obtained from pressure test and compression test of CFUT specimens are quite similar (Table 1). After getting the confining pressure, confined compressive strength of concrete (f_{cc}) and ε_{cc} is calculated using Eqs. (1) and (2) respectively. By having the value of f_{cc} and ε_{cc}, it is possible to draw stress–strain curve upto the peak. Now to obtain the post peak behavior of curve, the experimental results are matched with the numerical results by taking the various trial values of k_3 and k_4 with corresponding strains. The results of experiments and numerical simulations of concrete filled UPVC tubes are given in Table 2. It can be seen that error in the results obtained from simulation is within 1%. After getting the appropriate values of k_3 and k_4, an empirical relation between k_3 and D/t is obtained through regression analysis. A stress–strain model for concrete confined with UPVC tube is obtained by using the values of f_{cc}, ε_{cc}, k_3 and k_4. Figure 6 shows the curve between k_3 and D/t.

For the parameter k_3, following empirical relation is proposed

$$k_3 = 0.0001\left(\frac{D}{t}\right)^2 - 0.011\left(\frac{D}{t}\right) + 0.7635\ (20 \le D/t \le 40) \tag{5}$$

It was observed that the values of f_1 and k_3 decrease with the increase in diameter to thickness ratio. Confining pressure depends on the tensile strength of tube and the value of D/t. As a result the value of f_1 varies from 1.15 to 2.25 MPa (166.75–326.25 psi) when UPVC pipes with nominal pressure 0.6, 0.8 and 1.0 MPa (87, 116, 145 psi) were used. In the case of CFUT specimens, the degradation parameter (k_3) varies from 0.5 to 0.58 and k_4 is taken as $0.8k_3$.

The failure patterns of confined concrete specimens depend on the constraining factor (α) given by Wang and Yang (2010), Gupta (2013) and confinement ratio (β) given by Ansari and Li (1998). The constraining factor is defined as:

$$\alpha = A_p f_u / A_c f_c$$

where A_p and f_u are the cross-sectional area and tensile strength of UPVC tube respectively; A_c and f_c are the cross-sectional area and compressive strength of concrete cylinder respectively. The confinement ratio is defined as:

$$\beta = (f_l / f_c)$$

After calculating the parameters α and β, it was found that the value of β is twice the value of α. The value of β is given in Table 2. In the present case the value of α is less than 0.2 and β is less than 0.1 for all specimens. The failure pattern was observed as shear failure with slight bulging of tube. Wang and Yang (2010) and

Table 2 Experimental and numerical results

Specimen[a]	L (mm)	D/t	Ultimate load (kN)			Confin-ement ratio $\beta = (f_l/f_c)$	Mode of failure	Failure load (kN)	% strain at failure
			Experiment	simulation	From equation $P_u = f_c A_c + 2 f_u A_p$				
160P3PC30	800	40.10	759	753	724	0.0336	Shear cracks and slight bulge	348	1.69
160P4PC30		30.65	769	763	726	0.0448		351	1.50
160P5PC30		22.79	793	788	734	0.0616		372	1.49
200P3PC30		33.67	1123	1127	1131	0.0405		509	1.51
200P4PC30		24.66	1142	1135	1136	0.0567		657	1.49
200P5PC30		22.50	1182	1176	1135	0.0627		614	1.40
225P3PC30		26.01	1352	1354	1439	0.0533		607	1.50
225P4PC30		23.49	1420	1411	1440	0.0596		695	1.46
225P5PC30		22.57	1491	1482	1441	0.0624		772	1.38
160P3PC40		40.10	816	811	796	0.0302		325	1.55
160P4PC40		30.65	839	832	795	0.0402		384	1.42
160P5PC40		22.79	850	845	800	0.0553		456	1.32
200P3PC40		33.67	1231	1223	1241	0.0364		564	1.38
200P4PC40		24.66	1245	1239	1241	0.0510		599	1.38
200P5PC40		22.50	1280	1273	1237	0.0563		651	1.30
225P3PC40		26.01	1486	1471	1573	0.0479		694	1.43
225P4PC40		23.49	1550	1538	1572	0.0535		759	1.34

(continued)

Table 2 (continued)

Specimen[a]	L (mm)	D/t	Ultimate load (kN)			Confin-ement ratio $\beta = (f_l/f_c)$	Mode of failure	Failure load (kN)	% strain at failure
			Experiment	simulation	From equation $P_u = f_c A_c + 2 f_u A_p$				
225P5PC40		22.57	1610	1597	1571	0.0561	Macro cracks and slight bulge	806	1.28
75P3PC30-concrete loaded	1875	18.8	–	147	–	0.032	Buckling	69	0.536
75P3PC30-both loaded	1875	18.8	–	153	–	0.032		72	0.567
75P3PC30-concrete loaded	300	18.8	–	161	164	0.032	bulging	76	4.7

[a] 160P3PC30 represents the specimen composed of 160 mm diameter UPVC tube of class 3 (nominal pressure of 0.6 MPa) filled with plain concrete of grade M30. Other CFUT specimens are represented in the same manner. Similarly '75P3PC30' represents the specimens of 75 mm diameter tube with different lengths having same properties as the specimen '160P3PC30'

Note 1 in. 25.4 mm; 1 kN = 224.808 lbf

Fig. 7 Verification of mode of deformation

Gupta (2013) also found that specimens failed due to shear when the value of α was less than 0.204 and bulging when value of α was more than 0.204. The specimen (no. 114) of Richard et al. (1928) having the value of β equals to 0.07 showed the shear failure. Sfer et al. (2002) observed the failure due to several vertical and inclined cracks when confining pressure was less than 4.5 MPa (652.5 psi) or ($\beta < 0.15$). The typical deformed shape obtained from experiment is verified by the typical deformed shapes picked up from literature (Fig. 7).

The deformed shapes of few tested specimens and their corresponding load–displacement curves are plotted in Fig. 8a, b. The maximum displacement at failure was observed between 10 and 14 mm (0.394 and 0.551 in.). Strain at maximum load ε_{cc} was found to be 0.0043–0.0068. The angle of shear cracks from horizontal axis was measured. It was between 63° and 65°. The post peak displacement is important regarding ductile behavior of material. From the load–displacement curve, it was found that the displacement after the peak load was about 5–10 mm (0.197–0.394 in.). Load–displacement curves obtained from numerical analysis and experiments were found in good match (Fig. 9b). Figure 9c, d represent typical variation of load–displacement curves for the specimens made up of UPVC tubes with different diameter of same class and same diameter of different class respectively. It can be seen form Fig. 9c that the initial slope of curves increases as the diameter of tubes increases and the post peak slope are nearly same. While from Fig. 9d it may be observed that the initial slope is nearly equal and the post peak slope is decreasing with the higher class of tube due to enhancement of confinement. To see the slenderness effect on the capacity of CFUT; finite element analysis was carried out for long and short column, designated as "75P3PC30" with diameter 75 mm (2.95 in.). Lengths were kept as 1875 mm for long column and 300 mm (11.82 in.) for short column by keeping properties of tube and concrete same as in 160P3PC30. For long column, load was applied in two ways; only on concrete core and on concrete and tube both. For short column only concrete core was loaded. Buckling was occurred in long column in both the cases. The values of strains corresponding to failure stress ($k_4 fcc$) were 0.536% (concrete loaded) and 0.567% (both loaded) for long column

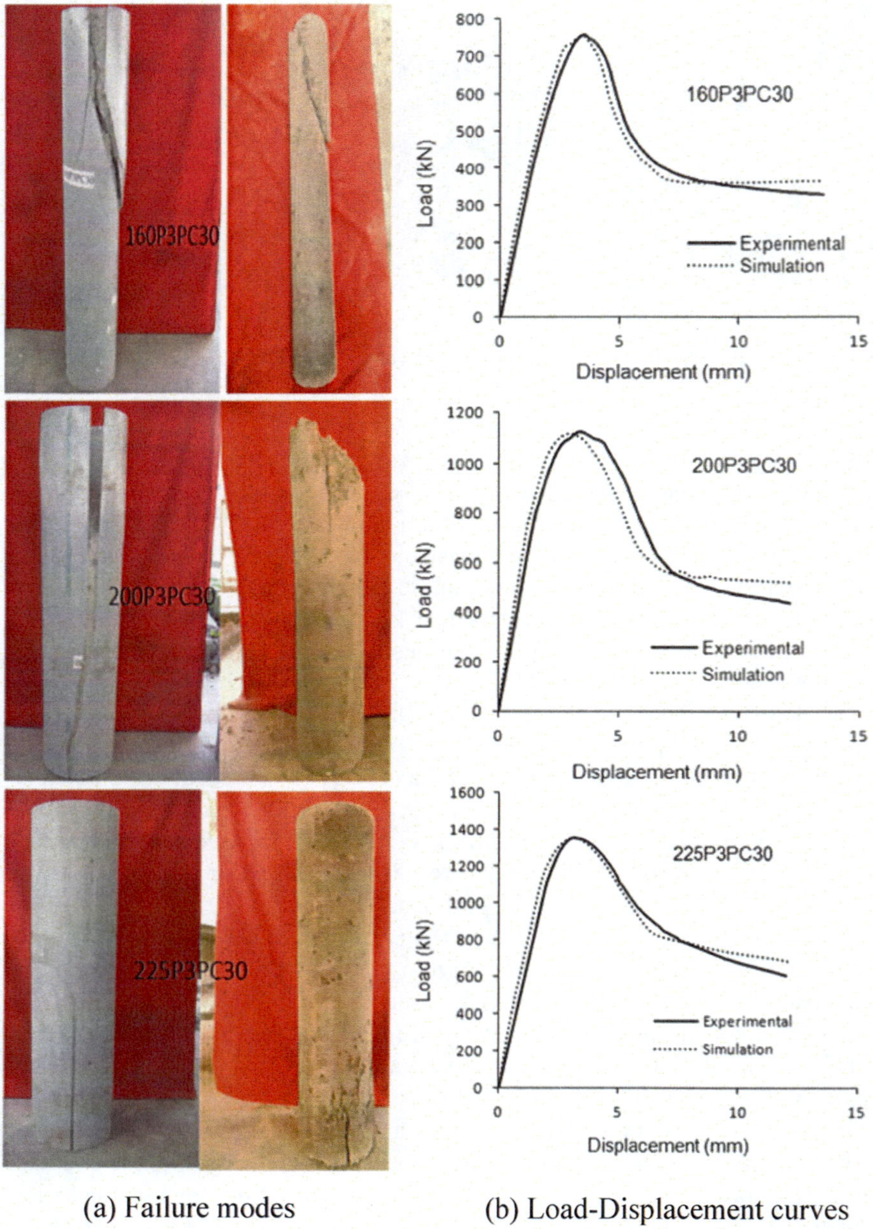

Fig. 8 **a** Failure modes, **b** load–displacement curves (*Note* 1 in. 25.4 mm; 1 kN = 224.808 lbf)

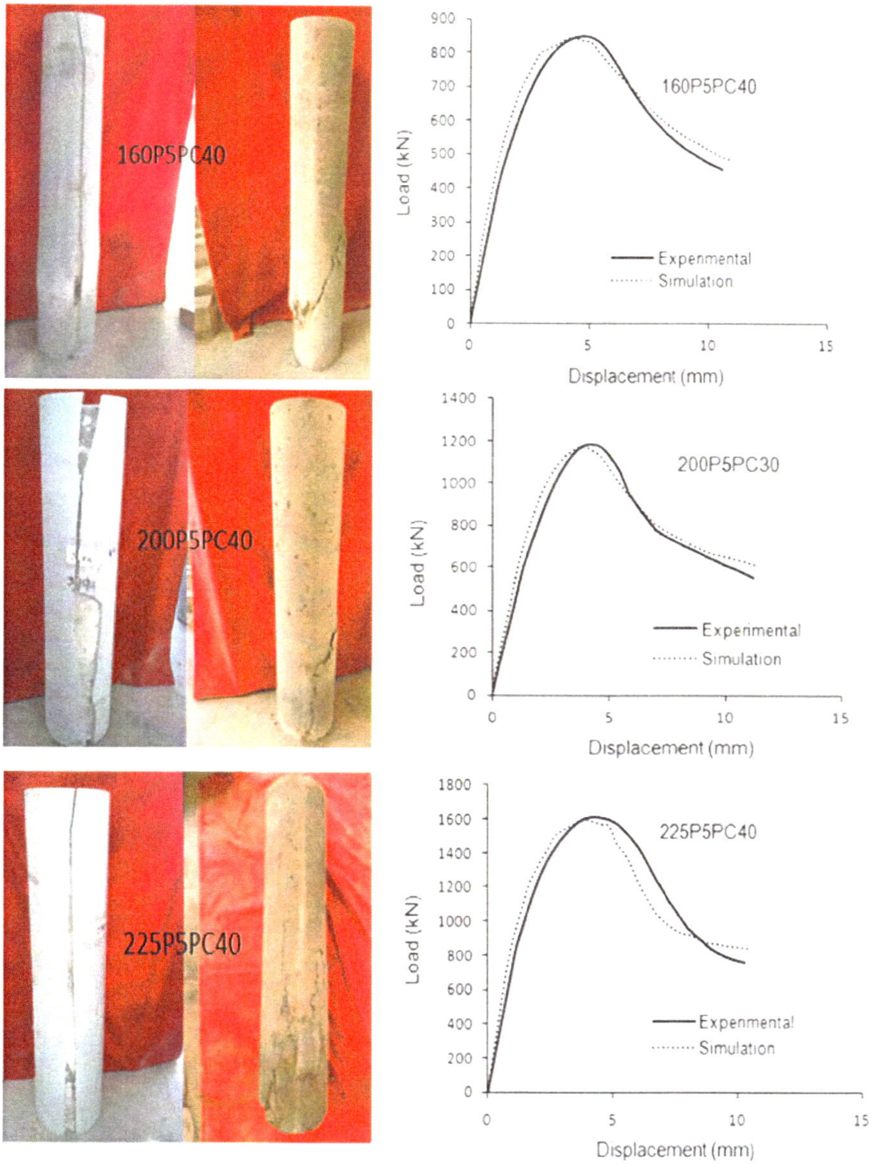

Fig. 9 **a** Failure modes, **b** load–displacement curves, **c** typical variation of load–displacement for different diameter and same class of tube, **d** typical variation of load–displacement for different class of same diameter tube, **e** deformed shapes of long column, **f** comparison of load–displacement curves for long and short columns (*Note* 1 in. 25.4 mm; 1 kN = 224.808 lbf)

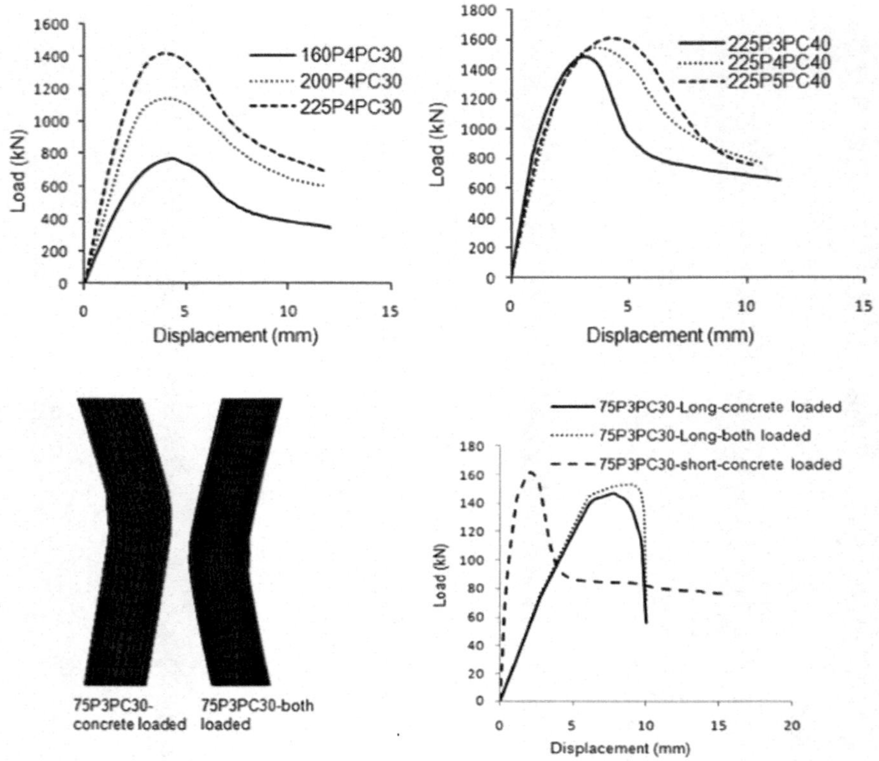

Fig. 9 (continued)

and 4.7% for short column (Table 2). The deformed shape of and load–displacement curves for the specimen "75P3PC30" are shown in Fig. 9e, f. Typical experimental and simulated deformed shapes are shown in Fig. 10. The deformed shapes obtained from experiments and simulations are quite similar.

For the design of concrete filled UPVC tubular column, load carrying capacity is required. An equation is proposed to get the load capacity of CFUT. Proposed equation is as follow:

$$P_u = f_c A_c + 2 f_u A_p \tag{6}$$

The ultimate load capacities of all the specimens are calculated using the above equation and presented in Table 2. The load calculated using above equation are within 0.35–7.5% higher and lower of from experimental load. To verify the above proposed equation, the ultimate load capacities of specimens from literature are calculated using equation no.6 and presented in Table 3. The calculated load capacities are within ±8.5% of experimental capacities. Experimental load capacities (from literature) versus load capacities computed from Eq. (6) are graphically presented

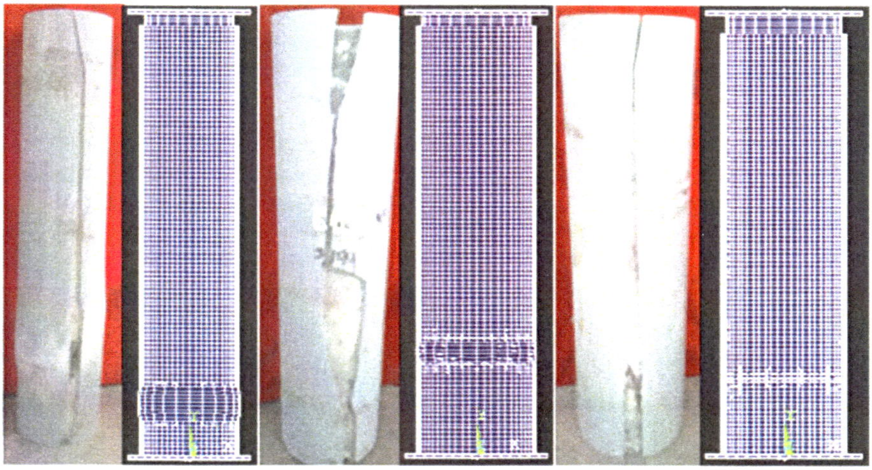

Fig. 10 Comparison of typical deformed shapes obtained from experiment and Simulation

Table 3 Verification of proposed equation (6)

References	Specimen	Ultimate load (kN)		
		Experiment	From equation $P_u = f_c A_c + 2f_u A_p$	% Variation
Kurt (1978)	2(B)	171	160	6.43
	7(C)	309	283	8.41
	9(B)	149	143	4.03
	14(C)	41	39	4.88
	18(C)	251	240	4.38
Gupta (2013)	T160M20PC-1	639	601	5.95
	T200M20PC-1	1026	958	6.63
	T140M25PC-1	530	533	0.57
	T140M40PC-1	714	738	3.36

Note 1 kN = 224.808 lbf

in Fig. 11. It can be concluded that proposed equation may be used for assessing the load capacity of concrete filled UPVC tubular columns. Table 4 presents the comparison of load capacity of specimens obtained from experiments and calculated using different models. It can be seen from Table 4 that models proposed by Richart et al. (1928) and Kurt (1978) underestimate the capacities. On the other hand the capacities calculated by Mander et al. (1988) and Saatcioglu and Razvi (1992) are on conservative side. The graphical representation of load capacities obtained from experiments and calculated by predicted models are depicted in Fig. 12.

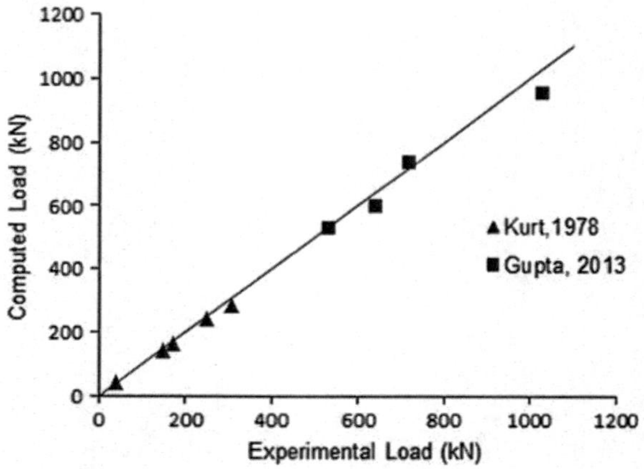

Fig. 11 Graphical representation of load capacities to verify proposed equation (6) (*Note* 1 kN = 224.808 lbf)

Table 4 Comparison of capacity (kN) obtained from experiments and calculated using different models

Specimens	Experiment	Richart et al. (1928)	Mander et al. (1988)	Kurt (1978)	Saatcioglu and Razvi (1992)
160P3PC30	759	726	776	709	778
160P4PC30	769	717	773	697	773
160P5PC30	793	717	782	692	777
200P3PC30	1123	1152	1246	1116	1243
200P4PC30	1142	1141	1247	1098	1239
200P5PC30	1182	1128	1235	1084	1226
225P3PC30	1352	1485	1628	1426	1614
225P4PC30	1420	1461	1603	1402	1589
225P5PC30	1491	1448	1588	1389	1575
160P3PC40	816	798	849	781	850
160P4PC40	839	787	844	767	842
160P5PC40	850	784	851	758	844
200P3PC40	1231	1262	1359	1226	1353
200P4PC40	1245	1246	1355	1203	1343
200P5PC40	1280	1231	1342	1187	1328
225P3PC40	1486	1619	1767	1560	1749
225P4PC40	1550	1592	1740	1533	1721
225P5PC40	1610	1578	1724	1520	1705
Mean	–	0.989	1.074	0.955	1.067
Standard Deviation	–	0.054	0.062	0.050	0.060
Coefficient of variance	–	5.43%	5.80%	5.24%	5.62%

Note 1 kN = 224.808 lbf

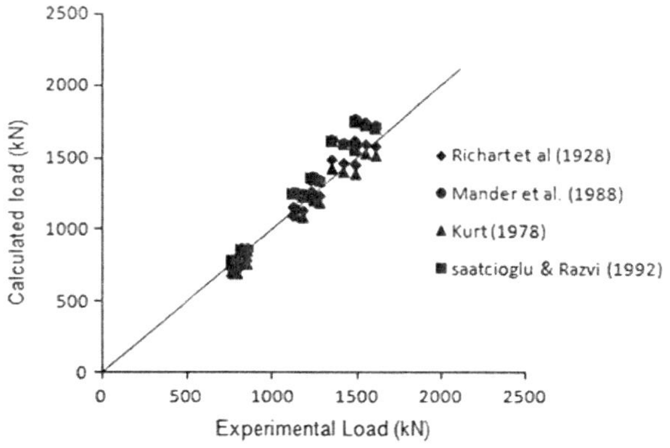

Fig. 12 Graphical representation of load capacities obtained from experiments and calculation (*Note* 1 kN = 224.808 lbf)

6 Conclusions

A new type of composite column, which can be obtained by filling concrete in unplasticised poly vinyl chloride (UPVC) tube is proposed and investigated in this study. On the basis of results obtained from experiments and nonlinear finite element analysis of concrete filled UPVC tubes, the following points can be concluded.

1. A stress–strain model for concrete confined with UPVC tube is proposed. This model is obtained by modifying the stress–strain model of concrete confined with steel tube. The post peak curve of concrete confined with steel tube is modified by introducing the parameters k_3 and k_4. The Material degradation parameter (k_3) depends on the value of D/t and decreases with the increase of D/t. The parameter meter k_4 depends upon value of k3 and equals to 0.8 k3. An empirical relation is proposed to calculate the value of and k_3.
2. The failure pattern of short specimens depends upon the constraining parameter (α) and confinement ratio (β). Failure of all specimens leads by development of shear crack and macrocracks with slight bulging.
3. An equation to calculate the load carrying capacity of concrete filled UPVC tubular columns is proposed. So this type of composite column may be easily designed using developed equation.
4. The post peak behavior of load-compression curve depends upon the confining pressure of tube and compressive strength of concrete. The absolute value of slope of curve increases with increase in concrete strength and decreases with increase in confining pressure.
5. The strain at failure for tested short specimens varies from 1.28 to 1.69%. It decreases with increases of compressive strength of concrete for the same class of tubes.

References

ACI-318 (2008) Building code requirements for structural concrete and commentary. American Concrete Institute, Detroit (USA)

Ansari F, Li Q (1998) High strength concrete subjected to triaxial compression. ACI Mater J 95(6):747–755

ASTM-D638-10. Standard test method for tensile properties of plastics. Annual book of ASTM Standards

Elwi AA, Murray DW (1979) A 3D hypoelastic concrete constitutive relationship. J Eng Mech 105(4):623–641

Fam AZ, Rizkalla SH (2001) Confinement model for axially loaded concrete confined by circular fiber-reinforced polymer tubes. ACI Mater J 98(4):451–461

Gupta PK, Verma VK (2014) Numerical study on axial compression of plastic tubes. The fourth International fib Congress on Improving Performance of Concrete Structures, Mumbai, India, pp 514–517

Gupta PK (2013) Confinement of concrete columns with unplasticized poly-vinyl chloride tubes. Int J Adv Struct Eng 5(9):1–8

Gupta PK, Verma VK, Nabam A, Ahuja AK (2012) A study on the behavior of axially loaded concrete filled unplasticised poly vinyl chloride (UPVC) tubes. In: Proceedings of fourth international conference on structural stability and dynamics, Jaipur, India, pp 662–667

Hu HT, Huang CS, Wu M-H, Wu Y-M (2003) Nonlinear analysis of axially loaded concrete-filled tube columns with confinement effect. J Struct Eng V 129(10):1322–1329

IS: 10262-2009. Concrete mix proportioning-guidelines. Bureau of Indian Standard. New Delhi, India

IS: 12235(Part 8)-1986. Methods of test for Unplasticized PVC pipes for potable water supplies. Bureau of Indian Standard, New Delhi, India

IS: 4985-2000 (Reaffirmed 2005) "Unplasticized PVC pipes for potable water supplies- specification. Bureau of Indian Standard, New Delhi, India

IS: 8112-1989 (Reaffirmed 2005) 43 grade ordinary Portland cement-specification. Bureau of Indian Standard, New Delhi, India

Kurt CE (1978) Concrete filled structural plastic columns. J Struct Div ASCE 104(1):55–63

Mander JB, Priestley MJN, Park R (1988) Theoretical stress-strain model for confined concrete. J Struct Eng. 114(8):1804–1830

Marzouck M, Sennah K (2002) Concrete filled PVC tubes as compression members. In: Proceedings of International Congress on Challenges of Concrete Construction at Scotland, UK

Ragab AR, Mahmoud MA, Khorshied SA (2001) Yielding of commercial poly (vinyl chloride) pipe material. J Polym Sci 81:991–999

Richart FE, Brandzaeg A, Brown RL (1928) A study of the failure of concrete under combined compressive stresses. Bull 185. Champaign; University of Illinois Engineering Experimental Station, IL, USA

Saatcioglu M, Razvi SR (1992) Strength and ductility of confined concrete. J Struct Eng 118(6):1590–1607

Saenz LP (1964) Discussion of equation for the stress-strain curve of concrete. ACI Struct J 61(3):1229–1235

Sfer D, Carol I, Gettu R, Etse G (2002) Study of the behavior of concrete under triaxial compression. J Eng Mech 128(2):156–163

Wang J, Yang Q (2010) Experimental study on mechanical properties of concrete confined with plastic pipes. ACI Mater J 107(2):132–137

Study of the Composite Action of FRP Floor Beams and RC Slab Under Flexural Loading

Himanshu Chawla, N. Chandramauli, and S. B. Singh

Abstract In this chapter, the effect of different types of connections on the flexural behavior of FRP-concrete composite slab is studied using finite element software ABAQUS. The connection between the FRP beams and RC slab is made using shear connectors and adhesive bonding. Web and flange elements of I-beams are connected with cohesive layer, because it is the weaker portion in the FRP I-beams and is highly prone to fail under three-point loading. Accuracy of the numerical model is verified by comparing the results with published experimental study. Further, the parametric study is performed on beams having different size and spacing of shear connectors. Along with, failure and service load of hybrid beams is determined for different geometric configuration of I-beams and width of RC slab. The flexural response of hybrid FRP-concrete beams obtained from finite element software shows good agreement with experimental testing. It is observed that shear connectors help in improving the strength of concrete-FRP hybrid beams. With increasing the length of shear connectors there is slight improvement in the strength of beam. As density of shear connectors decreases, consequently flexural stiffness of the beam (initial slope of load–deflection curve) decreases. It is also noted that strength and stiffness of the hybrid beam increases with increase in the length of the shear connectors. From the flexural study of the beams having different width-to-thickness (B/t) ratio, it is observed that some beams having different B/t ratio have same failure and service loads, due to the pre-mature failure of the web-flange junction of the I-beams. Hence, it is stated that addition of RC slab enhances the flexural stiffness, i.e., service load and also enhances the strength of web-flange junction. Addition of reinforced concrete slab over the FRP I-beams, reduces the chances of local buckling of the flange and web of the I-beam.

H. Chawla (✉) · N. Chandramauli
Civil Engineering Department, TIET, Patiala 147004, India
e-mail: himanshu.chawla11@gmail.com

S. B. Singh
Civil Engineering Department, BITS Pilani, Pilani 333031, India
e-mail: sbsingh@pilani.bits-pilani.ac.in

S. B. Singh et al. (eds.), *Emerging Trends of Advanced Composite Materials in Structural Applications*, Composites Science and Technology,
https://doi.org/10.1007/978-981-16-1688-4_6

1 Introduction

Fiber reinforced polymer (FRP) beams exhibit the brittle failure under three/four-point bending test. Reinforced concrete-FRP hybrid beams show pseudo-ductile characteristics, if the ideal bond formed by mechanical and/or adhesive connectors. Hence, it is important to study the flexural behavior of different connection of reinforced concrete-FRP hybrid beams. Hall and Mottram (1998) analyzed the combined FRP reinforcement and permanent formwork for concrete members. They have performed two tests such as 4-point bending test and Bond test. In FRP bond test, standard 150 mm cube mold for concrete was taken. The beam failed under shear and FRP panel failed at web-flange interface. Thus, they have concluded that reinforcement should be appropriate to optimize the use of materials. Correia et al. (2007) studied the flexural behaviour of GFRP concrete hybrid beams with interconnection slip. Alnahhal et al. (2008) studied the combined behaviour of concrete-FRP Bridge deck on steel beam. Authors have used hybrid deck system consisting of various units of GFRP trapezoidal cell and rectangular shell at the compression region and also concrete slab in the compression region, thus leading to the formation of deck of bridge. Shear keys were used and panels were connected for efficient transfer of bending moment and shear force. Strain gauges were used to measure slip between hybrid deck and girder. There were no traces of cracks or damage in shear connectors thus giving excellent performance of HFRP (hybrid FRP) concrete bridge decks under service loads. But these results are based on parameters used and can vary rapidly. Correia et al. (2009a) examined the concrete-GFRP hybrid sections for floors of the buildings. Authors provided a novel technique for connecting floors of building with GFRP pultruded profile using adhesive layer of epoxy and steel bolts. Flexure test was performed and it was concluded that in GFRP I-girder with lateral bracing helped in preventing the lateral torsional buckling of girders and the failure of the beam was due to the local buckling of the flange at top. In another study, Correia et al. (2009b) analyzed the behaviour of multi span concrete-GFRP hybrid beams in flexure. Two shear connection systems between GFRP pultruded and concrete layer was developed, one was stainless steel bolts fastened on either side and another is adhesive layer (epoxy) applied on the interface between two materials. Hybrid beams with 6 m length were tested in positive and negative bending. It was observed that failure was due to shear in web at supports and compression under the loading. The lateral torsional buckling prevents full exploitations of materials strength capacity so lateral bracing must be provided to avoid this failure.

Gonilha et al. (2013) have studied the creep, dynamic and static behaviour of a full scale GFRP-SFRSCC hybrid footbridge. GFRP pultruded profiles was considered as main girder and SFRSCC (steel fiber reinforced self-compacting concrete) slab was set over profile acts as extension bridge deck. An adhesive was provided between the girder and deck to proper connection with least slip between the materials. Chakrabortty et al. (2011) studied the flexural performance of outside filament wound on the hybrid FRP concrete beams. In the load–displacement curve, beams with steel fiber being reinforced with high strength concrete blocks shows linearity up to failure. The beams having no CFRP reinforcement failed by the lack of native bearing capability of the supports and the other beams failed by crushing of concrete. Kim and Shin (2011) have studied the flexural behaviour of reinforced concrete-FRP hybrid beams under bending test with sustained loading. Concrete layer was reinforced with steel bars and adhesive of constant thickness was used to attach the concrete layer to the FRP beams. Authors have concluded that concrete layer improved the strength and stiffness of FRP beams. Mendes et al. (2011) have developed the pedestrian bridge using fiber reinforced self-compacting concrete deck and GFRP profile. Creep tests were performed on two pedestrian bridge prototypes. The length was half of main span and transverse dimensions were four times smaller than main dimensions. Dagher et al. (2012) analyzed the bending behaviour of arches made from FRP tubes having in-filed concrete. Failure mode was rupture of FRP in the tensile region at the two-third of span. Zhang et al. (2015) studied the flexural behaviour of hybrid concrete beam reinforced with basalt fiber reinforced polymer bars (BFRP). From the experimental investigation, it was observed that BFRP bars have high bond strength and tensile strength but the large deflection was observed, which is due to the low elastic modulus. Qin et al. (2017) studied the effect of the reinforcement ratio on the flexural performance of concrete FRP hybrid beam. In order to enhance the strength and flexural ductility of FRP beams, it was reinforced with steel bars. It was observed that the hybrid reinforcement ratio b/w FRP and steel plays a pivotal role on the flexural performance of hybrid FRPRC beams. Authors have also examined the effect of reinforcement ratio on the bending performance of concrete beam in both over and under reinforced case using 3D finite element model. The result showed that appropriate hardness, toughness and ductility can be achieved by proper design of reinforcement ratio for hybrid beam.

Koaik et al. (2017) analyzed the behaviour of concrete GFRP hybrid beams under bending. Six I-beams were connected to a pair of blocks of concrete having dimension 100 × 200 × 200 mm. Two shear connector tests were performed to analyze the behaviour at the interface between GFRP beams and concrete layer. Digital image transducers were used to analyze the force-slip behavior. It was observed that the push out test overestimates the ultimate stress at the interface and underestimates the stress in composite beams. Cao et al. (2017) studied the flexural behaviour of expansive concrete beam reinforced with hybrid CFRP enclosure and steel bars. It was observed that flexural crack appeared at middle of span, which led to de-bonding between CFRP and concrete.

Araba and Ashour (2018) examined the flexural performance of concrete steel/GFRP reinforced hybrid continuous beam. The surface of GFRP beams was coated by sand to improve bond with concrete and transfer the load fully. Pull out test was conducted for FRP-concrete prism. During testing, failure modes were rupture of GFRP bars, failure by yielding of steel and crushing of concrete, shear failure and ductile flexural failure. Chellapandian and Prakash (2019) performed the three-point bending test on RC beams strengthened with HFRP laminates. Four specimens of square cross section of 230 mm with a total length of 2000 mm reinforced with 8 bars of 12 mm diameter providing a spacing of 100 mm c/c. It was noted that in the region having constant moment, first crack appeared on the bottom side and beyond the peak load there was crushing of concrete.

2 Numerical Model

In the present study, finite element software ABAQUS is used to investigate the flexural behaviour of FRP-concrete hybrid beams and shear strength of the concrete-FRP interface. The 3D deformable solid parts are used to model the concrete layer, web and flanges of FRP I-beam. Stiffeners of L and T-shaped were also made by 3D deformable solid, while the bearing plates were made up of non-deformable discrete rigid. The reason behind this is that the bearing plates adopted in the experiments were of steel and the flexural rigidity of plates was much larger than that of FRP beam.

2.1 Meshing

Meshing or discretization is generally carry out in FE analysis as it reduces the degree of freedom from infinite to finite and thus provide ease in solution. Meshing thus helps in breaking the domain into pieces where each piece represents an element which helps in applying the FE technique. In this study, combination of solid elements and shell elements were adopted from the ABAQUS library and utilized for modeling the three-point bend test specimen. Continuum 3-D eight node reduced integration elements C3D8R were used to model the concrete slab and the reinforcing steel re-bars. The bearing plates are modeled using 3 dimensional 4 nodes rigid element R3D4, i.e., four node element type. The stiffening elements (L and T) shapes are also modeled by C3D8R (Figs. 1 and 2).

In the FE model, non-linear material properties were provided to FRP I-section, bearing stiffeners and shear connectors. The brief explanation is given next.

Fig. 1 Discrete rigid bearing plate

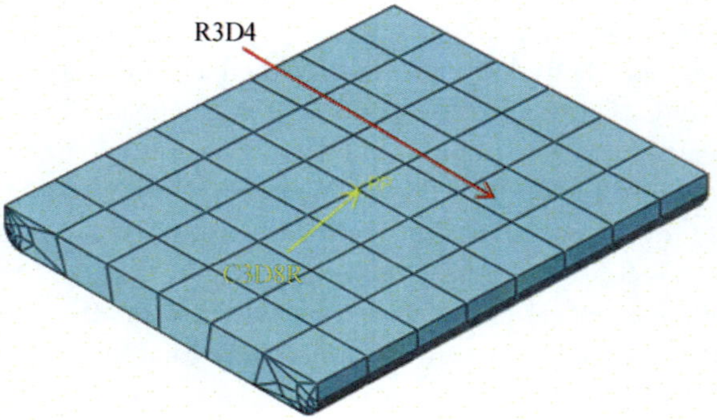

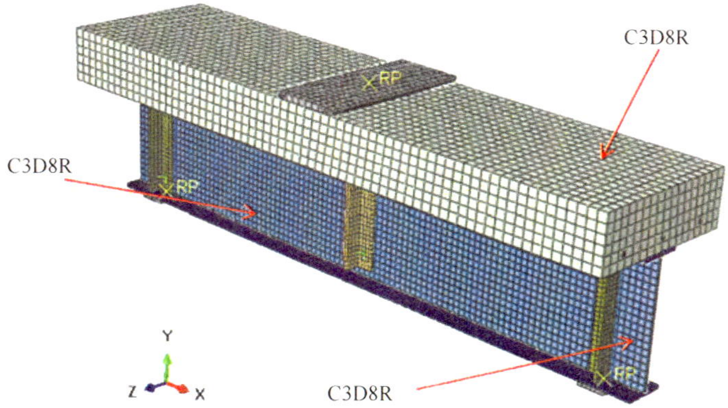

Fig. 2 Meshing of various parts of the model and different element type used

2.2 *Material Modeling*

2.2.1 Modeling of Concrete Layer

Material non-linearity in reinforced cement concrete slab is provided by adopting concrete damaged plasticity (CDP) model. Parameters of CDP analysis depends on the uni-axial compressive and tensile stress–strain curve and based on the flow plasticity theory. This theory helps in analyzing the plastic behavior by assuming a flow rule, which helps to describe the extent to which the material has been deformed. When concrete is subjected to severe state of inelastic stress, a significant change in volume occurs which is known as dilation.

A typical compressive stress–strain curve for concrete is shown in Fig. 3. Initially for the linear elastic range, stress is proportional to strain and the value of limit stress is taken as $0.4f_{cm}$ and for the parabolic portion, i.e., non-linear the stress ranges from $0.4f_{cm}$ to f_{cm}, where f_{cm} is the ultimate strength. The slope of the stress–strain curve,

Fig. 3 Compressive stress–strain curve of concrete

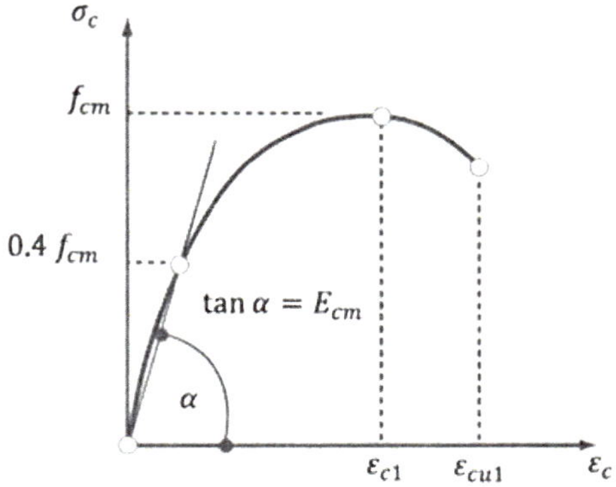

i.e., tanα represents the young modulus of concrete. The stress in non-linear region (σ_c), i.e., $0 < \varepsilon_c < \varepsilon_{cu1}$ can be determined by using Eq. (1)

$$\frac{\sigma_c}{f_{cm}} \approx \frac{\eta k - \eta^2}{1 + (k - 2)\eta} \tag{1}$$

$$\eta = \frac{\varepsilon_c}{\varepsilon_{c1}}, \varepsilon_{c1} = 0.7 f_{cm}^{0.31} \leq 2.8, k = 1.05(E_c \times |\varepsilon_{c1}|)/f_{cm} \tag{2}$$

The nominal ultimate strain is assumed to be 0.0035. After the peak stress, softening occurs and the stress in this region is calculated by Eq. (3)

$$\sigma_c = \left(\frac{2 + \gamma_c f_{cm} \varepsilon_c}{2 f_{cm}} + \gamma_c \varepsilon + \frac{\gamma_c}{2\varepsilon_c}\varepsilon^2\right)^{-1} \tag{3}$$

where γ_c is taken as 1.7. Failure of the concrete under tension and compression is calculated by damage variable d_c and d_t, respectively. The expression for damage variable in compression is given by the Eq. (4)

$$d_c = 1 - \frac{\sigma_c E_c^{-1}}{\varepsilon_c^{pl}\left(\frac{1}{b_c} - 1\right) + \sigma_c E_c^{-1}} \tag{4}$$

When these two variables of damage are equal to zero there is no damage in concrete otherwise when the variables value are equal to one it represents that the concrete has been fully damaged. The damage variable in case of compression is related to plastic strain (${\varepsilon_c}^{pl}$) which varies proportional to strain in inelastic zone (${\varepsilon_c}^{in}$), i.e.,

$$\varepsilon_c^{in} = \varepsilon_c - \sigma_c E_c^{-1} \tag{5}$$

The plastic strain can be calculated by

$$\varepsilon_c^{pl} = b_c \varepsilon_c^{in} \tag{6}$$

The above equation has been determined by using a factor b_c which is constant and ranging from $0 < b_c < 1$.

In the tension zone, the stress varies proportionally to strain until there is no emergence of cracks in concrete and after the crack starts to propagate the stress reduces non-linearly to zero value. When d_t is greater than zero and the maximum principle strain in plastic is positive and crack propagation in concrete starts. At the point of propagation of damage, the crack direction and the maximum principle strain in plastic region direction are orthogonal. As per the ABAQUS manual, at zero stress the value of strain is taken as ten times the strain at the point of failure. The damage variable in tension d_t depends on ${\varepsilon_t}^{pl}$. The value for constants b_c and b_t are

Table 1 Concrete plasticity damage properties

Dilation angle	Eccentricity	fb_0/fc_0	K	Viscosity parameter
13	0.1	1.16	0.7	0

Table 2 Compressive properties for concrete

Yield stress (N/mm^2)	Inelastic strain
16.75	0
20.00	0.0020
25.00	0.0035

Table 3 Tensile properties for concrete

Yield stress (N/mm^2)	Inelastic strain
3	0
4	0.012
5	0.020

assumed to be 0.7. The expression for damage is given below in the Eq. (7). The concrete damage properties is given in Tables 1, 2 and 3.

$$d_t = 1 - \frac{\sigma_t E_t^{-1}}{\varepsilon_t^{pl}\left(\frac{1}{b_t} - 1\right) + \sigma_t E_t^{-1}} \tag{7}$$

$$\varepsilon_t^{pl} = b_t \varepsilon_t^{in} \tag{8}$$

2.2.2 Bearing Plate

In order to avoid the deformation of bearing plate, the bearing plate was made with discrete rigid part. Due to its rigid nature no property was assigned. Stress concentration under the bearing plate was reduced by providing the high fillet radius of 0.006 at the edges.

2.2.3 FRP I-section and Stiffeners

Due to orthotropic nature of the laminates, composite lay-up technique has been selected to provide the material properties to I-sections and stiffeners. Equivalent material properties of FRP I-section and stiffeners are presented in Table 4. Hashin damage failure criteria is considered to investigate the failure of the laminate in

Table 4 Elastic properties (MPa) of FRP I-section and stiffeners

E_1	E_2	E_3	ν_{12}	ν_{13}	ν_{23}	G_{12}	G_{13}	G_{23}
27,950	8000	4000	0.28	0.28	0.21	2380	2380	1170

Table 5 Different strengths (N/mm^2) of flange and web elements of FRP I-section

Longitudinal tensile strength	Longitudinal compressive strength	Transverse tensile strength	Transverse compressive strength	Longitudinal shear strength	Transverse tensile strength
400,000	330,000	73,000	55,000	60,000	17,000

tension and/or compression under the flexural loading. Hashin failure criteria for matrix and fiber damage are given as follows (Table 5):

Matrix tension failure ($\sigma_{22} + \sigma_{33} > 0$):

$$(\sigma_{22} + \sigma_{33})^2/Y_t^2 + (\sigma_{23}^2 - \sigma_{22}\sigma_{33})/S_{23}^2 + (\sigma_{12}^2 - \sigma_{12}^2)/S_{12}^2 = 1 \quad (9)$$

Fibre tension failure ($\sigma_{11} > 0$):

$$\sigma_{11}^2/X_t^2 + (\sigma_{12}^2 + \sigma_{13}^2)/S_{12}^2 = 1 \quad (10)$$

Matrix compression failure ($\sigma_{22} + \sigma_{33} < 0$):

$$(\sigma_{22} + \sigma_{33})/Y_c \times [(Y_c/2s_{23})^2 - 1] + (\sigma_{22} + \sigma_{33})^2/4S_{23}^2 + (\sigma_{12} + \sigma_{13})^2/S_{12}^2 = 1 \quad (11)$$

Fibre compression failure ($\sigma_{11} < 0$):

$$\sigma_{11} = -X_c \quad (12)$$

Fibre-Matrix shear failure ($\sigma_{11} < 0$):

$$\sigma_{11}^2/X_c^2 + \sigma_{12}^2/S_{12}^2 + \sigma_{13}^2/S_{13}^2 = 1 \quad (13)$$

Damage evolution is used to reduce the stiffness of the laminate (flange and web), after the failure of the layers of laminate. Damage evolution helps in explaining the behaviour of the material after the post damage initiation, i.e., it explains the rate of degradation of the stiffness of the material after the criteria for damage initiation is satisfied. Stress in the layers after degradation is calculated by

$$\sigma = (1 - D)\sigma_d \quad (14)$$

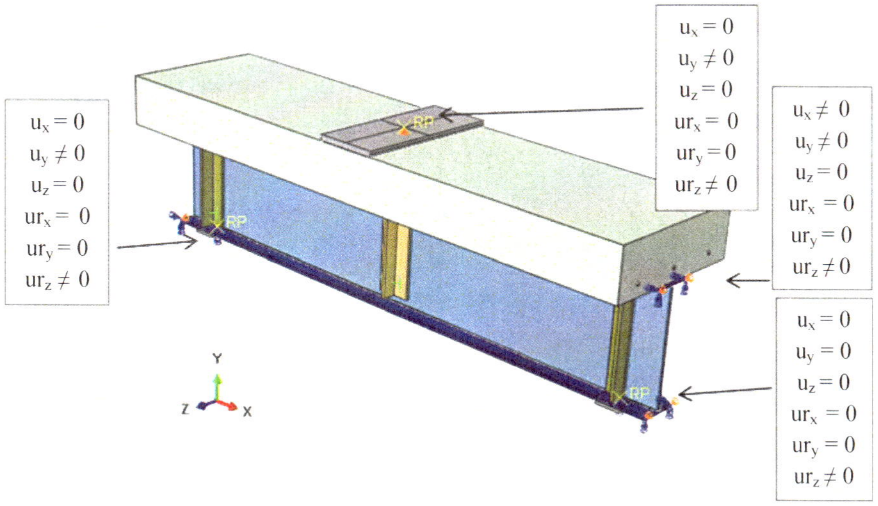

Fig. 4 Description of boundary conditions

where, σ_d is the stress generated due to undamaged response. Here, D is the overall damage variable, which captures the effect of all the combined active damage mechanism. The fracture occurs when $D = 1$, i.e., point of material has fully failed.

2.3 *Boundary Conditions*

Simply supported boundary conditions were applied at the ends of the beams as shown in Fig. 4. Along with, torsional boundary conditions were applied at the end of the compression flange (Fig. 4). As stated earlier, this study is performed under the displacement control mode. Therefore displacement of 50 mm is provided at the reference point of the bearing plate. Bearing plate is restrained to deflect in perpendicular to the direction of load applied and as well as torsion is also restrained as shown in Fig. 4.

2.4 *Interaction*

In the experimental investigation, it was observed that hybrid concrete-FRP I-beam fails due to the failure of shear connectors and/or web-flange junction failure of I-beam. Therefore, interaction of different parts plays the vital role in the prediction of the flexural strength of hybrid FRP-concrete I-sections. Different interaction technique provided in the simulation is explained in the following sections:

2.4.1 General Contact

General contact helps in providing interaction which allows you to make contact between many regions through a single interaction. Using this technique, interactions are provided between the bearing plates with concrete, and bottom flange surface with bottom bearing plate.

2.4.2 Tie Interaction

Tie interaction is provided in case of L- and T-shaped bearing stiffener with the flange of I-section. It is required to resist the penetration of the stiffeners in to the flange. In this interaction, bottom edge of stiffener is master surface and the surface of the laminate below the bearing stiffener is considered as slave surface.

2.4.3 Embedded Interaction

This type of interaction is used to provide the surface interaction of reinforcement bars and shear connectors with concrete. In this technique, embedded region is considered as bars and connectors, while the host region is considered as concrete FRP.

2.4.4 Surface to Surface Contact

These types of interactions are generally used for providing cohesive interaction, i.e., contact nonlinearity. At the web-flange junctions of FRP I-beam cohesive interaction is provided, it is because web-flange junctions of FRP I-beams are very prone to failure under the flexural loading. Surface of the flange is considered as master surface while the surface of web (thickness × length) is considered as slave surface. In the contact properties, cohesive behavior is chosen between flange and web element. Cohesive properties are presented in Table 6. In these properties, energy is chosen as the mode mix ratio with linear softening having Benzeggagh-Kennane as the specific mixed mode behavior. The value for power law/BK components is taken equal to 1.45.

Table 6 Properties of cohesive layer

K_{nn}(N/mm)^3	K_{ss}(N/mm)^3	K_{tt}(N/mm)^3	Normal only (MPa)	Shear mode-I (MPa)	Shear mode-II (MPa)	Normal fracture energy (N/mm)	Shear mode-I fracture energy (N/mm)	Shear mode-II fracture energy (N/mm)
1700	800	800	30	25	25	0.09	0.9	0.9

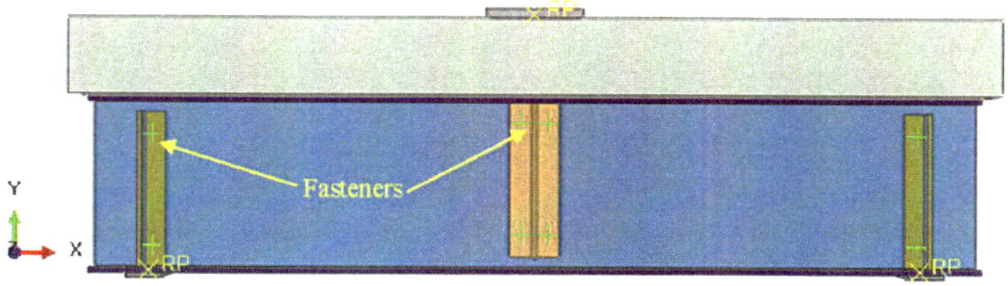

Fig. 5 Connection of stiffeners using fasteners

2.4.5 Fasteners

This is generally used to join the two or more surfaces. In the present case, they are used for joining L and T-shaded bearing stiffeners to the web of I-section. For this, firstly attachment points are created at particular offset from the edges of the stiffener. Further the fastener of radius 10 mm was provided at the attachment points as shown in Fig. 5.

3 Verification of the FE Model

This section presents the comparison of the flexural responses of the hybrid beams obtained from the proposed numerical model and published experimental study (Nguyen et al. 2015). Hybrid beam modeled in ABAQUS have fiber reinforced concrete layer over the compression flange of FRP I-section. Shear connectors are used to connect the concrete layer with flange of FRP I-section as shown in Fig. 6. Flexural response obtained from the numerical model is shown in Fig. 7 (Nguyen et al. 2015). It is observed that the proposed modeling technique shows the good agreement of numerical and experimental results Compression failure of the concrete layer was noticed under the bearing plate. The red region under the bearing plate in Fig. 8 shows the high stress concentration. Previous studies (Singh and Chawla 2019; Chawla and Singh 2019; Bank and Yin 1999) have shown that web-flange junction of the FRP I-section is the weaker portion of the beams and the first mode of failure of beams is the failure of web-flange junction. But the addition of the concrete layer over the compression flange of the FRP I-section enhances the flexural strength, and the failure mode of the beams is compression failure of concrete layer.

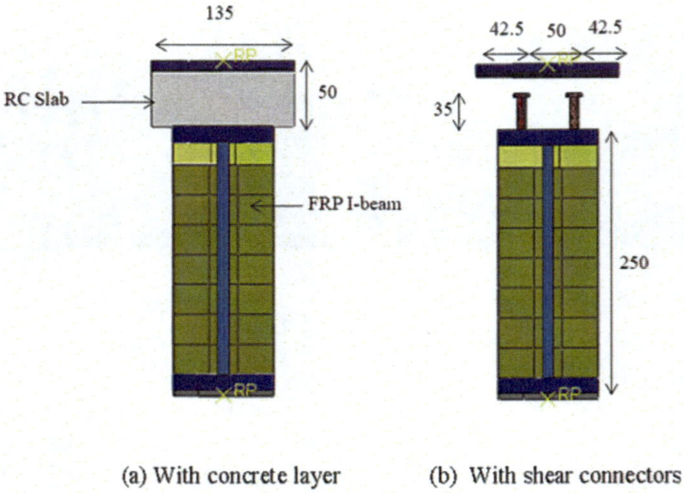

Fig. 6 Concrete-FRP hybrid beam modelled in ABAQUS (All dimensions are in mm)

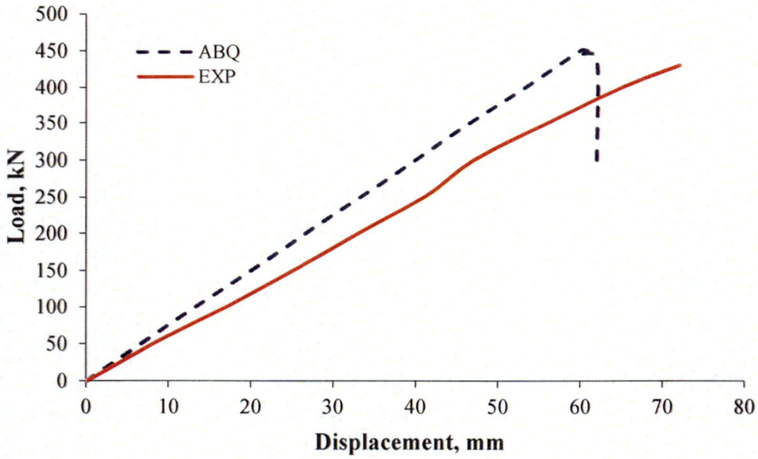

Fig. 7 Comparison of load–deflection response of the hybrid beam obtained in ABAQUS with published results (Nguyen et al. 2015)

Fig. 8 Stress variation in the concrete layer under three-point bending of hybrid beams

4 Parametric Study

After the verification of the numerical results with published results, a parametric study is performed on the beams having different geometric configurations and different degree and type of connections of RC slab and I-beam. Flexural study is conducted on the beams having different spacing and sizes of bolts. Furthermore, parametric study is performed on FRP beams having different length-to-depth (*L/d*) ratios, web depth-to-thickness (*D/t*) ratios and different width of reinforced concrete slab (b_c). In this analysis, influence of the different geometric properties of RC slab and I-section on the failure and service loads is evaluated. Service and failure loads is determined for beams having *L/d* ratios 3, 5 and 7; width of flange of I section-to-thickness ratios 11, 15 and 19. The ratio of depth of web-to-thickness of I-section is considered as 27, 31 and 35. All the above parametric values are analyzed with the different width of RC slab, i.e., 150, 300 and 450 mm. In the RC slab, steel bars of 10 mm diameter have center-to-center spacing of 100 mm, therefore 2, 3 and 5 longitudinal bars are provided in the slab of width 150 mm, 300 mm and 450 mm, respectively. Service load is calculated from the load–deflection response of the beam and it is the load corresponding to serviceable deflection (L/300). On the other hand, failure load is minimum of first ply failure load of the laminate and crushing load of the RC slab. The parametric study performed on the FRP-concrete hybrid beams is explained in the following sections:

4.1 Size and Spacing of Shear Connectors

This section presents the flexural response of the beams having different length and spacing of shear connectors. Figure 9 shows the flexural deformation of the hybrid beam with shear connectors. It is observed that end of shear connectors deformed due to the bending beam. From Figs. 10 and 11, it is noted that as the density of shear connectors decreases consequently flexural stiffness of the beam (slope of load–deflection curve) decreases. On the other hand, as the vertical distance of the shear connector from the edge of bearing plate decreases, i.e., length of the shear connectors increases the strength and stiffness of the hybrid beam also increases.

4.2 Influence of Variation of Length-to-Depth Ratio of I-beam

The FRP beams having low L/d ratio are prone to fail by local buckling. Therefore, failure and service loads are determined for beams having different length such as 600 mm, 900 mm and 1200 mm while keeping the same width and depth of the I-beams, i.e. 75 and 150 mm, respectively. In each beam, supports are provided at

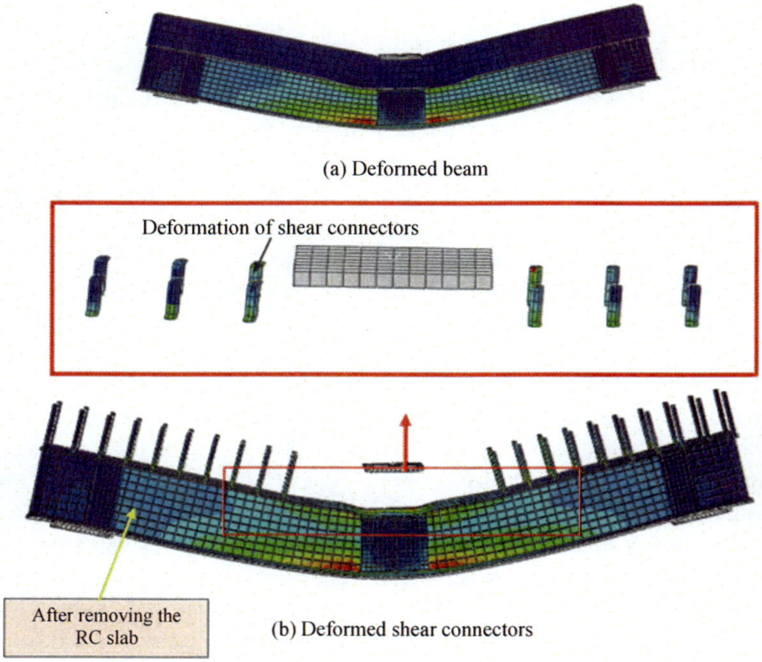

Fig. 9 Flexural behavior of FRP-concrete hybrid beam

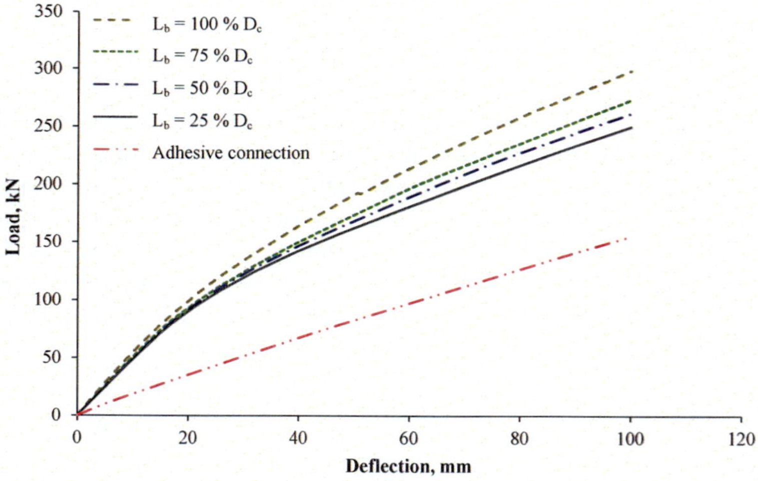

Fig. 10 Flexural response of the beams with different sizes of shear connectors

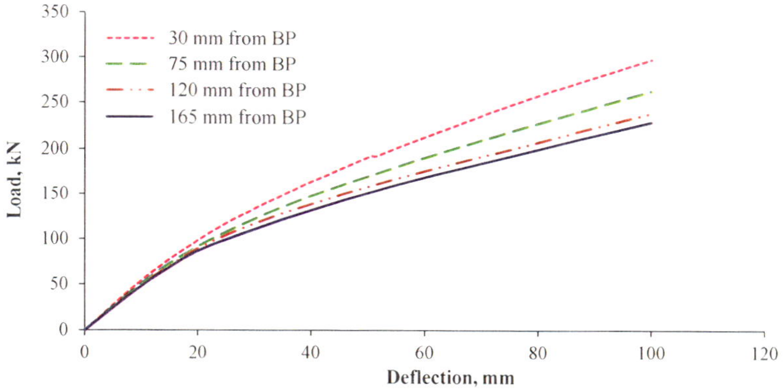

Fig. 11 Flexural response of the beams with different spacing of shear connectors

the distance of 75 mm from the ends, so that the L/d ratio is maintained as 3, 5 and 7. Along with the variation of L/d ratios, width-to-thickness (b_c/t) of concrete layer is also varied. Figure 12 show the flexural deformation of the beam having L/d ratio 7. The service and failure load of the beams having different L/d ratio is shown in Fig. 13. It is observed that with decrease in the length of the beam, flexural stiffness of the beam increases. The failure and service load is lower for the beams having higher L/d ratio as shown in Fig. 13. The failure mode in each beams was failure of web-flange junction.

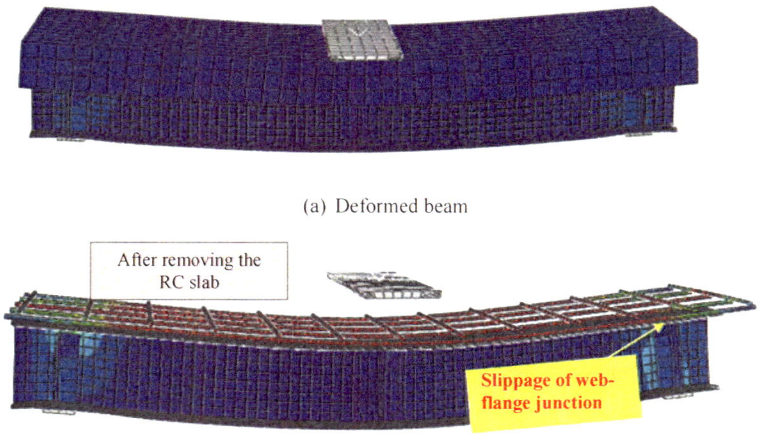

(a) Deformed beam

(b) Bars yielding and failure of web-flange junction

Fig. 12 Flexural deformation of the beam having L/d ratio 7 with RC slab of width 300 mm

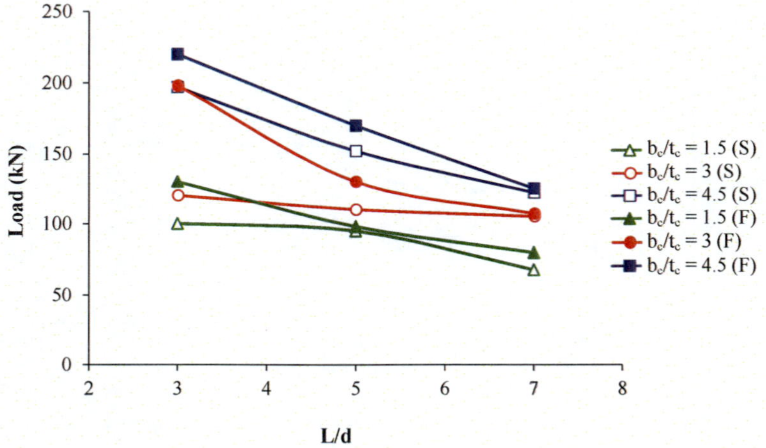

Fig. 13 Effect of L/d ratio on service and failure loads

4.3 Influence of Flange Width of I-beam

In another study, flange width (B) of I-beam is varied as 75, 100 and 125 mm, while the thickness (t) in each case is 6.5 mm and the B/t ratio is 11, 15 and 19. In order to investigate the effect of addition of RC slab on the flexural strength of FRP I-beam, the width of RC slab (b_c) is varied as 150, 300 and 450 mm. It is noted that in each beam failure of web-flange junction of FRP I-beam occurred as shown in Fig. 14. After failure, RC slab moves along with the flange element due to shear connectors between the RC slab and top flange. From Fig. 15, it is observed that some beams

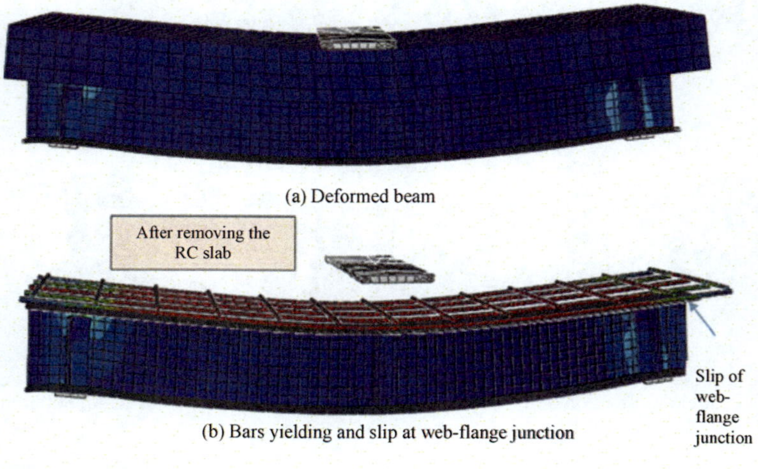

Fig. 14 Flexural deformation of the beam having B/t ratio 11 with RC slab of width 300 (mm)

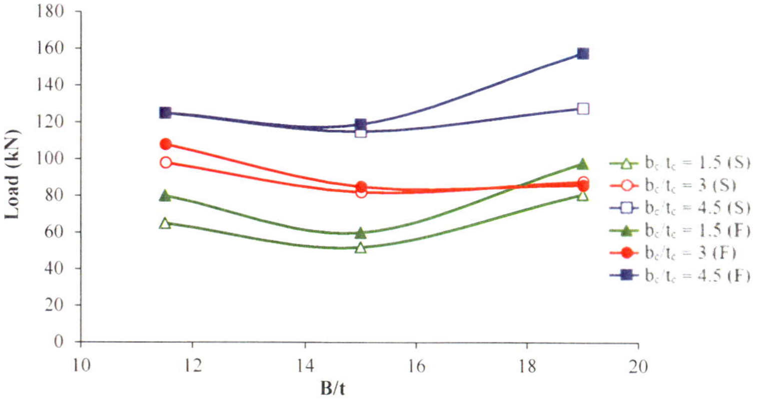

Fig. 15 Effect of B/t ratio on service and failure load

have same failure and service load, it is due to the pre-mature failure of the web-flange junction of the I-beams. Hence, it is stated that addition of RC slab enhances the flexural stiffness, i.e., service load and also enhance the strength of web-flange junction.

4.4 Effect of Depth I-beam

This section presents the influence of web depth and width of RC slab on the flexural response of the I-sections. The depth of the web is taken as 175, 200 and 225 mm with constant thickness 6.5 mm. Figure 16 shows that beams failed by failure of web-flange junction. It is noted that failure and service load increase with increase in the depth of the beams (Fig. 17). Alongwith, the lateral stability also increases with increase in the width of RC slab. Hence, the local buckling of the web is not observed and the failure of the web-flange junction occurs.

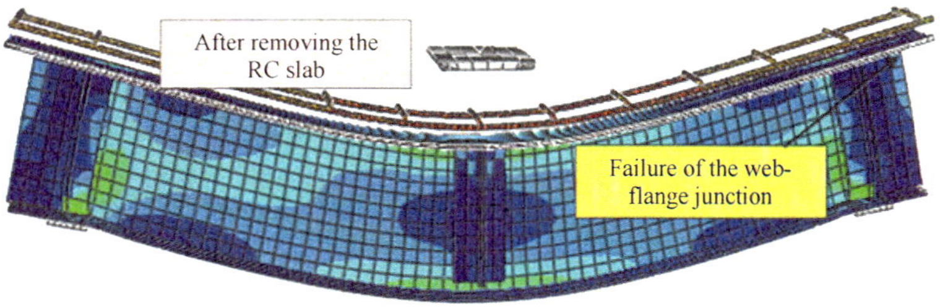

Fig. 16 Failure of the web-flange junction

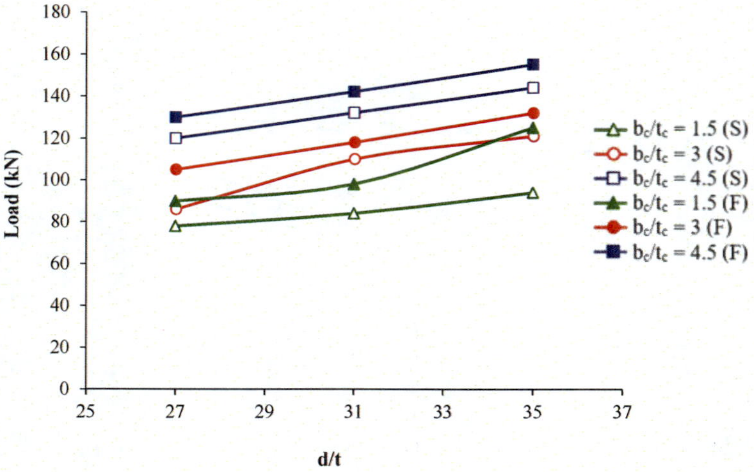

Fig. 17 Effect of D/t ratio on service and failure loads

5 Conclusions

In this chapter, flexural behavior of concrete-FRP hybrid beams is determined. Beams are modeled in finite element software ABAQUS. Shear connection between the FRP beams and RC slab is made using shear connectors and adhesive bonding. Web and flange elements of I-beams are connected with cohesive layer, because it is the weaker portion in the FRP I-beams and is highly prone to fail under three-point loading. Accuracy of the numerical model is verified by comparing the results with published experimental study. Further, the parametric study is performed on beams having different size and spacing of shear connectors. Alongwith, failure and service load of hybrid beams is determined for different geometric configurations of I-beams and width of RC slab. Based on the results obtained from numerical modelling, the following concluding remarks are made:

1. The proposed modeling techniques of hybrid FRP-concrete beams using finite element software ABAQUS is effective in producing the same flexural response and failure mode as obtained from experimental testing.
2. Shear connectors helps in improving the strength of concrete-FRP hybrid beams. With increasing the length of shear connectors, there is slight improvement in the strength of beam.
3. As density of shear connectors decreases consequently flexural stiffness of the beam (initial slope of load–deflection curve) decreases.
4. The strength and stiffness of the hybrid beam increases with increase in the length of the shear connectors.
5. With decrease in the length of the beam, flexural stiffness of the beam increases. The failure and service load is lower for the beams having higher L/d ratio.

6. From the flexural study of the beams having different B/t ratio, it is observed that some beams have same failure and service load, due to the pre-mature failure of the web-flange junction of the I-beams. Hence, it is stated that addition of RC slab enhances the flexural stiffness, i.e., service load and also enhances the strength of web-flange junction.
7. Addition of reinforced concrete slab over the FRP I-beams, reduces the chances of local buckling of the flange and web of the I-beam.

References

Alnahhal W, Aref A, Alampalli S (2008) Composite behavior of hybrid FRP-concrete bridge decks on steel girders. Compos Struct 84(1):29–43

Araba AM, Ashour AF (2018) Flexural performance of hybrid GFRP-steel reinforced concrete continuous beams. Compos B Eng 154:321–336

Bank LC, Yin J (1999) Failure of web-flange junction in postbuckled pultruded I-beams. J Compos Constr 3(4):177–184

Cao Q, Zhou J, Gao R, Ma ZJ (2017) Flexural behavior of expansive concrete beams reinforced with hybrid CFRP enclosure and steel rebars. Constr Build Mater 150:501–510

Chakrabortty A, Khennane A, Kayali O, Morozov E (2011) Performance of outside filament-wound hybrid FRP-concrete beams. Compos B Eng 42(4):907–915

Chawla H, Singh SB (2019) Stability and failure characterization of fiber reinforced pultruded beams with different stiffening elements, part 2: analytical and numerical studies. Thin-Walled Struct 141:606–626

Correia JR, Branco FA, Ferreira JG (2007) Flexural behaviour of GFRP–concrete hybrid beams with interconnection slip. Compos Struct 77(1):66–78

Correia JR, Branco FA, Ferreira J (2009a) GFRP–concrete hybrid cross-sections for floors of buildings. Eng Struct 31(6):1331–1343

Correia JR, Branco FA, Ferreira JG (2009b) Flexural behaviour of multi-span GFRP-concrete hybrid beams. Eng Struct 31(7):1369–1381

Dagher HJ, Bannon DJ, Davids WG, Lopez-Anido RA, Nagy E, Goslin K (2012) Bending behaviour of concrete-filled tubular FRP arches for bridge structures. Constr Build Mater 37:432–439

Gonilha JA, Correia JR, Branco FA, Caetano E, Cunha Á (2013) Modal identification of a GFRP-concrete hybrid footbridge prototype: experimental tests and analytical and numerical simulations. Compos Struct 106:724–733

Hall JE, Mottram JT (1998) Combined FRP reinforcement and permanent formwork for concrete members. J Compos Constr 2(2):78–86

Kim HS, Shin YS (2011) Flexural behavior of reinforced concrete (RC) beams retrofitted with hybrid fiber reinforced polymers (FRPs) under sustaining loads. Compos Struct 93(2):802–811

Koaik A, Bel S, Jurkiewiez B (2017) Experimental tests and analytical model of concrete-GFRP hybrid beams under flexure. Compos Struct 180:192–210

Mendes PJ, Barros JA, Sena-Cruz JM, Taheri M (2011) Development of a pedestrian bridge with GFRP profiles and fibre reinforced self-compacting concrete deck. Compos Struct 93(11):2969–2982

Nguyen H, Mutsuyoshi H, Zatar W (2015) Hybrid FRP-UHPFRC composite girders: part 1—experimental and numerical approach. Compos Struct 125:631–652

Qin R, Zhou A, Lau D (2017) Effect of reinforcement ratio on the flexural performance of hybrid FRP reinforced concrete beams. Compos B Eng 108:200–209

Singh SB, Chawla H (2019) Stability and failure characterization of fiber reinforced pultruded beams with different stiffening elements, part I: experimental investigation. Thin-Walled Struct 141:593–605

Zhang J, Cao D, Tu Y (2015) Flexural behaviors of hybrid concrete beams reinforced with BFRP bars and steel bars. Constr Build Mater 87:28–37

Large Scale Waste Utilisation in Sustainable Composite Materials for Structural Applications

Sanchit Gupta and Sandeep Chaudhary

Abstract Massive consumption of natural resources in composite materials for structural applications has resulted in an acute shortage of materials in the construction industry. In order to ensure sustainability, the reader is introduced to the concept of waste utilization in composite materials. Over the years, numerous wastes have been identified for large scale utilization in sustainable composite materials for structural applications. Utilization of waste primarily depends on the properties of waste and desired use in the composite matrix. The chapter demonstrates the potential of waste for large scale utilization in sustainable composite materials for structural applications through case studies. Processing of wastes to improve their large scale utilization in sustainable construction materials for structural application has also been discussed.

1 Introduction

Construction is a high volume industry, with concrete alone as the most consumed artificial material on earth. On account for this, even small amounts of materials consumed for a unit volume of construction can aggregate to a significant impact on the environment. The impact of high volume consumption in construction has accumulated over the years and has resulted in environmental concerns like scarcity of natural resources and heavy carbon foot-print associated with natural resources processing.

To address the growing environmental concern, the concept of waste utilization and circular economy has gained industry and academia's interest. In the circular economy (Fig. 1), waste from the construction industry is recycled, or other industries

S. Gupta · S. Chaudhary (✉)
Department of Civil Engineering, Indian Institute of Technology Indore, Simrol, Indore 453552, India
e-mail: schaudhary@iiti.ac.in

S. Gupta
e-mail: phd1801204004@iiti.ac.in

S. B. Singh et al. (eds.), *Emerging Trends of Advanced Composite Materials in Structural Applications*, Composites Science and Technology,
https://doi.org/10.1007/978-981-16-1688-4_7

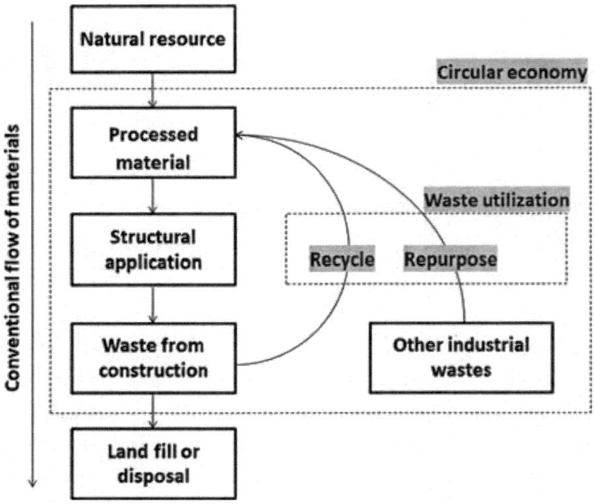

Fig. 1 Circular economy and waste utilization

are repurposed as a substitute material for natural resources. Waste utilization, on the one hand, lowers the dependency on natural resources and on the other, provides an efficient waste management strategy, making the construction more sustainable.

The construction industry holds potential for a broad spectrum of waste utilization in various capacities due to high volume consumption and the application of composite materials. Composite materials are a combination of two or more materials, engineered to produce a desired set of properties. Concrete, composite beams, and concrete-filled steel column tubes are a few of the composites commonly used in structural applications. Composite materials can be modified to utilize the desirable properties and compensate for the undesirable properties of its constituent elements, which simplifies the utilization of waste for the preparation of sustainable composite materials.

In order to address the growing environmental concerns, the chapter explores sustainable composite materials for structural applications prepared through large scale waste utilization. Subsequent sections will explore the general requirements of composite materials for selecting suitable wastes, processing waste, and existing studies on large-scale waste utilization to develop sustainable composite material for structural applications.

2 Composite Material and Selection of Waste

Figure 2 represents a broad classification of composites. The composite material consists of a matrix encompassing different reinforcements to obtain a desired uniform set of properties, like concrete. Further discussion on composite material is limited to the ceramic-based matrix and different forms of reinforcements for their wide structural applications. Structural elements are composite elements that

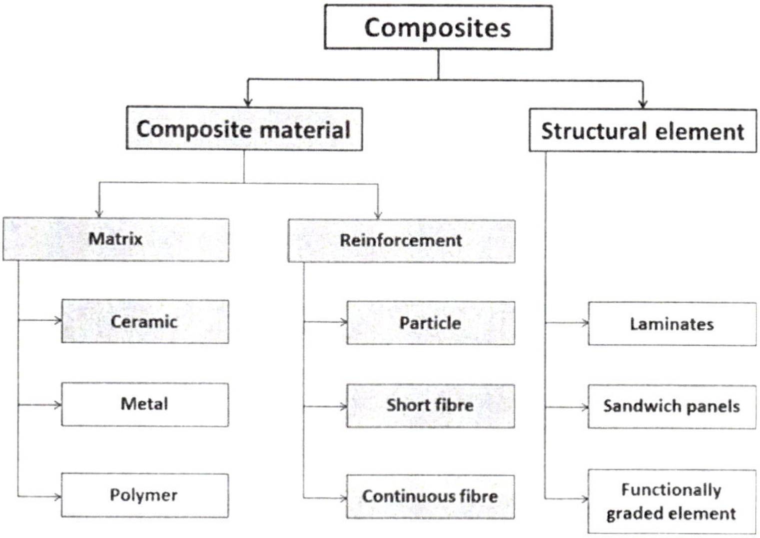

Fig. 2 Classification of composites

are designed to meet the specific requirements of structural applications, like laminates, sandwich panels, and steel–concrete composite beams. In structural elements, different layers are designed to have a specific property, limiting the scope of variation in material and potential of waste utilization.

2.1 *Ceramic Matrix Composite (CMC)*

CMCs are widely preferred for their durability and cost-effectiveness (Makul 2020). For most structural applications, CMCs are prepared using cement based binders and must have workability, cementing action, mechanical strength, and durability. The constituents of CMCs are typically chemically active, fine-sized particles, and can be substituted through wastes like fly ash, ground granulated blast furnace slag (GGBS), or limestone, also known as supplementary cementitious material (SCM) (Gupta and Chaudhary 2020).

2.2 *Particle Reinforced Composite (PRC)*

PRCs are the widest form of concrete. Particles, also known as aggregates, are embedded in a ceramic matrix to lower cost, provide strength and structural stability without hampering durability. In general, aggregates occupy the most massive volume in concrete composites and have the highest potential for waste utilization in terms of volume. Particles are required to be strong, dense, and inert; however, other forms of particles like hollow or lightweight aggregates have also found structural applications. Suitability of waste as particles typically depend on particle size distribution, density and durability. Various wastes have been identified for their use as particles, like recycled concrete aggregate (Prasad et al. 2020), and stone processing waste (Jain et al. 2019).

2.3 Short Fibre Reinforced Composites (S-FRC)

FRC are typically used for crack resistance, which improves the strength and durability of structures. Fibres provide a bridging action in concrete and are required to have a fibre like shape with large longitudinal length per unit lateral diameter (L/d ratio), good tensile strength, and sufficient bonding with the matrix. Wastes like shredded plastic can be used as S-FRC on account of similar characteristics. In general, fibres are added in small quantities to concrete, but on account of the size of the industry, small quantities of waste incorporation can result in large scale waste utilization.

2.4 Continuous Fibre Reinforced Composites (C-FRC)

Similar to S-FRC, C-FRC provides crack resistance and tension carrying capacity in the material. Continuous fibres needed for CFRC require good tensile strength and sufficient bonding with the matrix. Limited studies are available on C-FRC as typically waste is not available in terms of continuous fibres. One of the known applications is the use of continuous fibre produced from polyethylene terephthalate (PET) wastes, where up to 1% of fibre by volume was used to produce a beam with 2.8 times the flexural capacity of controlled concrete (Baldenebro-Lopez 2014).

3 Sustainable Composite Materials

Large scale waste utilization does not imply sustainability in its own; sustainable composite materials should also be technically viable and economically feasible. As composite materials are tailored to specific needs, the definition of technical viability and economic feasibility can change with variation in a structural application. To ensure better understanding, sustainable composite materials have been discussed through separate case studies.

3.1 Case Study of PRC—Repurposing of Fine Bone China Ceramic as Fine Aggregate

Discarded fine bone china ceramics conventionally end up as landfill due to a lack of recycling potential in the fired ceramic industry, but can be repurposed as fine bone china aggregate (FBA) for sustainable composite material (Siddique et al. 2019a). Properties of FBA differ from other ceramics on account of its higher CaO and lower SiO_2 content (Siddique et al. 2019a), and is shown to exhibit pozzolanic activity (Kannan et al. 2017). The effect of using FBA on the properties of concrete can be summarized through Table 1 (Siddique et al. 2017, 2018a, b, c, 2019a, b; Rajawat et al. 2018).

Literature shows that FBA improves the performance of concrete in all parameters except reduction in workability (Table 1). Addition of admixtures for compensating

Table 1 Effect of addition of FBA on the properties of concrete

Property	Observed trend	Trend up to (%)	At 100% replacement
Workability	Reduce	100	Lower
Porosity	Increase	100	Higher
28-day compressive strength	Increase	60	Higher
90-day compressive strength	Increase	100	Higher
180-day compressive strength	Increase	100	Higher
Flexural strength	Increase	80	Higher
Modulus of elasticity	Indefinite trend	–	Slightly higher
Impact resistance	Increase	40	Investigated up to 40%
Microstructure of interfacial transition zone (ITZ)	Improve	60	Denser than normal
Freeze–thaw resistance	Increase	100	Higher
Abrasion resistance	Increase	100	Higher
Drying wetting resistance	Increase	100	Higher
Resistance to chloride penetration	Increase	100	Higher
Corrosion resistance	Increase	100	Higher
Sulphate attack resistance	Decrease	100	Higher
Water penetration	Decrease	100	Lower
Residual strength after exposure to elevated temperature	Improved	100	Higher
Acid resistance	Increase	100	Higher

the loss in workability lowers the cost and environmental saving of an FBA, and maximum cost saving and environmental benefits appear at 60% utilization of FBA (Siddique et al. 2019a). It is also possible that by lowering cement or other content to account for improved performance by FBA may lower the environmental costs and make concrete more sustainable at 100% utilization of FBA, however literature lacks such investigation.

This discussed case highlights a waste from an alternative industry, which can be repurposed for structural applications. It shows that large scale utilization of wastes, like FBA, is possible for the development of sustainable composite materials.

3.2 Case Study of S-FRC—Shredded Rubber and Plastic Waste as Fibres in Concrete

The global annual generation of rubber and plastic waste amounts to approximately 6.5 billion tons (Li et al. 2020), which is greater than the annual global consumption of cement, approximately 4.0 billion tons (Shanks et al. 2019). An alternative to the conventional practice of incineration and landfilling, rubber and plastic wastes can be shredded in fibres to develop plastic fibre concrete (PFC) and rubber fibre concrete (RFC). Rubber and plastic provide a high tensile strength to weight ratio, and on account of their low density, they can provide a large volume alternative to conventional resources. A review of the properties of PFC and RFC is summarized in Table 2 (Gupta et al. 2017, 2020; Choudhary et al. 2020a, b; Jain et al. 2020).

Table 2 shows that the effect of fibre incorporation on concrete properties is different for rubber and plastic. The discussed S-FRC cases represent the typical scenario observed on large scale waste utilization, where one property improves while the other reduces. In the case of plastic waste, density reduction, and energy absorption are two positive effects of waste utilization, limiting the structural application of waste. Unlike plastic, rubber showed an additional improvement in flexural strength of concrete up to 15% utilization by volume of sand. Improved flexural characteristics can be utilized for structural applications, provided other technical characteristics meet the specifications. The discussed rubber fibre case shows large scale waste utilization in sustainable construction material with selective structural applications.

4 Processing for Large Scale Waste Utilization

Mineral and microstructural characterizations of waste have provided a better insight into the effect of waste utilization in structural applications. For example, optimum strength at 60% replacement of FBA can be understood by observing the improvement in ITZ (Siddique et al. 2017). Characterization provides a twofold benefit by

Table 2 Effect of addition of rubber and plastic waste fibres on the properties of concrete

Fibre	Property	Observed trend	Maximum utilization
Plastic bag	Workability	Decease	20% by weight of sand
	Density	Decrease	
	Compressive strength	Decrease	
	Flexural strength	Decrease	
	Modulus of elasticity	Decrease	
	Water permeability	Increase	
	Abrasion resistance	Increase	
	Energy absorption	Increase	
Rubber tyre	Workability	Reduce	15% by volume of sand
	Flexural strength	Increase	
	Water permeability	Decrease	
	Compressive strength	Decrease	25% by volume of sand
	Abrasion resistance	Increase	
	Water absorption	Increase	
	Carbonation resistance	Decrease	
	Acid resistance (mass loss)	Similar	
	Corrosion resistance	Decrease	
	Residual strength after exposure to elevated temperature	Decrease	
	Impact resistance	Improve	

optimizing waste utilization in sustainable composite materials and identifying the limits of given wastes.

Wastes can also be processed for overcoming the limitations identified through characterization. An excellent example of large scale utilization of processed waste is recycled concrete aggregate (RCA). RCA, on account of adhered mortar, shows a high variation in properties and generates weaker ITZ. Prasad et al. (2020) studied up to 40% incorporation of RCA after lime treatment and observed that treated RCA performed the same up to 20% utilization and lost their strength at 30% utilization, while untreated RCA started losing strength from 10% utilization alone.

Processing or treatment lowers the undesirable properties of waste and improves the large scale waste utilization in sustainable composite materials viable for structural applications. Other examples of waste treatment are the hydrophobic treatment of RCA (Mandolia et al. 2020), surface treatment of rubber for improved adhesion (Kashani et al. 2018), and calcination of clay for improving pozzolanic behavior (Ferreiro et al. 2019).

5 Conclusion

To suffice the growing need for structural constructions and the associated need for materials, sustainable strategies like waste utilization are gaining importance. Composite materials, prepared through a combination of different materials, can be engineered to desired properties different from their constituents. This makes the composite material ideal for waste utilization. The utilization of waste typically depends on its suitability as different constituents of a composite, and for structural applications, it can be classified in terms of ceramic matrix, particle reinforcement, short fibre reinforcement, and continuous fibre reinforcement. Different case studies demonstrate that waste can vary in sustainable composite materials based on the need for structural application. While, like FBA, few wastes improve the performance of concrete and have a good potential for large scale waste utilization, others, like RCA, reduce the performance of concrete, making it difficult for their use in structural applications. Advanced characterization techniques have resulted in a better understanding of waste and identifying the factors damaging composite materials' performance. Processing or suitable treatment of waste can lower the waste's negative characteristics and improve the large scale waste utilization in sustainable composite materials for structural applications.

References

Baldenebro-Lopez FJ (2014) Influence of continuous plastic fibers reinforcement arrangement in concrete strengthened. IOSR J Eng 4:15–23. https://doi.org/10.9790/3021-04411523

Choudhary S, Chaudhary S, Jain A, Gupta R (2020a) Assessment of effect of rubber tyre fiber on functionally graded concrete. Mater Today Proc 28:1496–1502. https://doi.org/10.1016/j.matpr.2020.04.830

Choudhary S, Chaudhary S, Jain A, Gupta R (2020b) Valorization of waste rubber tyre fiber in functionally graded concrete. Mater Today Proc 3–8. https://doi.org/10.1016/j.matpr.2020.03.122

Ferreiro S, Canut MMC, Lund J, Herfort D (2019) Influence of fineness of raw clay and calcination temperature on the performance of calcined clay-limestone blended cements. Appl Clay Sci 169:81–90. https://doi.org/10.1016/j.clay.2018.12.021

Gupta S, Chaudhary S (2020) State of the art review on supplementary cementitious materials in India—I: an overview of legal perspective, governing organizations, and development patterns. J Clean Prod 261:121203. https://doi.org/10.1016/j.jclepro.2020.121203

Gupta T, Siddique S, Sharma RK, Chaudhary S (2017) Effect of elevated temperature and cooling regimes on mechanical and durability properties of concrete containing waste rubber fiber. Constr Build Mater 137:35–45. https://doi.org/10.1016/j.conbuildmat.2017.01.065

Gupta T, Siddique S, Sharma RK, Chaudhary S (2020) Effect of aggressive environment on durability of concrete containing fibrous rubber shreds and silica fume. Struct Concr 18–23. https://doi.org/10.1002/suco.202000043

Jain A, Gupta R, Chaudhary S (2019) Performance of self-compacting concrete comprising granite cutting waste as fine aggregate. Constr Build Mater 221:539–552. https://doi.org/10.1016/j.conbuildmat.2019.06.104

Jain A, Siddique S, Gupta T et al (2020) Utilization of shredded waste plastic bags to improve impact and abrasion resistance of concrete. Environ Dev Sustain 22:337–362. https://doi.org/10.1007/s10668-018-0204-1

Kannan DM, Aboubakr SH, EL-Dieb AS, Reda Taha MM (2017) High performance concrete incorporating ceramic waste powder as large partial replacement of Portland cement. Constr Build Mater 144:35–41. https://doi.org/10.1016/j.conbuildmat.2017.03.115

Kashani A, Ngo TD, Hemachandra P, Hajimohammadi A (2018) Effects of surface treatments of recycled tyre crumb on cement-rubber bonding in concrete composite foam. Constr Build Mater 171:467–473. https://doi.org/10.1016/j.conbuildmat.2018.03.163

Li X, Ling TC, Hung Mo K (2020) Functions and impacts of plastic/rubber wastes as eco-friendly aggregate in concrete—a review. Constr Build Mater 240:117869. https://doi.org/10.1016/j.conbuildmat.2019.117869

Makul N (2020) Modern sustainable cement and concrete composites: review of current status, challenges and guidelines. Sustain Mater Technol 25:e00155. https://doi.org/10.1016/j.susmat.2020.e00155

Mandolia R, Siddique S, Chaudhary S (2020) Effect of different hydrophobic treatments on properties of recycled aggregate concrete. In: Lecture notes in civil engineering. Springer, Berlin, pp 121–130

Prasad D, Pandey A, Kumar B (2020) Sustainable production of recycled concrete aggregates by lime treatment and mechanical abrasion for M40 grade concrete. Constr Build Mater 121119. https://doi.org/10.1016/j.conbuildmat.2020.121119

Rajawat D, Siddique S, Shrivastava S et al (2018) Influence of fine ceramic aggregates on the residual properties of concrete subjected to elevated temperature. Fire Mater 42:834–842. https://doi.org/10.1002/fam.2639

Shanks W, Dunant CF, Drewniok MP et al (2019) How much cement can we do without? Lessons from cement material flows in the UK. Resour Conserv Recycl 141:441–454. https://doi.org/10.1016/j.resconrec.2018.11.002

Siddique S, Shrivastava S, Chaudhary S (2017) Lateral force microscopic examination of interfacial transition zone in ceramic concrete. Constr Build Mater 155:688–725. https://doi.org/10.1016/j.conbuildmat.2017.08.080

Siddique S, Shrivastava S, Chaudhary S (2018a) Evaluating resistance of fine bone china ceramic aggregate concrete to sulphate attack. Constr Build Mater 186:826–832. https://doi.org/10.1016/j.conbuildmat.2018.07.138

Siddique S, Shrivastava S, Chaudhary S (2018b) Durability properties of bone china ceramic fine aggregate concrete. Constr Build Mater 173:323–331. https://doi.org/10.1016/j.conbuildmat.2018.03.262

Siddique S, Shrivastava S, Chaudhary S, Gupta T (2018c) Strength and impact resistance properties of concrete containing fine bone china ceramic aggregate. Constr Build Mater 169:289–298. https://doi.org/10.1016/j.conbuildmat.2018.02.213

Siddique S, Chaudhary S, Shrivastava S, Gupta T (2019a) Sustainable utilisation of ceramic waste in concrete: exposure to adverse conditions. J Clean Prod 210:246–255. https://doi.org/10.1016/j.jclepro.2018.10.231

Siddique S, Gupta T, Thakare AA et al (2019b) Acid resistance of fine bone china ceramic aggregate concrete. Eur J Environ Civ Eng 1–14. https://doi.org/10.1080/19648189.2019.1572543

Stress Block Parameters of Confined Fibrous Recycled Self Compacting Concrete

Pancharathi Rathish Kumar, M. L. V. Prasad, and K. L. Radhika

Abstract In present day construction, utilisation of sustainable materials and technologies has become inevitable. One such idea is the use of recycled aggregate for rapidly depleting natural aggregates. Self Compacting Concrete (SCC) is normally used in problematic casting conditions where congestion of reinforcement is encountered. This is a technically and economically viable solution. With better performance in terms of strength and durability, it is fast becoming a good alternative method of making an Environmental friendly (Eco) concrete. Using steel fibers in concrete is known to enhance the performance of concrete in the hardened state. In order for designers to utilise Fibrous Recycled Cycled Aggregate Based Self Compacting Concrete (RASCC), it is important that the stress–strain relationship and the stress block parameters of such a novel material need to be investigated. This will enable to have a comprehensive outlook on the constitutive behaviour of this material. An analytical model is developed based on the results of experimentation and compared with the Euro Code provisions.

Keywords Waste material · Fibers · SCC · Aggregate recycled · Stress block parameters · Micro structure · Stress–strain analytical model

1 Introduction

There is paradigm shift in thinking from a strength based design to a sustainable based design. With natural materials becoming scarce one is forced to thing in the direction of sustainability without compromising on the important aspects like strength

P. Rathish Kumar (✉)
Civil Engineering Department, NIT Warangal, Warangal, Telangana, India
e-mail: rateeshp@gmail.com; rateeshp@nitw.ac.in

M. L. V. Prasad
Civil Engineering Department, NIT Silchar, Silchar, Assam, India

K. L. Radhika
Civil Engineering Department, Osmania University, Hyderabad, Telangana, India

S. B. Singh et al. (eds.), *Emerging Trends of Advanced Composite Materials in Structural Applications*, Composites Science and Technology,
https://doi.org/10.1007/978-981-16-1688-4_8

and durability. In fact, a quadruple bottom line approach including aspects such as social, economical, environmental and technological is not to be forgotten. This should become a governing criterion in the material selection too. Utilising concrete demolished waste (CDW) as aggregate in concrete making is one such alternative solutions to the above problem (Yang et al. 2008; Nwakaire et al. 2020). The CDW material can be treated and used as a secondary raw material and can further solve the problems of waste disposal without any compromise on the quality of the final product (Mc Neil and Kang 2013). Concrete is a generic material which has very good applications in construction field. Inspite of its versatility, concrete has one major drawback in terms of ductility and this is more so when recycled aggregate is used as a constituent material (Kumar 2017).

Self Compacting Concrete (SCC) has important qualities as flowability and passing ability and is cohesive enough to overcome the problem of segregation or bleeding (Ahmad et al. 2017). The ability of self consolidation and improved density increases the efficiency and quality of concrete. SCC is noise free as there is no requirement of vibration. This improves the durability and thus safety, improving the working environment (Siad et al. 2014; Ahmad et al. 2017). The property of filling narrow areas and confining the reinforcement in concrete members, enhances the structural performance (Martínez-Garcia et al. 2020). The brittleness of concrete can also be overcome by introduction of discrete fibers which take part in imparting the tensile load and also prevent propagation of cracks (Ahmad et al. 2017). The addition of fibers thus improves the ductility of the cementitious composite materials by delaying the formation and extension of the crack. Finally, the strength properties of concrete using CDW as aggregate depend on the quality of the raw material used and the replacement ratio.

2 Scope of the Work

The congested reinforcement in high rise building elements such as columns and beams makes quite problematic in placing and compacting the concrete to a required compaction level. The advantage of SCC flowing into the congested reinforcement, filling all the formwork without any voids and avoiding problem of bleeding and segregation is making researchers think towards the strong and durable structures (Dhiyaneshwaran et al. 2013). The utilization of recycled material as aggregate opens a way to reuse the waste materials from building industry (Güneyisi et al. 2016). The deformation of concrete along with integrity can be archived by the lateral ties along with the fiber addition. The present paper deals with the study of Fibrous Self Compacting Concrete utilising Recycled Concrete as Aggregates and the stress–strain behaviour is predicted for the above developed FSCC made with RA. In the present paper, M20 and M40 grades are considered for understanding the behavior of Fibrous Self Compacting Concrete (FSCC) made with Recycled Aggregate (RA).

3 Methodology

A major volume of concrete is occupied by the aggregates and their physical and chemical characteristics have greater effect on properties of concrete. In this paper, the work focuses on the characteristics of recycled aggregate made of demolished waste. The basic properties of utilising recycled aggregate were found before using it as an ingredient in concrete. The important properties of recycled aggregate like density, specific gravity, absorption of water, % of voids, impact value, crushing value are determined and compared with the natural aggregate as per IS: 2386-1970.

3.1 Recycled Aggregate

The utilization of CDW aggregate as an ingredient in preparation of structural and non-structural concrete elements is economical and technically a viable solution to the problem of disposal of used concrete for landfills (Prasad and Rathish Kumar 2007). This will ensure sustainability and paves the way for disposal problems of huge volume of concrete waste from demolished of old building and samples of concrete blocks from concrete laboratories (Saravana Kumar and Dhinakaran 2015; Zhou and Chen 2017). The aggregate demand is increasing and reached to 21 billion tones in the developed countries like China, India etc. The demand for natural aggregate after 2010 is approximately 10 billion tons annually (Yang et al. 2008; Mastali et al. 2016) for the construction industry. This forced the use of recycled aggregate which is both an economical and a sustainable alternative.

3.2 Fibrous Self Compacting Concrete with Recycled Aggregate

Use of Self-Compacting Concrete (SCC) is increasing in congested reinforcement because of its easy placement in difficult casting conditions. For the application of difficult casting conditions, the concrete must have high flowability, passing ability along with fluidity along with good cohesiveness (Okamura and Ouchi 2003). The placement rate increases rapidly with use of SCC enabling time consumption for construction and ensures flow easily around congested reinforcement. SCC satisfies three important requirements-Passing ability, filling ability along with resistance to segregation as per EFNARC specifications-2005 (EFNARC 2005). The property of concrete and fibers influences the performance of Fiber Reinforced Concrete (FRC). The orientation of fibers, geometry, distribution and orientation are the main characteristics of a fiber (Prasad and Rathish Kumar 2009; Ahmad et al. 2017). The usage of fibers in concrete as crack arrester avoids initiation and propagation of crack by bridging the bond between materials. Use of fiber in concrete enhances ductility of concrete member and delays the crack propagation (Aslani and Nejadi 2013; Ahmad et al. 2017). The fibers commonly used in the concrete are natural fiber, synthetic fiber, polypropylene fiber, carbon fiber and glass fiber. In the present study glass fibers were used.

4 Experimental Program

The program includes preparation of total of 150 specimens of size 150 × 150 × 450 mm and testing them for finding the stress block parameters of M20 and M40 grades of concrete. Of these 150 specimens, 75 prisms are of M20 grade and the remaining 75 prisms are of M40. The specimens in each group comprise of fibrous and recycled aggregate based specimens. In each set, an average of three specimens were cast and tested after 28 days of curing. For the prediction of compressive strength some additional 150 × 150 mm cubes were cast as companion specimens. Sampling for microstructure analysis is done based on specimens of size 1.0 × 1.0 × 0.5 cm size from these companion samples.

5 Materials Used

In this work OPC (Ordinary Portland Cement confirming to IS: 12269-1987) was used and part of it was replaced with locally available fly ash. River sand confirming to Zone-II (IS-2386-1997) was used as fine aggregate (F.A). The aggregate obtained from recycled concrete was used as coarse aggregate (C.A) and is confirming to IS: 383-2016 grading. Glass fiber (GF) was used in the present study. The fibers were of 24 mm length, specific gravity 2.6 and specific surface area 105 m^2/kg. The fresh properties of SCC Viz./passing ability, flowability and segregation resistance were modified by adding Polycarboxylic ether based water reducing admixture superplasticizer and Viscosity Modifying Agent (VMA) Glenium Stream-II were added in optimum dosages for better workability of SCC (Shah et al. 2017). The SCC mixes were developed using Nansu mix design method (Su et al. 2001) and the details were shown in Table 1. The fresh and hardened properties of SCC without and with glass fiber and with RA are shown in Table 2.

5.1 Curing and Testing

The prisms cured for 28 days were tested under axial compression for developing stress–strain cures as per Standards. The details of specimens cast are shown in Tables 3 and 4. The specimens were tested under axial compression in a servo controlled computerized machine of 1000 kN capacity. The entire testing was done

Table 1 Mix proportion details of recycled aggregate SCC with and without GF

S. No.	Grade of concrete	Cement (kg)	Coarse aggregate (kg)	Fine aggregate (kg)	Flyash (kg)	Glass fiber (kg)	Water (L)	Dosage of S.P (L)	Dosage of VMA (L)
1	M20-NF	275	760	970	165	–	198	2.50	0.25
2	M20-GF	275	760	970	170	1.00	198	3.00	0.35
3	M40-NF	425	740	900	165	–	195	3.50	0.40
4	M40-GF	425	740	900	165	1.00	195	4.00	0.45

NF No Fiber; *GF* Glass Fiber

Table 2 Fresh and hardened properties of RA based SCC with and without GF

S. No.	Grade of concrete	Slump cone test		V funnel test		L box test			Compressive strength (MPa)
		H-flow (mm)	T50 (time in sec)	Time for complete discharge (s)	T 5 min in (s)	Time for 0–200 mm spread	Time for 0–400 mm spread	H2/H1	
1	M20-NF	675	5.10	11.08	14.21	3.10	6.34	0.85	28.45
2	M20-GF	650	5.16	11.25	14.56	3.25	6.61	0.80	29.12
3	M40-NF	680	5.12	12.04	15.0	4.40	6.60	0.82	47.26
4	M40-GF	660	5.23	12.23	15.51	4.68	6.79	0.80	47.41

NF No Fiber; *GF* Glass Fiber

Table 3 Details of M20 grade FSCC with RA prism specimens

Designation	Longitudinal steel		Lateral steel		F_i	M20 grade cube strength (MPa)	M20 grade plain prism strength (MPa)
	No (mm)	Dia (mm)	Dia (mm)	Spacing (mm)			
XP00	–	–	–	–	0.00	28.74	20.5
XG01	4	6.00	8	150	0.00		
XG02	4	6.00	8	100	0.00		
XG03	4	6.00	8	75	0.00		
XG04	4	6.00	8	50	0.00		
XP10	–	–	–	–	0.09	28.91	21.15
XG11	4	6.00	8	150	0.09		
XG12	4	6.00	8	100	0.09		
XG13	4	6.00	8	75	0.09		
XG14	4	6.00	8	50	0.09		
XP20	–	–	–	–	0.18	29.05	21.2
XG21	4	6.00	8	150	0.18		
XG22	4	6.00	8	100	0.18		
XG23	4	6.00	8	75	0.18		
XG24	4	6.00	8	50	0.18		
XP30	–	–	–	–	0.27	29.17	21.35
XG31	4	6.00	8	150	0.27		
XG32	4	6.00	8	100	0.27		
XG33	4	6.00	8	75	0.27		
XG34	4	6.00	8	50	0.27		
XP40	–	–	–	–	0.36	29.23	21.55
XG41	4	6.00	8	150	0.36		
XG42	4	6.00	8	100	0.36		
XG43	4	6.00	8	75	0.36		
XG44	4	6.00	8	50	0.36		

Note X M20 grade SCC; *P* Plain prism; *G* Glass fiber prisms
XG01 M20 grade SCC without fiber with 150 mm spacing of lateral ties
XG12 M20 grade SCC with fiber (Fi = 0.089) and with 150 mm spacing of lateral ties

based on strain rate control. The load was increased gradually in the initial stages and was increased slowly until the ultimate load is reached. After attaining the peak load, the increase in strains was at a slightly faster rate and was accompanied with a decline in the load carrying capacity of the prisms. The details of testing along with test setup are shown in Fig. 1.

Table 4 Details of M40 grade FRSCRAC prism specimens

Designation	Longitudinal steel		Lateral steel		Fi	M40 grade cube strength (MPa)	M40 grade plain prism strength (MPa)
	No (mm)	Dia (mm)	Dia (mm)	Spacing (mm)			
YP00	–	–	–	–	0.00	45.83	34.59
YG01	4	6.00	7.96	150	0.00		
YG02	4	6.00	7.96	100	0.00		
YG03	4	6.00	7.96	75	0.00		
YG04	4	6.00	7.96	50	0.00		
YP10	–	–	–	–	0.09	46.02	34.98
YG11	4	6.00	7.96	150	0.09		
YG12	4	6.00	7.96	100	0.09		
YG13	4	6.00	7.96	75	0.09		
YG14	4	6.00	7.96	50	0.09		
YP20	–	–	–	–	0.18	46.16	35.2
YG21	4	6.00	7.96	150	0.18		
YG22	4	6.00	7.96	100	0.18		
YG23	4	6.00	7.96	75	0.18		
YG24	4	6.00	7.96	50	0.18		
YP30	–	–	–	–	0.27	46.28	35.48
YG31	4	6.00	7.96	150	0.27		
YG32	4	6.00	7.96	100	0.27		
YG33	4	6.00	7.96	75	0.27		
YG34	4	6.00	7.96	50	0.27		
YP40	–	–	–	–	0.36	46.39	35.78

Note Y M40 grade SCC; *P* Plain prism; *G* Glass fiber prisms
YG01 M40 grade SCC without fiber with 150 mm spacing of lateral ties
YG44 M40 grade SCC with Fi = 0.357 fiber and 50 mm spacing of lateral ties

5.2 *Microstructure Analysis of Concrete*

Scanning Electron Microscopy (SEM) (Fig. 2) and energy dispersive X-ray analyzer (EDXA), do not fall under the realm of any standard procedure. SEM was used for understanding the cement hydration (Chatterji and Jeffery 1962). The energy dispersive X-ray analyzer with the scanning electron microscopy attachment was used for the microanalysis of concrete in the present work.

The cubes were tested for strength under compression, after 28 days of age of curing. A representative sample from the core of the cube was taken. The sample was grinded on all the surfaces using bench grinder except the surface to be tested. The sample was reduced to 1.0 × 1.0 × 0.5 cm. Since concrete sample is nonconductive,

Fig. 1 Test setup and specimen under test

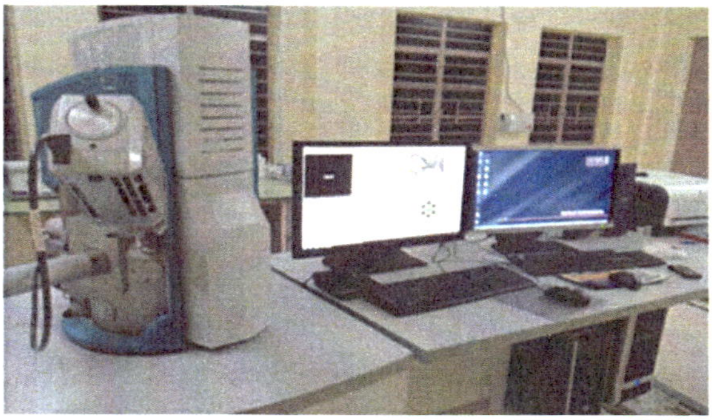

Fig. 2 SEM and energy dispersive X-ray analyser

in order to make it conductive, gold coating was given prior to the examination using SEM. The sample was prepared in such a way that it should contain aggregate mortar interfacial transition zone (ITZ) and also mortar content.

5.3 SEM Analysis of SCC with Recycled Aggregate

The microstructure investigation was done using Scanning electron microscope (SEM). The poor strength of recycled concrete aggregate compared to natural coarse aggregate necessitated to investigate the microstructure features to understand the void distribution and densification of the matrix. It is possible to use filler materials and make RASCC mixes suitable for construction purpose. The studies on M20 and M40 grades of concrete were carried out and the images are examined and compared.

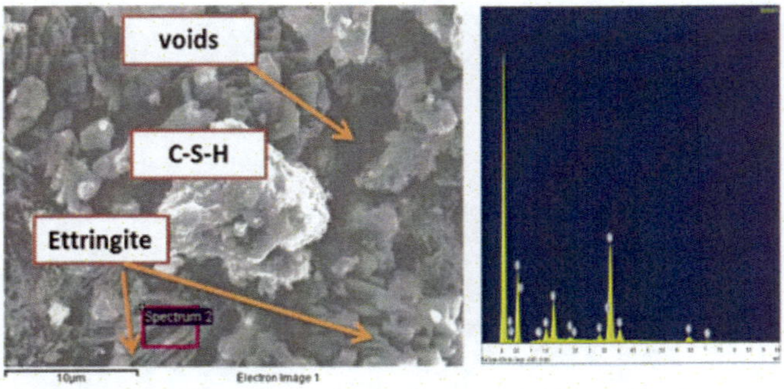

Element	Weight%	Atomic%
O	40.67	59.81
Al	1.18	1.05
Si	12.72	6.83
S	1.18	0.88
Fe	1.30	3.79
Ca	42.96	25.64
Totals	100.00	

Fig. 3 SEM image and EDXA spectrum and elemental composition of M20 grade SCC-RA

5.4 M20 Grade SCC with RA

Figure 3 shows the SEM image and the EDAX spectrum and elemental composition of M20 grade SCC with RA. The SEM image shows the presence of porous CSH gel with maximum amount of voids at ITZ zone. Some traces of Ettringite (AFm) needles were also observed. Through the elemental composition, the observed ratio Ca/Si was 3.75 (Al + Fe)/Ca was 0.19 while S/Ca is 0.04. These ratios indicate the presence of higher Ca amount which depicts the presence of weaker CSH gel and some amounts of calcium hydroxide (Ca $(OH))_2$.

5.5 M40 Grade SCC with RA

The SEM image of M40 RASCC along with the EDAX spectrum images for evaluating elemental composition are shown in Fig. 4. The SEM image shown for M40 grade concrete is quite different from low grade concrete (M20) due to less number of voids and better microstructure. A denser microstructure was observed in M40 grade

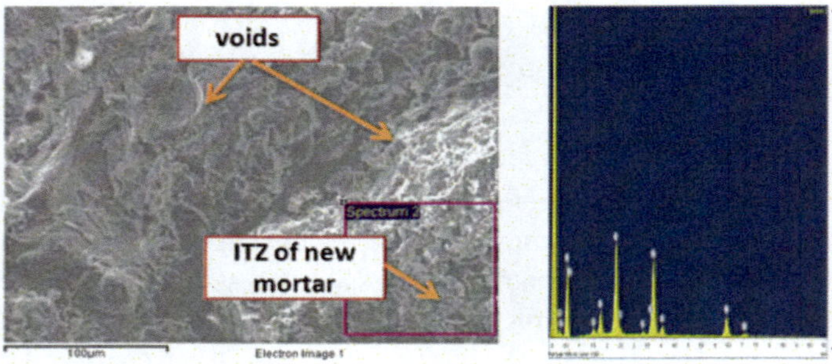

Element	Weight%	Atomic%
O	52.40	72.39
Mg	0.41	0.37
Al	1.90	1.56
Si	6.39	5.03
S	0.53	0.37
K	1.60	0.90
Ca	31.03	17.11
Fe	5.75	2.28
Totals	100.00	

Fig. 4 SEM image and EDXA spectrum and elemental composition of M40 grade SCC-RA

SCC with RA as it contains more filler materials and added admixtures resulting in the formation of dense CSH gel. The elemental composition shown in Fig. 4 gave the ratios of Ca/Si, (Al + Fe)/Ca and S/Ca as 3.4, 0.22 and 0.02 respectively. The higher presence of calcium shown in the spectrum indicates less presence of CSH and more contents of AFm and CH. The image observed shows new mortar sticking on the surface.

5.6 Stress–Strain Behavior of FSCC with Recycled Aggregate

For developing the stress–strain behaviour of FSCC with RA, the main parameters involved in this study strength of concrete, spacing of lateral ties, fiber volume, and grade of concrete and core size of the test specimen were considered. These parameters influence the behaviour of tie confined FSCC with RA. Confinement index (C_{iR}) and Fiber index (F_{iR}) two non-dimensional parameters are considered linking all the parameters influencing the performance of FSCC with RA.

5.7 The Confinement Index

The term confinement index is defined as

$$C_{iR} = \left(P_b - \bar{P_b}\right)\left(f_v / f_{c'}\right)\sqrt{\frac{b}{s}} \tag{1}$$

where,

P_b the ratio of the volume of ties to the volume of concrete.
$\overline{P}_b$ the ratio of the volume of ties to the volume of concrete corresponding to limiting pitch.
b breadth of the prism.
s spacing of ties.

The Stress in steel binder is given by $f_v = \varepsilon_v . E_s$ (for maximum yield strength).

ε_v strain and
E_s modulus of elasticity.

5.8 Fiber Index

The degree of confinement provided by fiber can be expressed with Fiber index (F_{iR}). There will be good bonding between cover and core because of the passive confinement of the fibers in core concrete (Ramesh et al. 2003). The addition of fibers is going to improve the resistance against crack propagation, strength and energy absorption of concrete, rather than other improvements in the properties of the concrete.

$$\text{Fiber index } F_{iR} = (w_f.)\left({}^{l}/_{d}\right) \tag{2}$$

$(w_f.)$ Weight Fraction of fiber.
(l/d) Aspect Ratio of fiber.

If the stress–strain behaviour of a material is known, then it is easy to analyze the member theoretically (Prasad and Rathish Kumar 2009). In case of steel, the Stress–Strain relationship is not complicated as there is very little variation of material compared to that of concrete. But in case of confined and unconfined concrete behaviour, there is a much variation. In regular practice, we use stress–strain relation from the codal specifications for the analysis and design, but these are not suitable for new concretes like self compacting concrete. With the advantages of the use of recycled aggregate in FSCC, the stress–strain relation for this concrete material needs to be developed. Generally, confinement can be provided by the use of lateral ties along with addition of fibers in the concrete. Confined specimens show better performance and there will be no spalling of cover and improvement in the deformation along with integrity of concrete.

For vibrated concrete, to predict the stress–strain relation with lateral ties, many empirical models have been reported in the literature during the last four decades. In the present work, to predict the stress–strain relation of FSCC with RA confined with lateral ties, an empirical confinement model is developed from experimental work. The grading of concrete (M20 and M40), variation in the confinement indices (0, 2, 3, 5, 7 ties) and spacing of ties (0, 150 mm, 100 mm, 75 mm, 50 mm) were taken as the parameters for the study. The stress–strain curves for the above parameters are drawn and relationships obtained. The diameter of the lateral ties was 8 mm. From the experimentation, it was noticed that there is a decreases the spalling of core concrete with the use of fibers in addition to producing a passive confinement to the core concrete. The addition of glass fiber along with the confinement improved the deformation characteristics and maintained the integrity of concrete. To study the effect of fiber additions in SCC with RA, Fiber Index is taken as the parameter. Stress–strain curves are developed for M20 grade of concrete and for M40 grade of concrete for different confinement indices and for variation in the fiber indices (Figs. 5a–6c).

5.9 *Confinement Index, Fiber Index Versus Stress Ratio and Strain Ratio*

The experimental stress–strain curves for all the specimens are drawn by taking the average of three values. Since the parameters that characterize the stress–strain curve

Fig. 5 **a** M20 grade FSCC with RA (Ci = 0.16). **b** M20 grade FSCC with RA (Ci = 0.32). **c** M20 grade FSCC with RA (Ci = 0.72)

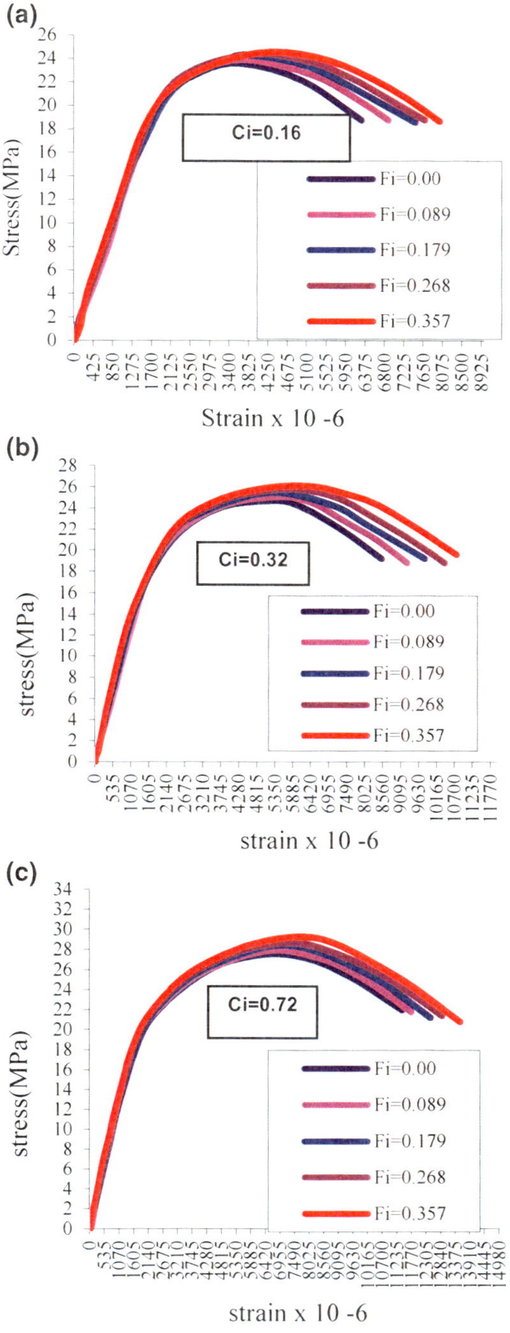

Fig. 6 M40 grade FSCC with RA (Ci = 0.09). **b** M40 grade FSCC with RA (Ci = 0.19). **c** M40 grade FSCC with RA (Ci = 0.43)

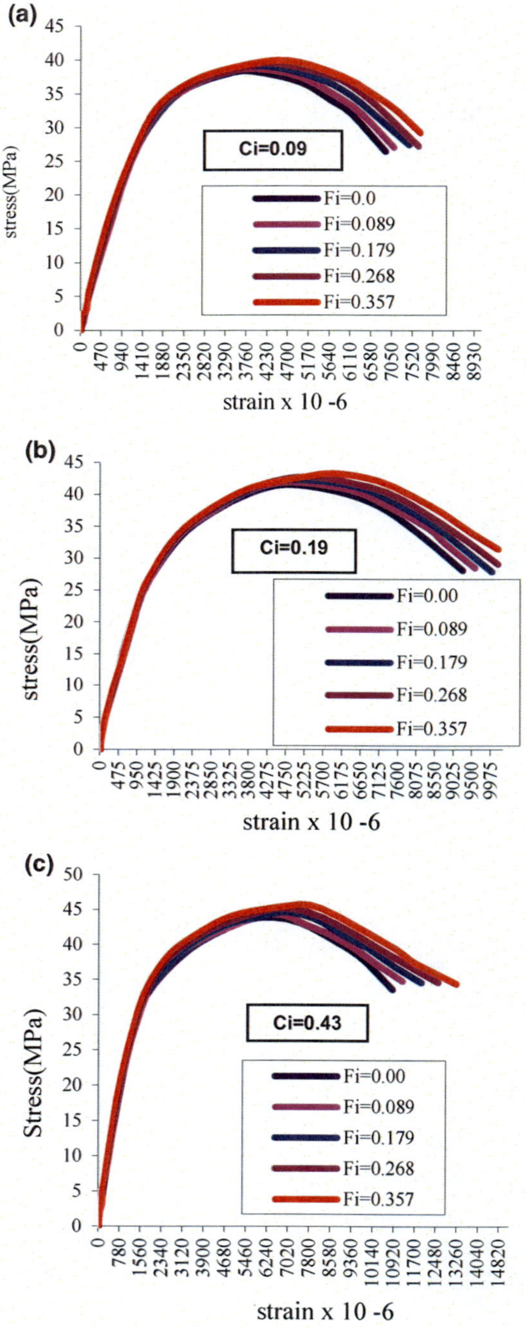

are the peak stress and the strain corresponding to peak stress, the confining effects of ties and fiber on these parameters are evaluated. The ratio of the peak stress of the tie confined fiber based specimen (f_{cb}) to that of a plain concrete companion specimen (fc') is determined. Similarly, the ratio of the strains ($\varepsilon_{cb}/\varepsilon_c'$) at maximum stress is determined. The relationship between the confinement index (Ci), fiber index (Fi) and the improvement in the ultimate strength and strain at ultimate are established.

The relationship between stress ratio and Confinement Index, similarly relationship between Strain ratio and Confinement Index for FSCC with RA is shown in Eqs. 3 and 4. Similarly, the relationship between Stress Ratio versus Fiber Index and Strain Ratio versus Fiber Index for FSCC with RA is shown in Figs. 7a–8b.

The confined Concrete Strength:

$$\frac{\varepsilon_u}{\varepsilon'} = (1 + 4.142C_{iR}) + (1 + 0.838F_{iR}) \tag{3}$$

The confined Concrete Strain:

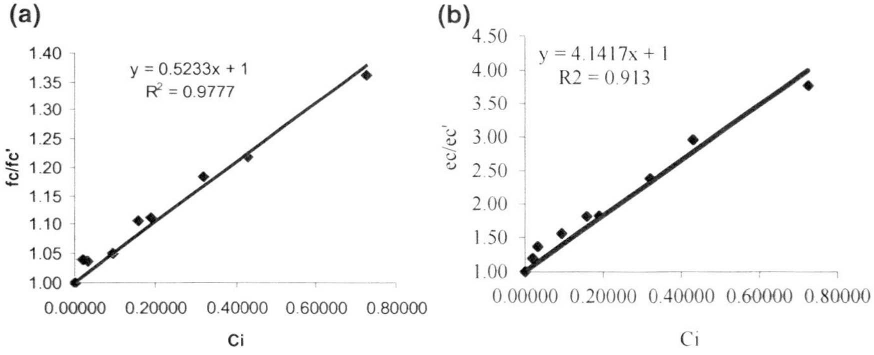

Fig. 7 **a** Stress ratio versus Ci. **b** Strain ratio versus Ci

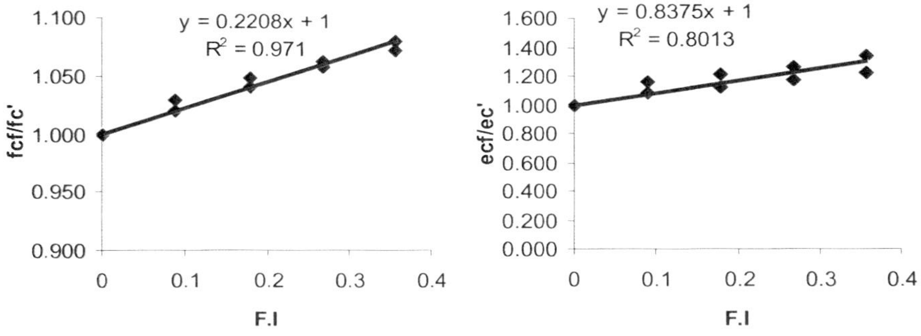

Fig. 8 **a** Stress ratio versus Fi. **b** Strain ratio versus Fi

$$\frac{f_u}{f'} = (1 + 0.523C_{iR}) + (1 + 0.221F_{iR}) \tag{4}$$

The use of ties at closer spacing provide larger volumetric ratio of confining steel and hence the more confinement index.

5.10 Non-dimensionalised Stress–Strain Curve

From the experimentation, it was observed that upto 80% of the ultimate, the stress–strain behaviour is linear and beyond this it becomes non-linear. After the ultimate value, the stress–strain response for all the FSCC with RA prisms is gradually drooping and appears to have a consistent and steady gradient. From this similarity we can give a conclusion that the stress ratio and the strain ratio plot falls into the same pattern, where, stress ratio is the stress at any level to the corresponding stress at ultimate, while, strain ratio is the strain at any level to the corresponding strain at ultimate strength. From the non-dimensionalised stress ratio and strain ratio as explained above, the Confinement Index (C_i) effect can be neutralized.

The non dimensionalised values of stress and strain are plotted on the y-axis (ordinate) and x-axis (abscissa) as shown in Fig. 9. The comparison between the characteristic values and the analytical values are shown in the Fig. 10. The stress block represents the stress–strain behavior and to develop the parameters of the stress

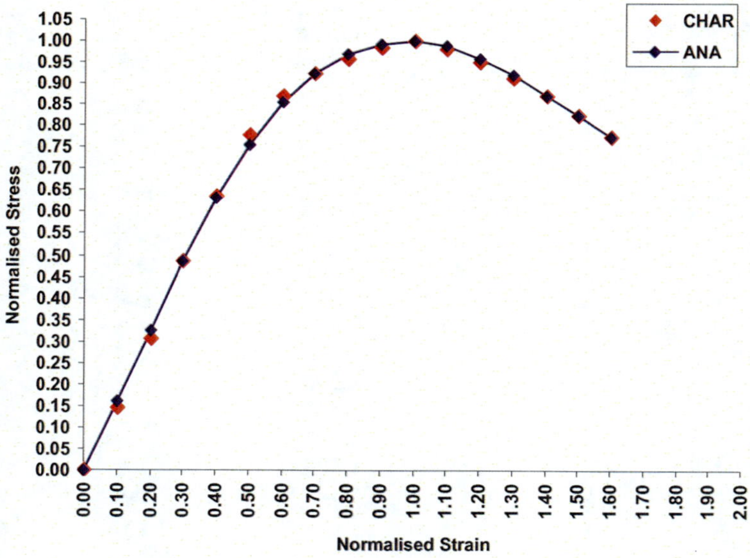

Fig. 9 Stress ratio versus strain ratio

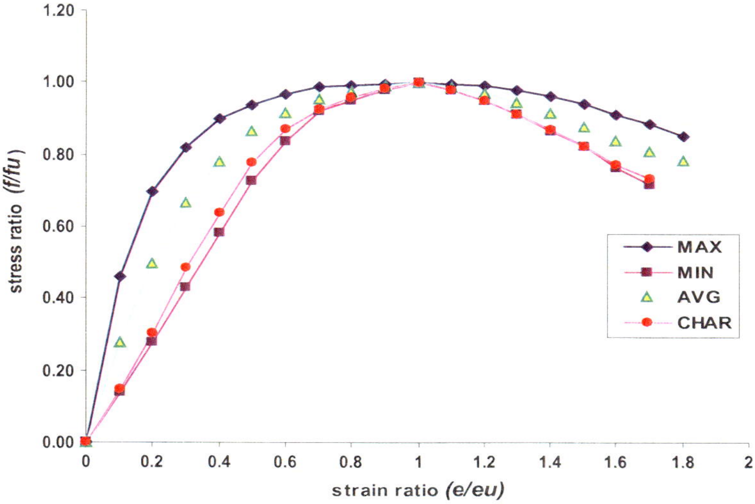

Fig. 10 Normalised stress versus strain

block, Saenz has proposed a polynomial equation and the same is used in this present study (Saenz 1995).

For predicting the stress–strain behaviour of normal vibrated concrete in compression Saenz proposed the following equation.

$$f = \frac{A\varepsilon + D}{1 + B\varepsilon + C\varepsilon^2} \tag{5}$$

where,

f stress corresponding to strain and
ε strain in concrete.

The normalized values of strain and stress ie., strain ratio and stress ratios are taken as independent and dependent variables, to express the non-dimensional behavior of FSCC with RA stress–strain curves, instead of taking the proposed values of Saenz.

$$\frac{f}{f_u} = \frac{A^1\left({}^{\varepsilon}/_{\varepsilon_u}\right)}{1 + B^1\left({}^{\varepsilon}/_{\varepsilon_u}\right) + C^1\left(\frac{\varepsilon}{\varepsilon_u}\right)^2} \tag{6}$$

where

$$A = A^1\left({}^{f_u}/_{\varepsilon_u}\right), B = B^1\left({}^{1}/_{\varepsilon_u}\right), C = C^1\left({}^{1}/_{\varepsilon_u^2}\right)$$

To predict the ascending and descending portions of the stress–strain curve for FSCC with RA, two sets of constants (A1, B1, C1) are proposed. One set represents the ascending portion and other set represents the descending portion. The details of the boundary conditions used for both ascending and descending portions of Stress–Strain Curve of FSCC with RA are as given below:

1. At $\frac{\varepsilon}{\varepsilon_u} = 1; \frac{f}{f_u} = 1;$
2. At $\frac{\varepsilon}{\varepsilon_u} = 1; d\left({}^{f}/_{f_u}\right)/d\left({}^{\varepsilon}/_{\varepsilon_u}\right) = 0;$

Here, because of use of fibers, the additional boundary conditions for the ascending portion of stress- strain curve used are

3. At $\left({}^{\varepsilon}/_{\varepsilon_u}\right) = 0.3; \left({}^{f}/_{f_u}\right) = 0.487;$

Similarly, for the descending portion of stress- strain curve, additional boundary required condition used are

4. At $\left({}^{\varepsilon}/_{\varepsilon_u}\right) = 1.445; \left({}^{f}/_{f_u}\right) = 0.85;$

The above (3) and (4) conditions are obtained from the experimental results.

At $\left({}^{\varepsilon}/_{\varepsilon_u}\right) = 0.3$, the curve deviates from the initial tangent.

At $\left({}^{\varepsilon}/_{\varepsilon_u}\right) = 1.445$ the reduction in strength is 15%. Satisfying the boundary conditions,

The constants for the ascending and descending portions of the curve are obtained:

For Ascending Portion A1 = 1.55, B1 = −0.45, C1 = 1.0 and For Descending Portion A1 = 0.777, B1 = −1.22, C1 = 1.0

Thus, the stress–strain equation for the ascending portion of fibrous self compacting concrete with recycled aggregate is

$$\frac{f}{f_u} = \frac{1.55\left({}^{\varepsilon}/_{\varepsilon_u}\right)}{1 - 0.45\left({}^{\varepsilon}/_{\varepsilon_u}\right) + 1.0\left(\frac{\varepsilon}{\varepsilon_u}\right)^2} \tag{7}$$

while for the descending portion of fibrous self compacting concrete with recycled aggregate the stress–strain equation is

$$\frac{f}{f_u} = \frac{0.78\left({}^{\varepsilon}/_{\varepsilon_u}\right)}{1 - 1.22\left({}^{\varepsilon}/_{\varepsilon_u}\right) + 1.0\left(\frac{\varepsilon}{\varepsilon_u}\right)^2} \tag{8}$$

The generalized stress–strain equation for fibers self compacting concrete with recycled aggregate can be written as

$$f = \frac{A\varepsilon}{1 + B\varepsilon + C\varepsilon^2} \tag{9}$$

where $A = A^1\left({}^{f_u}/_{\varepsilon_u}\right), B = B^1\left({}^{1}/_{\varepsilon_u}\right), C = C^1\left({}^{1}/_{\varepsilon_u^2}\right)$.

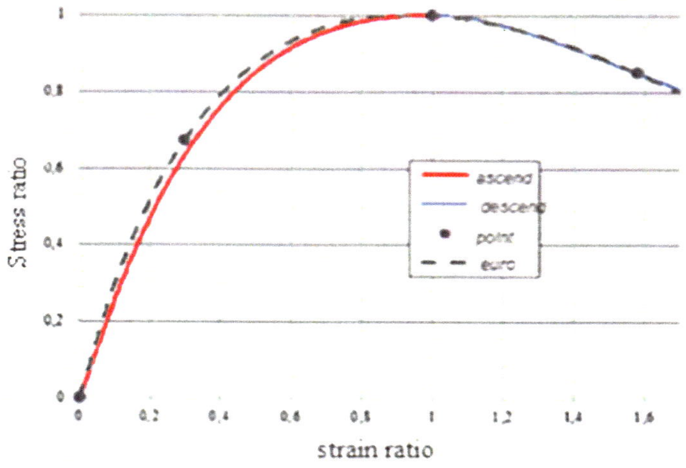

Fig. 11 Comparison with Euro-code

From the proposed stress–strain equation, the stress blocks parameters useful in predicting the ultimate moment of resistance and curvature of FSCC with RA is obtained. The variation in the theoretical and experimentally predicted stress–strain curves is less than 5%. Also, it can be observed from Fig. 11 that the Recycled Aggregate based FSCC stress–strain curve has shown similar behavior as that proposed by the Euro code 2 (EN 1992-1-1:2005 and EN 1992-1-2:2005).

6 Conclusions

From the properties of recycled aggregate, it can be concluded that the coarse aggregate obtained from crushing demolished waste and laboratories can be used for structural concrete works. The strength properties confirm the fact that recycled aggregate is in no way inferior to natural aggregate.

- From the SEM images, it can be understood that the micro cracks at the ITZ zones are interlinked and only tiny length cracks were observed. In SCC with RA specimens, more number of voids with larger diameter was observed and this may be due to presence of adhered old mortar on the surfaces of recycled coarse aggregates.
- From a comparison of EDXA Spectrums of different grades of concrete, it can be observed that the Ca/Si values decreased with increase in the grade of concrete indicating denser presence of C–S–H gel formation in higher grades of concretes.
- Ettringite formation was observed in recycled aggregate based concretes at the ITZ zone. The cracking at the ITZ in SCC with RA specimens is clear and the cracks are interconnected.

- From the confinement studies on Fibrous Self Compacting Concrete made with Recycled Aggregate, with increase in confinement index and fiber index, it is proved that there is an increase in peak stress, strain at peak stress and strain at 85% of peak stress.
- The ties and fiber contributes to the improvement in the confinement of FSCC with RA. The influence of lateral ties was brought out through Confinement Index, while the influence of fiber was brought out through Fiber Index.

$$\text{Confinement Index } C_{iR} = \left(P_b - \bar{P}_b\right)\left(f_v / f_{c'}\right)\sqrt{\frac{b}{s}}$$

$$\text{Fiber Index } F_{iR} = \left(w_f\right)\left(l/d\right)$$

- The effect of fiber on ascending portion of load–deflection curve is nominal whereas in the descending portion it is dominant and the curve became less steep because of the enhanced ductility and toughness.
- The confined strength and strain at peak stress due to lateral ties and fiber index were developed from the experimental study and the relationship with respect to fiber index and confinement and are as below:

$$\frac{f_u}{f'} = (1 + 0.523C_{iR}) + (1 + 0.221F_{iR})$$

$$\frac{\varepsilon_u}{\varepsilon'} = (1 + 4.142C_{iR}) + (1 + 0.838F_{iR})$$

- The equation proposed (non-dimensional) in this investigation to predict the constitutive behaviour of FSCC with RA in axial compression is very close to the experimental values and hence prediction is accurate.
- From the developed stress block parameters, the ultimate moment and curvature of the FSCC with RA section can be determined and which will be very useful to the practical design engineers using this type of concrete in the field.

Acknowledgements I would like to express my thanks to the Director, National Institute of Technology Warangal, for extending the technical and financial support for carrying out this research.

References

Ahmad S, Umar A, Masood A (2017) Properties of normal concrete, self-compacting concrete and glass fibre-reinforced self-compacting concrete: an experimental study. Procedia Eng 173:807–813

Aslani F, Nejadi S (2013) Self-compacting concrete incorporating steel and polypropylene fibres: compressive and tensile strengths, moduli of elasticity and rupture, compressive stress–strain curve and energy dissipated under compression. Compos B Eng 53:121–133

Chatterji S, Jeffery JW (1962) Studies of early stages of paste hydration of cement compounds, I. J Am Ceram Soc 45(11):536–543

Dhiyaneshwaran S, Ramanathan P, Baskar I, Venkatasubramani R (2013) Study on durability characteristics of self-compacting concrete with fly ash. Jordan J Civ Eng 7(3):342–353

EFNARC (2005) Specifications and guidelines for self-compacting concrete. European Federation of Producers and Applicators of Specialist Products for Structures, United Kingdom. https://www.efnarc.org/pdf/SCCGuidelinesMay2005.pdf

EN 1992-1-1:2005. Eurocode 2: design of concrete structures. Part 1–1: general rules and rules for buildings. CEN

EN 1992-1-2:2005. Eurocode 2: design of concrete structures. Part 1–2: general rules—structural fire design. CEN

Güneyisi E, Gesoglu M, Algın Z, Yazıcı H (2016) Rheological and fresh properties of self-compacting concretes containing coarse and fine recycled concrete aggregates. Constr Build Mater 113:622–630

Indian Standard Code IS: 12269 (1987) Specifications for 53 grade ordinary Portland cement. Bureau of Indian Standards, New Delhi, India

Indian Standard Code IS: 2386 (1997) Methods of test for aggregates for concrete. Reprinted. Bureau of Indian Standards, New Delhi, India

Indian Standard Code IS: 383 (2016) Specification for coarse and fine aggregates from natural sources for concrete. Bureau of Indian Standards, New Delhi, India

Kumar R (2017) Influence of recycled coarse aggregate derived from construction and demolition waste (CDW) on abrasion resistance of pavement concrete. Constr Build Mater 142:248–255

Martínez-Garcia R, Ignacio Guerra-Romero M, Morán-del Pozo JM, de Brito J, Juan-Valdés A (2020) Recycling aggregates for self-compacting concrete production: a feasible option. Materials 13:868. https://doi.org/10.3390/ma13040868

Mastali M, Dalvand A, Sattarifard AR (2016) The impact resistance and mechanical properties of self compacting concrete reinforced with recycled glass fibre-reinforced polymers. J Clean Prod 124:312–324

Mc Neil K, Kang T-K (2013) Recycled concrete aggregates: a review. Int J Concr Struct Mater 7(1):61–69. https://doi.org/10.1007/s40069-013-0032-5

Nwakaire CM, Yap SP, Onn CC, Yuen CW, Ibrahim HA (2020) Utilisation of recycled concrete aggregates for sustainable highway pavement applications: a review. Constr Build Mater 235:117444

Okamura H, Ouchi M (2003) Self-compacting concrete. J Adv Concr Technol 1(1):5–15

Prasad MLV, Rathish Kumar P (2007) Strength studies on glass fiber reinforced recycled aggregate concrete. Asian J Civ Eng (Build Hous) 8(6):679–690

Prasad MLV, Rathish Kumar P (2009) Development of analytical stress-strain model for glass fiber reinforced self compacting concrete. Int J Mech Solids Res India Publ 4(1):25–37. ISSN 0973-1881

Ramesh K, Seshu DR, Prabhakar M (2003) Constitutive behaviour of confined fibre reinforced concrete under axial compression. Cem Concr Compos 25(3):343–350(8)

Saenz LP (1995) Discussion of the paper, equation for the stress-strain curve of concrete, by P. Desayi and S. Krishnan. J ACI 61(9):1229–1235

Saravana Kumar P, Dhinakaran G (2015) Mechanical and durability properties of slag based recycled aggregate concrete. IJST Trans Civil Eng 39(C2):271–282

Siad H, Mesbah HA, Mouli M, Escadeillas G, Khelafi H (2014) Influence of mineral admixtures on the permeation properties of self-compacting concrete at different ages. Arab J Sci Eng 39(5):3641–3649

Su N, Hsu KC, Chai HW (2001) A simple mix design methods for self compacting concrete. Cem Concr Res 31(12):1799–1807

Yang K-H, Chung H-S, Ashour AF (2008) Influence of type and replacement level of recycled aggregate on concrete properties. ACI Mater J 105(3):289–296

Zhou C, Chen Z (2017) Mechanical properties of recycled concrete made with different types of coarse aggregate, concrete aggregates. Constr Build Mater 134:497–506

Applications of Fabric Reinforced Cementitious Mortar (FRCM) in Structural Strengthening

M. Chellapandian and S. Suriya Prakash

Abstract Over the past two decades, the use of advanced composite materials such as Fiber reinforced Polymers (FRP) and Fabric Reinforced Cementitious Matrix (FRCM) has been widely adopted for the strengthening of critical infrastructural assets such as bridges, building, dams, and tunnels. This chapter presents a state-of-the-art review on the use of FRCM for strengthening of reinforced concrete (RC) structures under different load combinations. Initially, the material characterisation of FRCM through the tensile and bond test are described. Then, a detailed overview on the overall behavior and failure mode of FRCM strengthened RC members under compression, shear, flexure, torsion and seismic loads are discussed. Moreover, few studies highlighting the FRCM strengthening of un-reinforced masonry structures are also discussed. In addition, few case studies where the use of FRCM strengthening is preferred over the other techniques are discussed and the key parameters are analysed.

Keywords Strengthening · Fabric reinforced cementitious matrix · Fiber reinforced polymers

1 Introduction

Maintenance of critical infrastructural assets such as bridges, building, dams, and tunnels has become a global issue and requires immediate attention. Some of them have deteriorated and need to be strengthened to ensure safe performance. Most of the strengthening work focus on the deteriorated concrete structures either due to natural calamities (earthquake, fire, etc.) or corrosion, as shown in Fig. 1. As per the statistics from Indian Bridge Management System (IBMS) in 2016, more than

M. Chellapandian
Department of Civil Engineering, Mepco Schlenk Engineering College, Mepco Nagar, Sivakasi, Tamil Nadu 626005, India

S. Suriya Prakash (✉)
Structural Engineering Division, Department of Civil Engineering, Indian Institute of Technology Hyderabad, Kandi, Sangareddy, Telangana 502285, India
e-mail: suriyap@ce.iith.ac.in

S. B. Singh et al. (eds.), *Emerging Trends of Advanced Composite Materials in Structural Applications*, Composites Science and Technology,
https://doi.org/10.1007/978-981-16-1688-4_9

(a) (b)

Fig. 1 Damage of RC members. **a** Shear failure due to earthquake and **b** damaged RC beam due to corrosion (*Courtesy* https://civildigital.com)

40,000 bridges constructed before 60 years are close to complete their service life and require immediate strengthening. In addition, 37,000 railway bridges in India require immediate strengthening (Singh et al. 2002). It is worth mentioning that most of the structures requiring strengthening are under service, which complicates the strengthening process. Moreover, the present status of the structures and their capacity are unpredictable, which further complicates the optimistic strengthening design.

Various strengthening such as steel jacketing, concrete enlargement, FRP retrofitting is commonly used to improve the performance of existing structures. However, these conventional techniques have various disadvantages due to their heavy dead loads, increased installation time and labour requirements. Use of fibre-reinforced polymer (FRP) for structural strengthening of RC elements has become popular since the last two decades. The main advantages of FRP strengthening, when compared to the conventional strengthening techniques, includes (a) Ease of handling, (b) Lightweight, (c) Corrosion resistance, (d) High specific strength and stiffness (strength/stiffness to weight ratio), (e) High impact resistance, (h) High dielectric capability with a nonconductive property, (g) Good insulator with low thermal conductivity, (h) Lower installation cost, lesser maintenance cost, and longer service life. Strengthening of concrete members with different FRP techniques such as near-surface mounting, external confinement and their hybrid combinations were successfully used for improving the overall behaviour under different load combinations (Pachalla and Prakash 2017; Kuntal et al. 2017; Chellapandian et al. 2017, 2019; Jain et al. 2017; Kankeri et al. 2018; Chinthapalli et al. 2020). Despite having numerous advantages, FRP strengthening has a shortcoming which includes the use of organic resin to bond the parent member and fibre. The performance of organic resins is compromised when exposed to high temperature and extreme weather conditions leading to reduced durability of the strengthened systems. Due to the limitations of existing techniques, there is a need for the development of innovative strengthening materials and techniques and fibre reinforced cementitious mortar (FRCM) is one of them.

1.1 Background on FRCM

Textile-based composite materials known as fabric reinforced cementitious mortar (FRCM) or textile reinforced mortar (TRM) comprise of fabric meshes and cementitious mortar. The fabric meshes are fabricated with long-woven, knitted or even unwoven fibre roving in two orthogonal directions (bi-directional). The fabric meshes are bonded to the surface of concrete members using cementitious mortar. Different types of textiles used as reinforcement in the FRCM system includes steel, polyphenylene bezobisoxazole (PBO), glass, carbon and basalt. Among them, PBO, glass and carbon-based fibre textiles showed in Fig. 2 are predominantly used in the structural strengthening of RC members. Steel fibre textile is commercially available in the uni-directional form with several twisted cords (Density = 1–10 cords/cm). Each cord consists of five wires with a total sectional area (A_{cord}) of 0.538 mm^2. Three straight wires from the core area of the cord and two wires are twisted around them in a helical manner (Zou and Sneed 2020). Other non-metallic fibre textiles such as PBO, glass and carbon are commercially available with the mesh size varying between 8 and 30 mm. The weight of these non-metallic textiles varies between 150 and 600 g/m^2 (Koutas and Bournas 2017).

The effectiveness of FRCM application usually depends on the mechanical properties of textiles, spacing of roving in each direction and the degree of penetration of mortar matrix. The spacing of roving and their quantity can be controlled based on the independent requirement to allow the mesh formation. Moreover, this property makes it effective for various strengthening applications. The mechanical interlock, which is the essential characteristics of FRCM strengthened members, can be achieved by providing perforations between the fibre roving. This helps in developing mechanical

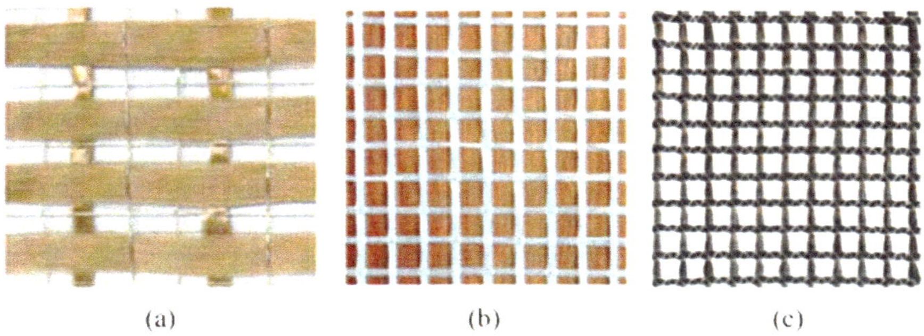

Fig. 2 Different types of fibres used for strengthening. **a** PBO textile fibre; **b** glass textile fibre and **c** carbon textile fibre

interlock between the matrix and FRCM mesh. However, this increased stiffness can limit its application for complex strengthening procedures, i.e., shear strengthening using full wrapping or U-wrapping techniques.

1.2 Advantages of FRCM Over Conventional Strengthening Methods

Use of inorganic cement-based bonding materials as a replacement for organic resin can help in overcoming the issues that concern the use of FRP composites which includes: (a) improved ultraviolet (UV) resistance, (b) better resistance against high temperatures, (c) handling and use are more straightforward as the inorganic resins are water-based, (d) emit no toxic smells during strengthening, and (e) comparable chemical bonding with concrete. However, the direct application of inorganic cement-based mortar over the FRP sheet would result in poor bond performance due to two significant reasons such as (a) the granularity of the mortar, (b) penetration and impregnation of fibre sheets will be difficult. Replacing textiles instead of continuous fibre sheets can help in improving the fibre–matrix interactions. Thus, the use of fabric/textiles with the cement matrix is termed as fabric reinforced cementitious mortar (FRCM). FRCM based strengthening techniques are extensively investigated over the last twenty-five years. FRCM can be used for a wide range of structural strengthening similar to the FRP composites. Moreover, the stability of textile material and the mechanical interlock between the matrix and textile can be significantly improved by coating the non-metallic textiles with polymers. Figure 3 highlights the advantages of using FRCM over FRP for structural strengthening. The various aspects of FRCM, which includes mechanical characterization, strengthening procedure, behaviour under different load combinations and selected case studies are discussed in the following sections.

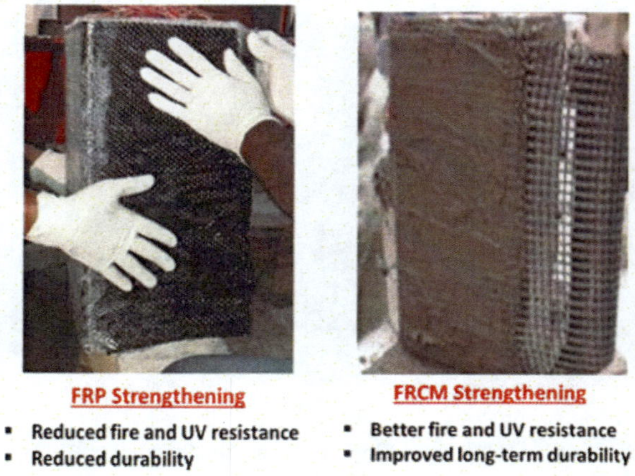

Fig. 3 Schematic representation of the advantages of FRCM over FRP

2 Mechanical Characterization of FRCM Systems

The composition of the mortar used as the matrix in FRCM systems significantly affects its response as a composite material, because the impregnation of fibres with mortar is quite crucial for achieving a good bond between the fibres and the matrix. The mortar shall include fine granules and have a plastic consistency. It shall be workable with low viscosity (for easy application to vertical or steep surfaces). Minimum shear strength shall be ensured to prevent the debonding of the composite material from the substrate. Due to the shear strength requirements, cement-based mortars are widely used as a matrix in FRCM.

2.1 Tension Test

It is essential to characterize the FRCM composites used for structural strengthening so that the mechanical properties such as tensile strength, ultimate strain and the modulus of elasticity can be determined. Many researchers in the past have developed the procedure for performing the tensile tests on FRCM based composites (Contamine et al. 2011; Ascione et al. 2015; Arboleda et al. 2016; D'Antino and Papanicolaou 2018). Some in the past have also investigated the dynamic tensile behaviour of FRCM coupons (Zhu et al. 2011). The accuracy of results from the tensile tests rely mainly on two factors which include, (a) shape of FRCM coupon (flat plate or dumbbell), (b) gripping methods (Hartig et al. 2012; De Santis et al. 2017; D'Antino and Papanicolaou 2017, 2018). Figure 4a shows the dimensions of the standard FRCM coupon and the setup for direct tension test (ASTM 1996). The specimens were loaded in a displacement-controlled manner (0.5–1.0 mm/min) and the measured displacements were used for calculating the ultimate strain level.

Table 1 shows the comparison of different mechanical properties of FRP and FRCM composites obtained from the material characterization (direct tension test). The values were obtained from the standard coupon samples prepared as per standards and tested under tension. Figure 4b shows the failure mode of FRCM coupons under direct tension. The failure mode of FRCM relies strongly on the gripping method adopted. The failure mode by rupture of textile in the mid-span of the coupon is desirable than the other two failure modes such as rupture of textile at the end or slippage within the matrix (De Santis et al. 2017; D'Antino and Papanicolaou 2018). Moreover, the failure can occur at the grips due to the insufficient grip pressure as a result of which the tensile load is transferred by shear stresses. A minimum of five FRCM coupons needs to be tested for ensuring the reliability of results from a direct tensile test.

Figure 4c shows the stress–strain behaviour of FRCM coupons under tension. The tri-linear stress–strain curve usually characterizes the tensile response with the three major stages such as (a) Stage I—un-cracked stage with the linear response until the formation of the first crack, (b) Stage II—formation of multiple cracks in the matrix

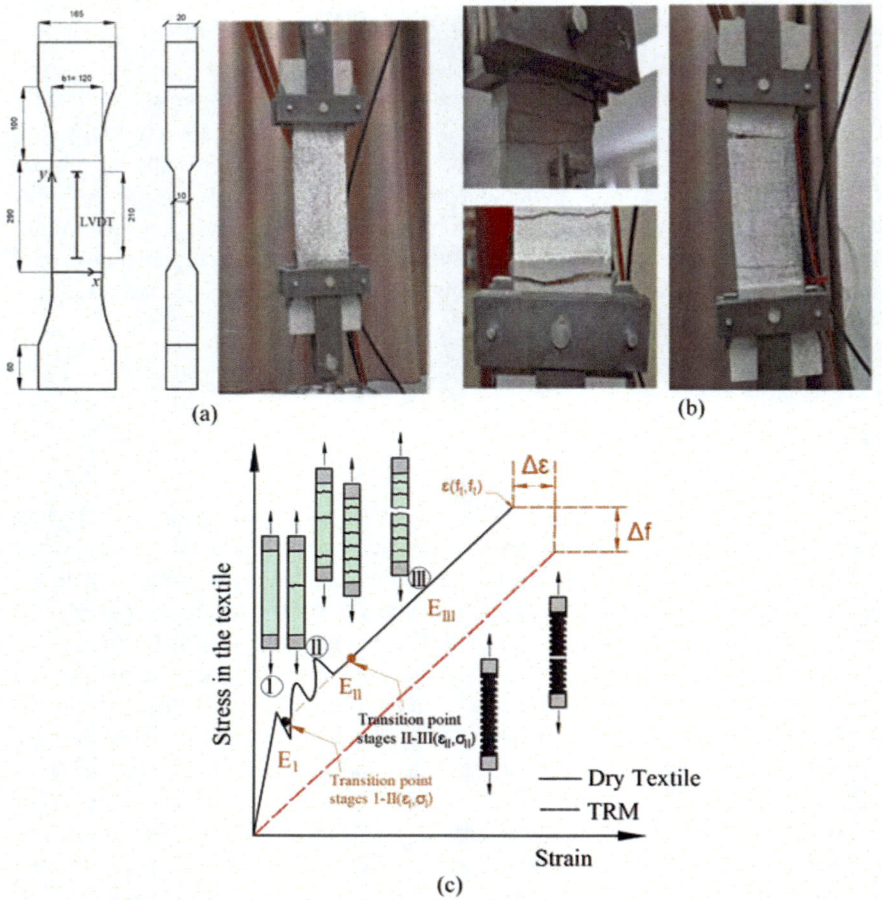

Fig. 4 FRCM under tension. **a** Test setup; **b** failure modes and **c** tensile behavior of FRCM (*Picture Courtesy* D'Antino and Papanicolaou 2018). *Note* All dimensions are in "mm"

Table 1 Comparison of mechanical properties—FRP versus FRCM

Material	Ultimate strength (MPa)		Elastic modulus (GPa)		Ultimate strain (%)	
	FRP	FRCM	FRP	FRCM	FRP	FRCM
Glass	500–1200	600–850	30–65	50–65	2.0–4.5	1.5–5.0
Carbon	1200–2400	800–1350	120–165	80–110	1.0–1.5	0.8–1.5
PBO	–	1250–3100	–	105–210	–	1.5–2.5

and (c) Stage III—end of crack formation with a stiffer response until the rupture of fibres. The transition from Stage-I to II where there are few kinks in the stress–strain graph was due to the development of multiple cracks in the fibre-mortar interface, and the stiffness of the composites was significantly reduced. The extent of cracking relies largely on the properties of textile and mortar used, and the curve stabilises after the cracking.

2.2 *Bond Test*

Understanding the bond performance of FRCM and the substrate is essential to evaluate the efficacy of strengthening technique. Either single-lap or double-lap shear bond test is performed to evaluate the bond of FRCM system with the concrete or masonry unit. In the single lap test, the FRCM reinforcement is attached to one side of the block and kept fixed. A portion of the textile system is left unbonded and pulled through a displacement controlled mode (ASTM 2013). In the double-lap shear test, the surface is bonded with a U-shaped textile wrap, i.e., on two sides of the substrate. Figure 5 shows the different failure modes under shear bond test. They include (a) debonding with cohesive failure in the substrate, (b) failure at the reinforcement-to-substrate interface, (c) failure at the textile-to-matrix interface, (d) sliding of the textile within the reinforcement thickness, (e) tensile rupture of the textile in the unbonded portion and (f) rupture of the textile inside the mortar matrix (Ascione et al. 2015). Bond tests show the efficiency of load transfer by the substrate-reinforcement interface allowing the weakest failure mechanism to activate. The failure mode usually depends on the following parameters such as (i) shear strength of the mortar matrix, (ii) the tensile strength of the textile, and (iii) the textile-to-matrix bond/interlocking. The first three failure modes are highlighted in Fig. 5. Force-slip diagram of first three failure modes shows a nonlinear curve followed by flat post-peak behaviour and a sudden brittle failure (Fig. 6a). For failure mode 'D', a

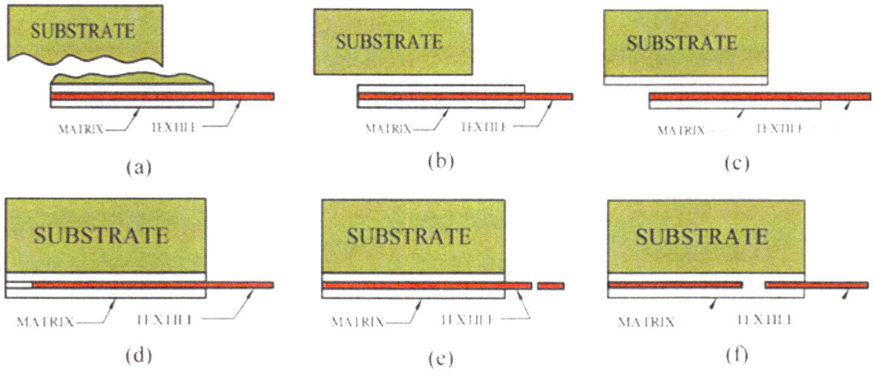

Fig. 5 Comparison of different failure modes under shear bond test

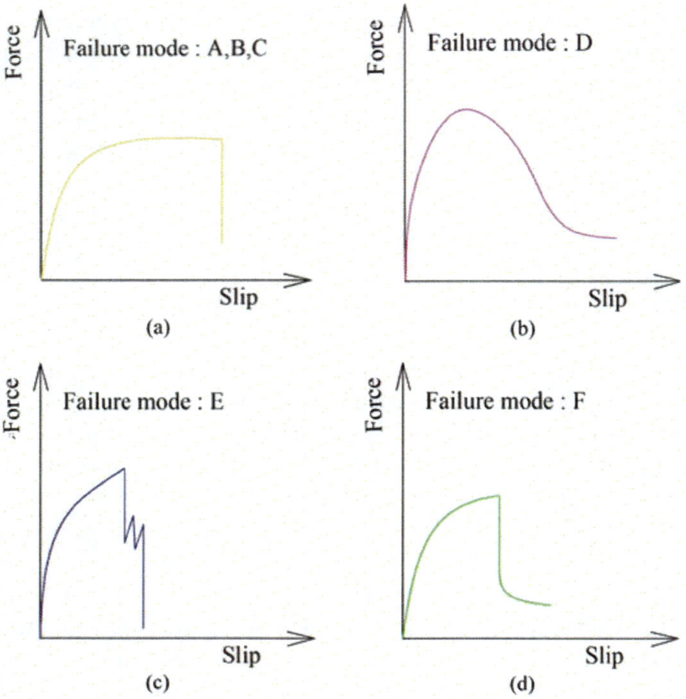

Fig. 6 Overall load–slip behaviour in shear bond test

gradual load reduction (parabolic shape curve) occurs due to the progressive friction loss of the textile sliding within the mortar (Fig. 6b). For the failure mode 'E', an immediate load reduction may occur before the flat branch of the curve is reached (Fig. 6c). Finally, the load-slip behaviour related to failure mode 'F' shows a sudden reduction (Fig. 6d).

3 Strengthening Using FRCM—Procedure

Depending on the type of loading, the strengthening scheme using FRCM system is designed. For example, strengthening with a complete wrapping scheme can be used for enhancing the shear resistance and confinement of RC columns. Similarly, bonding FRCM plates at the soffit of the RC beam can help in enhancing its flexural capacity. The strengthening scheme will consist of four essential steps, as shown in Fig. 7.

(a) providing a corner radius to avoid possible stress concentrations followed by surface roughening and pressurizing with a water jet (Fig. 7a).

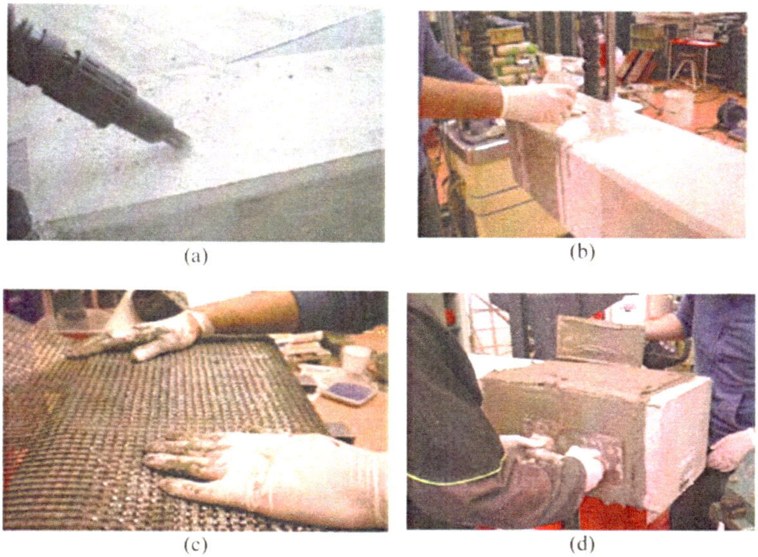

Fig. 7 Strengthening using FRCM system. **a** Surface preparation; **b** surface wetting; **c** FRCM application and **d** finishing (*Picture Courtesy* Escrig et al. 2015)

(b) Surface wetting using water spray to ensure consistency during the application of inorganic based cementitious matrix (Fig. 7b).
(c) Application of the first layer of the inorganic cementitious matrix (4–5 mm thick) and providing externally bonded FRCM system over it (Fig. 7c).
(d) Filling the textiles with the final layer of inorganic based cementitious matrix and surface finishing (Fig. 7d).

4 Different Applications of FRCM

Table 2 summarises the details of different studies which have used FRCM for strengthening RC and masonry structures under different load combinations. FRCM can be used for strengthening under different load combinations which includes:

- Column confinement
- Flexural strengthening
- Shear strengthening
- Strengthening under torsion and combined loading
- Seismic strengthening for RC/masonry structures.

Table 2 Summary of literature for FRCM strengthened members under different loads

Type of loading	Authors investigated	Type of strengthening	Failure modes
Axial load	Triantafillou et al. (2006), Bournas et al. (2007), Peled (2007), Triantafillou (2007), Garcia et al. (2010), Colajanni et al. (2014), Ombres (2014), Trapko (2014) and Thermou et al. (2015)	Full wrapping of FRCM jacket	(a) Debonding of FRCM at the lap end (b) Rupture of FRCM Jacket
Flexure	Bruckner et al. (2006), Papanicolaou and Triantafillou (2006), Triantafillou (2007), Papanicolaou et al. (2009), Ombres (2011), Schladitz et al. (2012), Elsanadedy (2013), Loreto et al. (2013), Babaeidarabad et al. (2014), Napoli and Realfonzo (2015), Ebead et al. (2017), Sneed et al. (2016), Koutas and Bournas (2017) and Raoof and Bournas (2017)	Bonding of FRCM sheets or laminates at the tension side	(a) Slippage of fibres within the matrix (b) Debonding at the concrete-matrix interface (c) Debonding with peeling off of cover concrete (d) Rupture of fibres
Shear	Triantafillou and Papanicolaou (2006), Bruckner et al. (2008), Blanksvard et al. (2009), Al-Salloum et al. (2012), Contamine et al. (2013), Azam and Soudki (2014), Jung et al. (2015), Loreto et al. (2015), Ombres (2015), Tetta et al. (2015), Awani et al. (2016), Tzoura and Triantafillou (2016), Aljazaeri and Myres (2017) and Tetta et al. (2018)	Full wrapping and U-wrapping	(a) Debonding with peeling off of concrete cover (b) Fracture of FRCM jacket (c) Localized FRCM jacket damage (d) Flexure failure

(continued)

Table 2 (continued)

Type of loading	Authors investigated	Type of strengthening	Failure modes
Torsion	Schladitz and Curbach (2012), Alabdulhady et al. (2017) and Alabdulhady and Sneed (2018, 2019)	Full wrapping and U-wrapping	(a) Concrete crushing close to the restrained end (b) Rupture of fibres in the textile
Seismic loading—RC structures	Bournas et al. (2007, 2009), Bournas and Triantafillou (2011a, b) and Bournas (2016)	Full wrapping of FRCM jacket; hybrid combination of NSM + FRCM jacketing	(a) Relocation of the plastic hinge (b) Expansion at the column base
Masonry structures	Prota et al. (2006), Papanicolaou et al. (2007), Papanicolaou et al. (2008), Harajli et al. (2010), Augenti et al. (2011), Faella et al. (2011), Parisi et al. (2011), Koutas et al. (2014), Cascardi et al. (2017), Mezrea et al. (2017) and Akhoundi et al. (2018)	Full coverage of FRCM jackets for the entire wall (in-plane loading) Full coverage of FRCM jackets for the top portion (out-of-plane loading)	(a) Tensile rupture (b) Debonding (c) Crushing (d) Shear cracking (d) Flexure cracking

4.1 Lateral Confinement

Wrapping of compression members using FRCM system can improve the performance under axial compression through increased passive confinement. The of FRCM confinement is to increase both the axial strength and deformation capacity of columns through lateral confinement. The fibres in the hoop direction can improve the confinement and help in increasing the ultimate strain of concrete. However, the confinement effectiveness is better in circular columns than rectangular sections. Due to the presence of sharp corners, the rectangular elements are more susceptible to premature rupture of fibres resulting in reduced peak strength and strain when compared to circular members (Fig. 8). The use of FRCM has been explored extensively in the past as an alternative for FRP based systems (Triantafillou et al. 2006; Bournas et al. 2007; Colajanni et al. 2014; Trapko 2014).

Figure 9a shows the procedure the applying confinement using FRCM. The number of layers required for increasing the lateral confinement usually depends on demand to increase the capacity of the member. It is essential to maintain the minimum corner radius of 25 mm to prevent any possible premature failure at corners. Moreover, a minimum overlap must be provided to prevent the possible early debonding. The important parameters which affect the confinement of FRCM strengthened RC members include (a) FRCM reinforcement ratio, i.e. the number of layers used for confinement, (b) geometry of the section (circular or rectangular), (c) unconfined concrete strength (low strength concrete or high strength concrete),

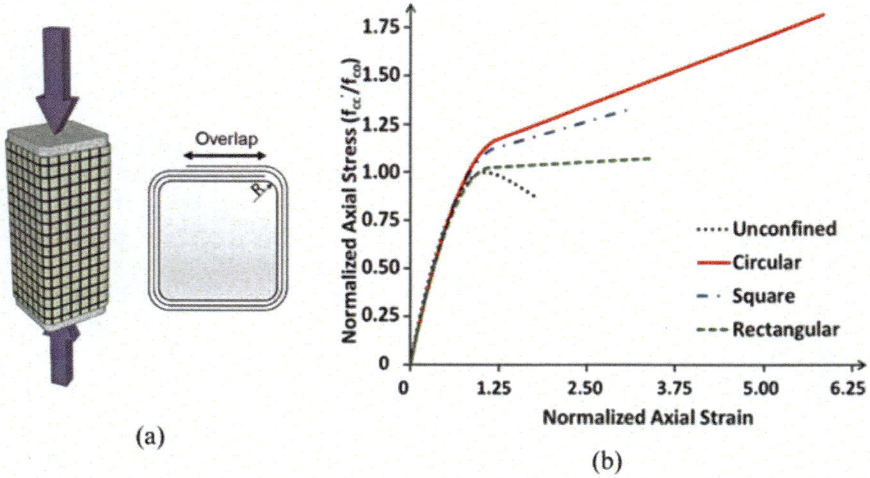

Fig. 8 Confinement of RC columns using FRCM. **a** Overview and **b** typical behaviour

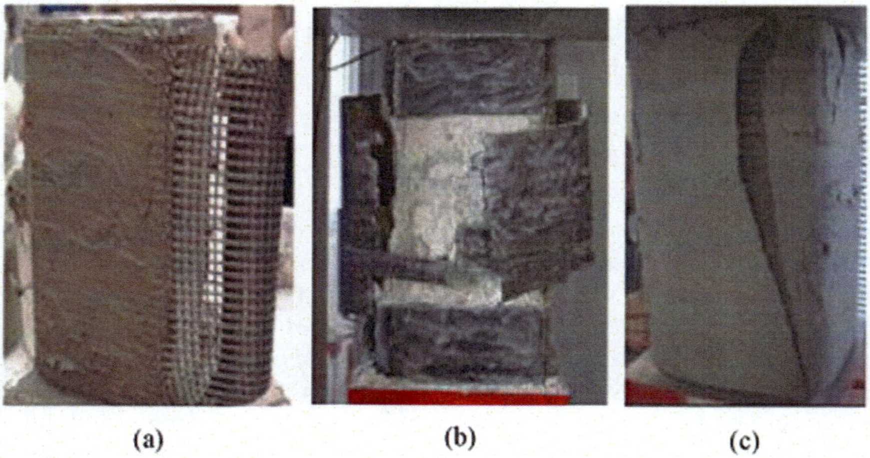

Fig. 9 Failure of columns under compression. **a** Strengthening procedure and failure; **b** textile rupture and **c** debonding (*Picture Courtesy* Bournas et al. 2007)

(d) strength of the inorganic cementitious matrix used, (e) effect of corner radius and overlap length and (f) effect of fibre orientation.

RC columns confined with FRCM can have two possible failure modes, namely:

(a) **Rupture of FRCM jacket** (Fig. 9b)—When the overlap provided is sufficient, the failure of FRCM strengthened members occurs by rupture of FRCM system, i.e., the fracture of fibres in the hoop direction (Bournas et al. 2007; Peled 2007). Similarly, providing a large number of FRCM layers for rectangular members can result in brittle failure at their corners due to stress localization.

(b) **Debonding at the end of the lap** (Fig. 9c)—Due to the use of low strength mortar, the debonding failure occurs where the lap terminates (Triantafillou et al. 2006). The similar failure mode is also possible due to the shorter overlap length and dense mesh with good mortar impregnation (Thermou et al. 2015).

Few researchers have compared the effectiveness of column strengthening using both FRCM and FRP (Triantafillou et al. 2006; Bournas et al. 2007). The researchers concluded that confinement using FRCM is only 10–20% less effective than the FRP confinement. ACI 549.4R-13 provides an idealized bilinear constitutive law for FRCM confined concrete which is very much similar to the FRP confined concrete. The tensile strain in the FRCM composite is limited to 1.2%

4.2 *Strengthening Under Flexure*

The design of reinforced concrete beams and slabs are usually governed by flexure and may require strengthening due to various reasons. Flexural strengthening of RC members through the FRCM system on the soffit (Bruckner et al. 2006; Papanicolaou and Triantafillou 2006; Triantafillou 2007; Ombres 2011; Schladitz et al. 2012; Elsanadedy et al. 2013; Loreto et al. 2013; Ebead et al. 2017; Sneed et al. 2016; Koutas and Bournas 2017; Raoof and Bournas 2017). FRCM can be used to enhance the additional moment capacity of RC members. Most of the FRCM system have bi-directional textiles, i.e., fibre roving in both directions. The fibres present parallel to the axis of the member helps in developing tensile stress, whereas the fibres in the perpendicular direction help in achieving mechanical interlock.

Figure 10a shows the representation of the load–displacement behaviour of RC members under flexure. Control specimens (under-reinforced sections) with no strengthening typically have a ductile behaviour, i.e., large ultimate displacement before failure. When strengthened with low reinforcement ratio of FRCM, negligible increase in the initial stiffness can be witnessed followed by good improvement in the post-cracking stiffness. However, the specimens undergo failure once after reaching the peak load. The failure mode observed will be similar to the control specimen due to the slippage of fibres within the matrix. With the increase in FRCM reinforcement ratio, i.e., the number of plies more than two, the post-cracking stiffness and peak load increase significantly. However, the ultimate strain reduces significantly, and the member undergoes sudden failure due to debonding with or without peeling off of cover concrete. Similar results were reported by Napoli and Realfonzo (2015). As shown in Fig. 10b, SRP refers to Steel Reinforced Polymer (SRP). LD, MD and HD refer to density of polymer tape used. Example: SRP-1LD denotes the slab strengthened with a single layer of low density (LD) steel tape impregnated with a matrix.

Four failure modes have been reported in the literature which includes (a) Intermediate crack debonding, (b) Fibre slippage through the mortar, (c) Debonding with the peeling off of concrete cover and (d) textile rupture (Ombres 2011; Schladitz et al.

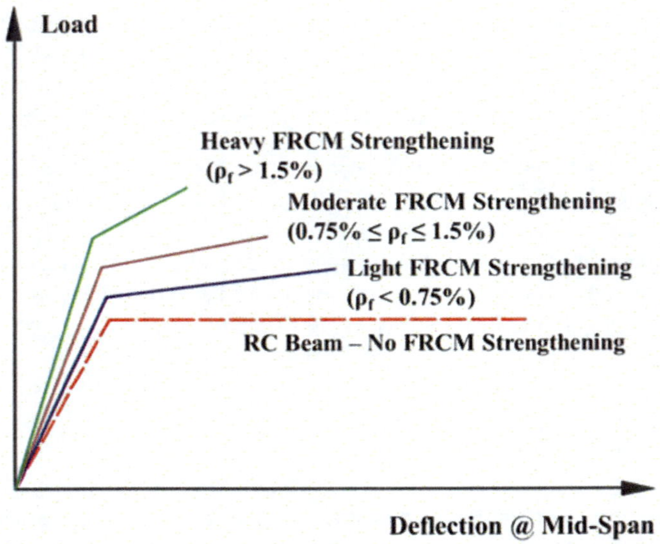

(a)

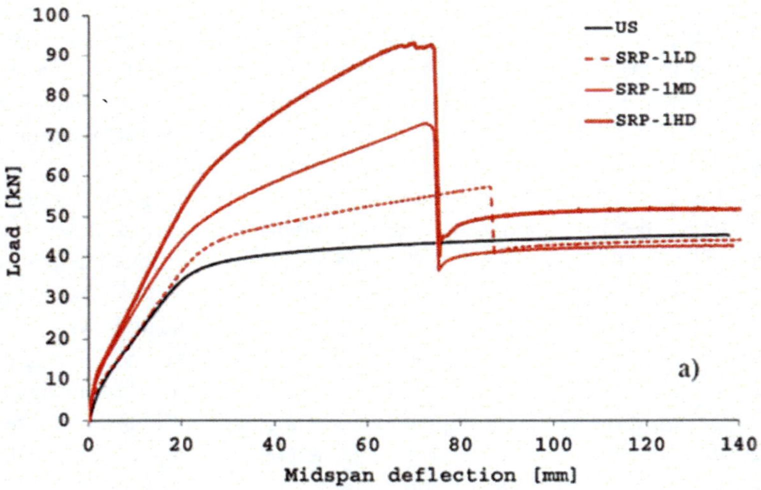

(b)

Fig. 10 Schematic representation of flexural members with different FRCM ratios. **a** Strengthening with different FRCM ratios and **b** typical load–displacement behavior

2012; Loreto et al. 2013; Sneed et al. 2016; Ebead et al. 2017; Raoof et al. 2017). When the specimens are strengthened with more than two plies of FRCM, there is a possibility of failure occurring through debonding mode (Fig. 11a). The debonding at the concrete-matrix interface can occur through two possible modes called as (a) intermediate debonding, i.e., initiation of debonding from the region of maximum bending moment and (b) end debonding, i.e., detachment of FRCM initiates from the end and propagates towards the mid-span. When the specimens are strengthened with less number of FRCM plies (i.e., mild FRCM reinforcement ratio), the failure tends to occur through slippage of fibres within the matrix (Fig. 11b). Similarly, when the bond between the concrete –matrix is strong enough, the debonding of FRCM reinforcement becomes less likely and the entire FRCM peels of along with the cover

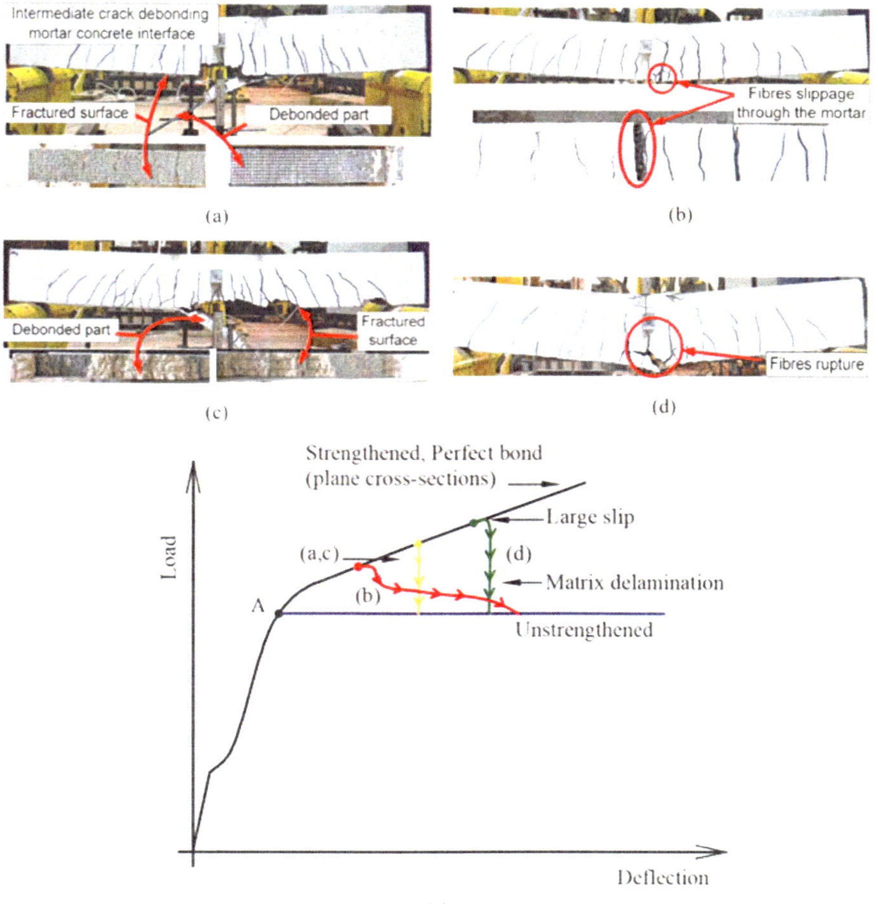

Fig. 11 Failure modes of FRCM strengthened flexural members. **a**–**d** Different failure modes under flexural loading and **e** typical load–displacement behavior with the corresponding failure (*Picture Courtesy* Raoof et al. 2017)

concrete (Fig. 11c). The debonding can initiate either from the mid-span (due to flexure) or in shear span (due to flexure-shear crack). Another possible failure mode under flexure may occur due to the rupture of fibres when the maximum moment region is subjected to high tensile stress (Fig. 11d). Here, the fibre rupture occurs in a single section. Moreover, the failure mechanism is brittle, and the specimen experiences a sudden load drop after reaching a peak load.

Different factors that affect the flexure strengthening of RC members using FRCM includes:

a. FRCM reinforcement ratio—increasing the number of plies or layers of FRCM can help in increasing the overall moment capacity. However, the failure mode may change from one to the other. Example: Application of more than two layers of FRCM may convert the failure mode to debonding with or without peeling off of cover concrete.
b. Steel reinforcement ratio—the efficiency of FRCM strengthening improves significantly for under-reinforced RC members.
c. Influence of inorganic cementitious mortar—Use of polymer modified cementitious mortar can enhance the chemical bond compared to conventional mortar type (Elsanadedy et al. 2013). Similarly, the strength of mortar determines the failure mode of FRCM strengthened members. Use of low strength mortar results in failure due to fracture of the jacket, whereas the use of high strength mortar results in debonding failure.
d. Geometry of textile–spacing and orientation of fibres in the textile.

4.3 Strengthening Under Shear

RC beams/ girders often require strengthening under shear due to various causes such as (a) corrosion issues in existing reinforcement, (b) insufficient shear reinforcement, (c) low strength concrete, (d) increase in demand and (e) convert the failure mode from brittle shear to ductile flexure mode. FRCM may be used to enhance the shear capacity of RC members by applying in the form of either U-wrap or complete wrapping. U-wrap can be used when the strengthening under shear is essential for beams; i.e., the top portion of the member is not accessible due to the presence of the slab (Fig. 12). Complete wrapping is preferred for columns which require strengthening under shear (Fig. 13). A number of studies in the past have investigated the feasibility of FRCM application in shear strengthening of RC members (Triantafillou and Papanicolaou 2006; Bruckner et al. 2008; Blanksvard et al. 2009; Al-Salloum et al. 2012; Contamine et al. 2013; Jung et al. 2015; Ombres 2015; Tetta et al. 2015; Tzoura and Triantafillou 2016; Aljazaeri and Myres 2017; Tetta et al. 2018). The shear behaviour of FRCM strengthened members were studied through testing of beams under static three or four-point bending loads at different shear span to depth ratios.

The schematic representation of the load–deflection behaviour of un-strengthened and FRCM strengthened RC beams is shown in Fig. 14. The un-strengthened speci-

Fig. 12 Typical failure mode and strengthening under shear. **a** Diagonal-shear failure in RC beams and **b** retrofitting solutions using U-Wrap FRCM

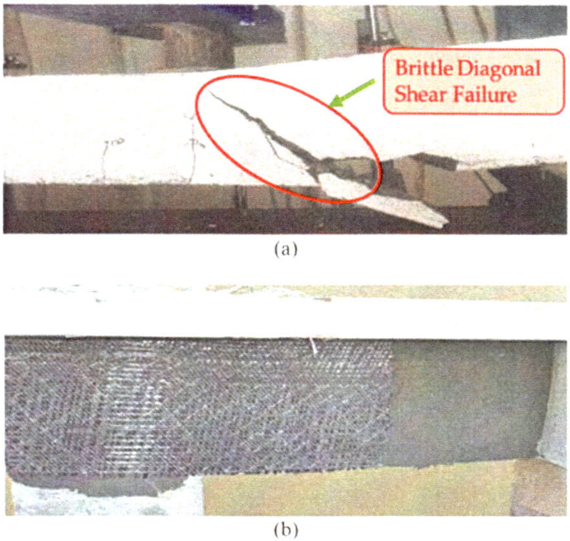

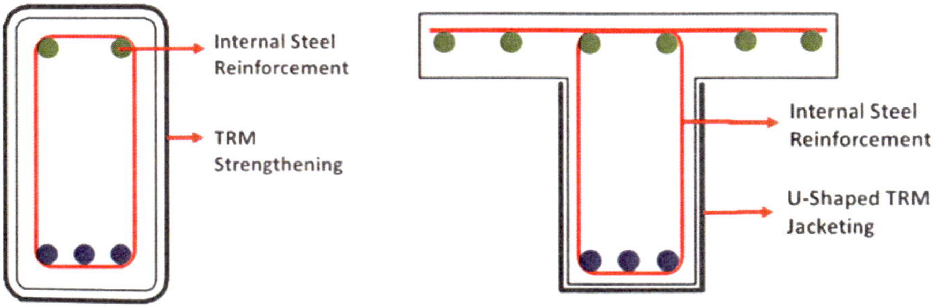

Fig. 13 Types of FRCM shear strengthening of RC members

Fig. 14 Load–deflection behaviour comparison under shear

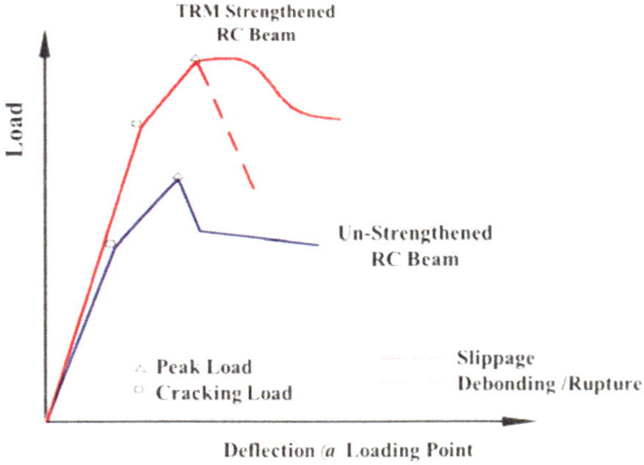

mens show a brittle failure once after reaching the peak load. The behaviour of the FRCM strengthened specimen is similar to the control one until the cracking load. After cracking, FRCM is activated and shows a significant improvement in the post-cracking behaviour. After reaching the peak loads, two types of failure modes are possible namely (a) debonding or rupture of fibres which are very brittle and show a sudden load drop and (b) localized damage of jacket, i.e., slippage of fibres through the inorganic matrix. The latter one exhibits a pseudo-ductile failure mode, i.e., the degradation is not sudden and is gradual.

Figure 15 shows the different failure modes of FRCM strengthened RC members under shear loads. The un-strengthened control beam, which is shear deficient, can fail due to shear-tension mode, as shown in Fig. 15a. Strengthening will aim to exploit the full capacity of the member and convert the failure mode from shear dominant to bending dominant. Four major failure types under shear are possible for FRCM strengthened members which includes.

a. Debonding with peeling off of concrete cover (Fig. 15b)—This type of failure occurs predominantly for U-wrapped members. This failure type helps in exploiting the full shear capacity of the members.
b. Fracture of FRCM jacket—Mainly occurs in members applied with full wrapping scheme. However, this failure mode does not utilize the full capacity and makes the member less effective (Tzoura and Triantafillou 2016).
c. Localized FRCM jacket damage—One of the predominant failure modes observed in most of the FRCM shear strengthened specimens with dry textile

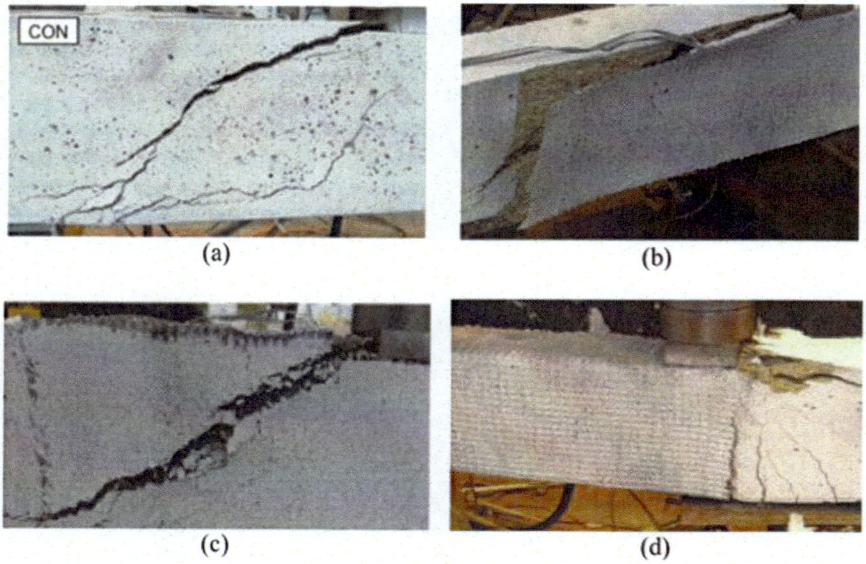

Fig. 15 Possible failure modes under shear. **a** Shear–tension failure; **b** fracture of FRP jacket; **c** damage of FRCM jacket and **d** flexure failure (*Picture Courtesy* Tetta et al. 2018)

materials (Loreto et al. 2015; Tetta et al. 2016). Here, the failure occurs due to the slippage of the mortar through the matrix.

d. Flexure failure (Fig. 15d)—development of shear crack is stopped due to the effectiveness of FRCM jacket, and the failure occurs due to the opening of flexure crack.

4.4 Strengthening Under Torsion or Combined Loads

Reinforced Concrete (RC) bridges, buildings and other structures are subjected to torsional loading in addition to other load combinations. Several studies have reported the effect of torsion on the behaviour of RC members strengthened with FRCM composites (Alabdulhady et al. 2017; Alabdulhady and Sneed 2018, 2019). The wrapping scheme used for torsional strengthening was similar to the one used for shear dominant ones, i.e., full wrapping (with and without spacing along the length) and U-wrapping schemes (T-beams where the top portion is not accessible) may be employed.

Figure 16a shows the test setup used for applying torsional load (Alabdulhady et al. 2017). The torque will be generated to the specimen using the loading arm with some eccentricity relative to the centroid of the section. Threaded steel rods which are anchored to the strong floor were used to support the reaction arm (Fig. 16a). The control specimens without any strengthening had failure initiation with the formation of two complete spiral cracks at an angle of 45° to the longitudinal axis (Fig. 16b). The final failure was due to the concrete crushing in the mid-span of the test specimen. The specimens strengthened with U-wrapping system had a similar failure mode to the control specimen expect the location of crushing, which was close to the restrained ends. In addition, slippage of fibres into the matrix was also observed. It is worth mentioning that the 3-sided configuration shows only a marginal increase in torsional strength enhancement. For specimens strengthened with complete wrapping scheme, the failure initiated due to rupture of fibres in the textile. The specimen also had concrete crushing after the loss of confinement at the mid-span and near the reaction end. The full-wrapping with FRCM helps in complete utilization of torsional capacity (torque increase more than 100%) and the failure surface is distributed throughout the member.

4.5 Seismic Strengthening of RC Structures

FRCM jackets can be used for enhancing the deformation and energy dissipation capacity of reinforced concrete (RC) columns (Bournas et al. 2007, 2009; Bournas and Triantafillou 2011a, b). RC columns require strengthening for avoiding global failure under seismic actions. The reasons for strengthening includes: (a) better confine the plastic hinge locations, (b) improve the cyclic deformability of columns

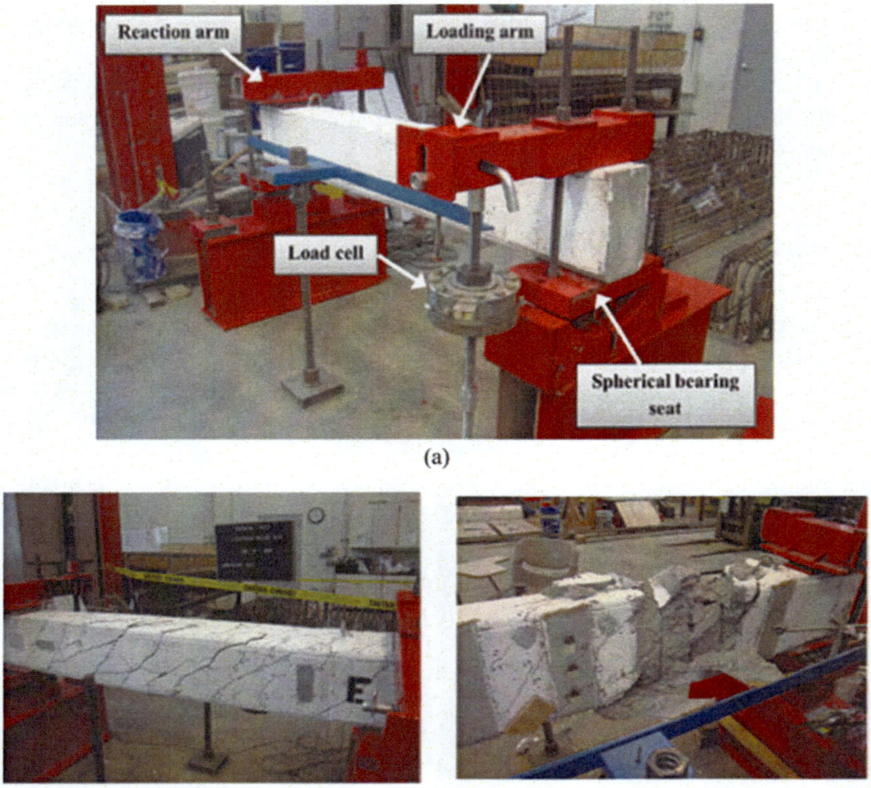

(a)

(b)

(c)

Fig. 16 Test setup and failure mode comparison under torsion. **a** Typical test setup; **b** crack pattern at initial stages and **c** ultimate failure (*Picture Courtesy* Alabdulhady et al. 2017)

by increasing their energy dissipation, (c) improve the behaviour of poorly detailed old RC columns, (d) prevent buckling of longitudinal reinforcements and (e) protect the locations of lap-splices.

Earlier, researchers used FRP as a useful material for improving the deformability of RC columns under seismic loads. However, recent studies have shown that FRCM increased the energy dissipation and cyclic deformation of old type RC columns with poor detailing. Moreover, the effectiveness of FRCM was found similar to the FRP strengthened RC columns under seismic loads.

Figure 17a shows the load–drift response of columns with and without FRCM strengthening. In the graph, L0_M4 and L0_M4G correspond to carbon and glass FRCM used for seismic strengthening of columns. Both the FRCM strengthened specimens showed superior deformability and energy dissipation capacity of poorly detailed RC columns under seismic actions. Moreover, the control specimen failed due to the buckling of longitudinal reinforcement and disintegration of concrete at the base. In the case of FRCM strengthened RC columns, both the carbon and glass

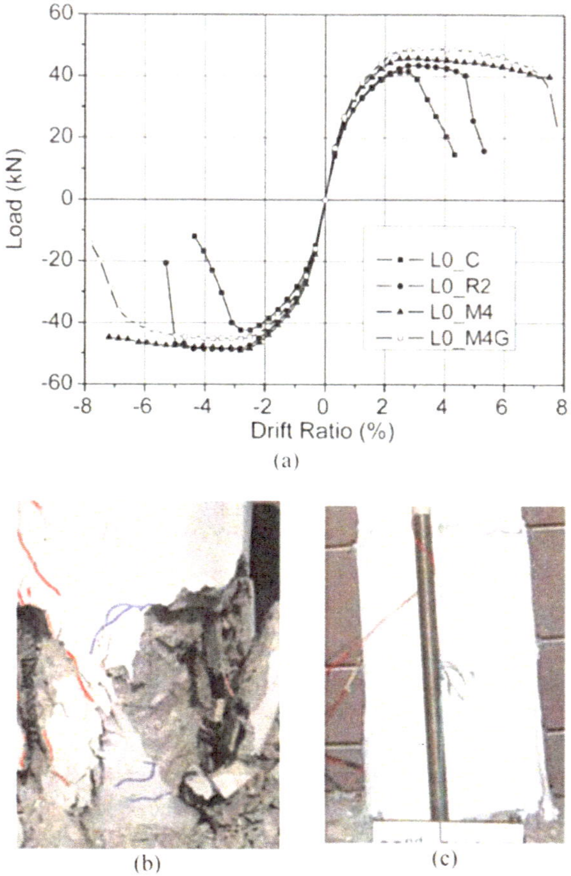

Fig. 17 Seismic behaviour of un-retrofitted and FRCM retrofitted RC columns. **a** Load–drift ratio behaviour; **b** ultimate failure of control specimen and **c** ultimate failure of FRCM retrofitted specimen (*Picture Courtesy* Bournas and Triantafillou 2011a, b)

fibres in textiles developed excellent composite action and were able to resist loads without any early fibre rupture. The ultimate failure was due to the lateral expansion at the base of the column without rupture of fibres in the FRCM jacket (Fig. 17c).

4.6 Strengthening of Masonry Structures

Un-reinforced masonry structures when subjected to in-plane seismic loading can undergo failure due to diagonal cracking (cracking along the bed joints) or shear sliding. Therefore, the strengthening of un-reinforced masonry structures is of great importance. FRCM can be used for strengthening masonry frames subjected to different loading cases such as (a) confinement of masonry columns, (b) strengthening against in-plane bending and shear demand of masonry spandrels and piers, (c) improving the out-of-plane bending capacity of bridge piers and spandrels, (d) strengthening of other masonry structures like arches, domes and vaults. Strengthening masonry structures by confinement is usually not required as they possess sufficient axial capacity. However, different factors such as ageing, deterioration, eccentricity during loading reduces the capacity significantly. The original axial capacity of masonry structures can be restored by providing FRCM confinement in the circumferential direction to increase its peak strength and deformation levels.

Several studies have investigated the effect of FRCM confinement on the behaviour of masonry structures (Faella et al. 2011; Cascardi et al. 2017; Mezrea et al. 2017). Moreover, the models which are used for predicting the axial capacity of FRP strengthened masonry members can be used for FRCM strengthened as well (Cascardi et al. 2017; Mezrea et al. 2017).

Another failure of masonry walls is due to the in-plane bending and shear occurring as a result of seismic action parallel to the wall. The combination of shear and bending occurs with or without the influence of axial loads. FRCM composites are provided with the vertical fibres at the extremities as close as possible to the highly stressed areas for protecting the masonry walls from the in-plane bending stresses, (Fig. 18a). However, if the walls are to subjected to a combination of stress, full coverage of wrapping should be carried out, as shown in Fig. 18b. Effect of in-plane loading on the behaviour of masonry walls have been extensively investigated (Prota et al. 2006; Papanicolaou et al. 2007; Augenti et al. 2011; Parisi et al. 2011; Koutas et al. 2014).

Out-of-plane failure of masonry walls is also possible due to the inertial forces acting in the horizontal direction. Figure 18c shows the overturning of masonry wall due to the plastic hinge formation at the base or certain height from the base. This can be due to the different parameters like slenderness of the wall and boundary conditions. For this loading combination, FRCM can be applied to the top portion of the wall and anchored correctly to the orthogonal walls like a belt, as shown in Fig. 18d. FRCM based strengthening of masonry walls subjected to out of plane bending (cyclic loading) showed good improvement in strength and deformation capacity (Papanicolaou et al. 2008; Harajli et al. 2010).

Koutas et al. (2015) investigated three-storey full-scale masonry infill frames strengthened with FRCM jacketing (Fig. 19). They wanted to understand the contribution of masonry infills to the in-plane seismic loading on the existing RC structures when strengthened using FRCM. The test results revealed that the masonry infills strengthened using FRCM scheme exhibited more than 50% increase in lateral strength and deformation capacity when compared to the un-strengthened specimen. Moreover, the energy dissipation capacity increased by more than 25% when compared to the un-strengthened one.

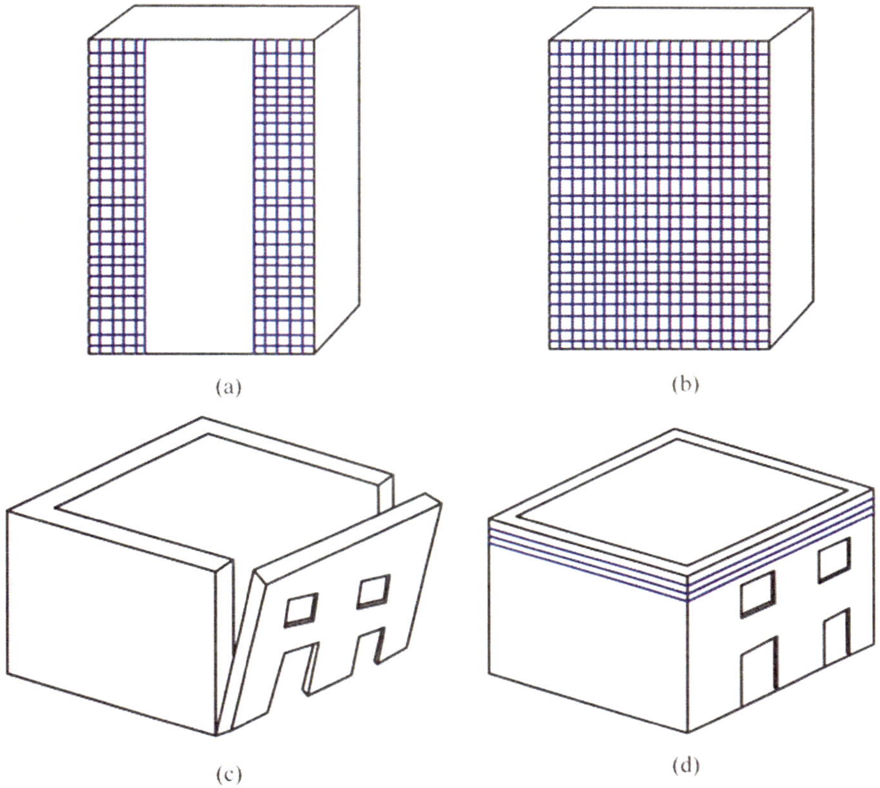

Fig. 18 Seismic strengthening of masonry walls. **a** Partial wrapping scheme against in-plane bending stresses; **b** full wrapping scheme against in-plane bending stresses; **c** out of plane failure and **d** strengthening against out of plane failure

5 Case Studies and Special Applications

5.1 Strengthening San Siro Stadium, Italy

San Siro stadium, located in Milan, Italy, was constructed in the year 1925. The bearing structure of the lower portion of the stadium is an RC frame, as shown in Fig. 20a. The building was renovated in the year 2002, which was new steel structures but left unconnected with the existing RC structure (Trimboli and Mantegazza 2004). A new foundation system was laid for transferring the additional load to the ground. During preliminary condition assessment, and a significant deterioration in the RC beam was detected (Fig. 20b). U-shaped FRCM wrapping is used often for enhancing the flexural and shear capacity of the RC beam (Fig. 20c).

Other strengthening techniques such as concrete jacketing, steel plate bonding and FRP strengthening were preferred earlier. However, those techniques could not

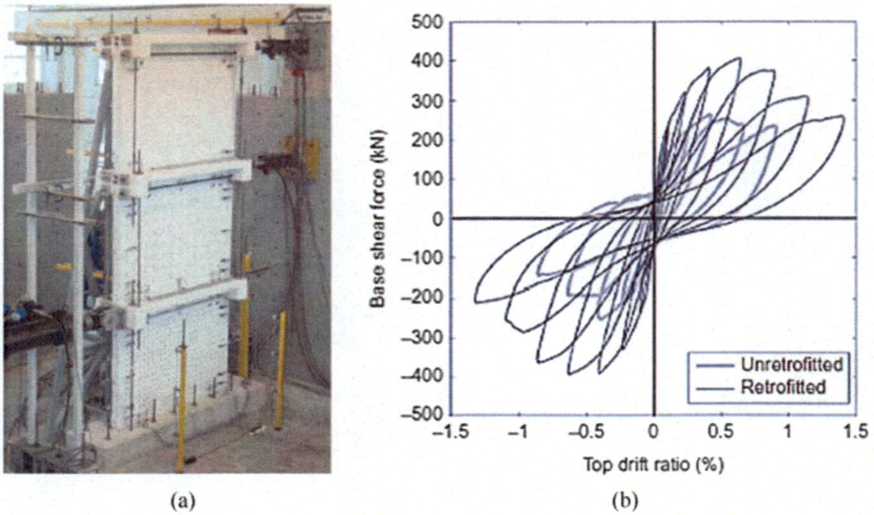

Fig. 19 Seismic strengthening of infilled masonry walls. **a** Failure mode and **b** overall behaviour comparison (*Picture Courtesy* Koutas et al. 2015)

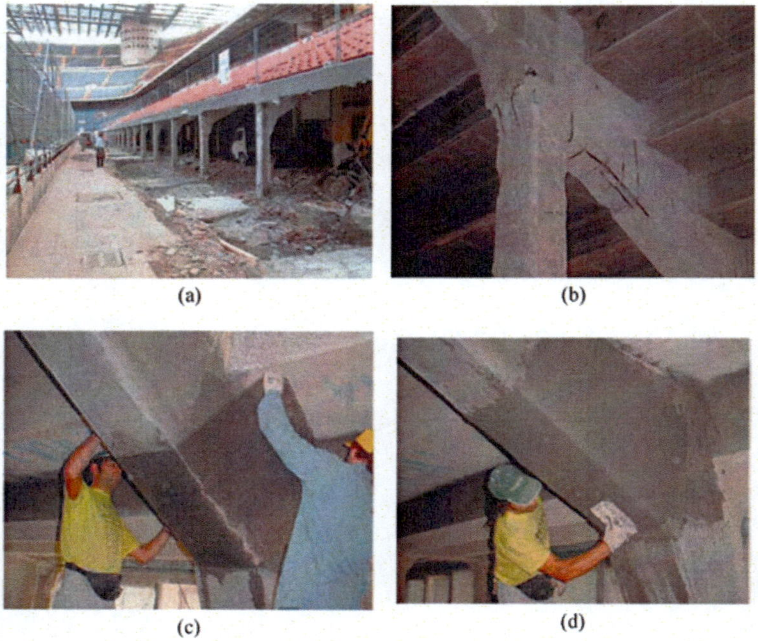

Fig. 20 Strengthening San Siro stadium. **a** Site view; **b** preliminary assessment; **c** strengthening by U-shaped FRCM and **d** surface finishing with cementitious grout (*Picture Courtesy* Trimboli and Mantegazza 2004)

be chosen due to several site constraints and shorter time required to complete the project. Moreover, FRCM jacketing turned out to be an ideal solution for this repair as the no preparation for the damaged surface was essential. The damaged RC beam was first applied with the inorganic cementitious matrix followed by the installation of FRCM jackets. The surface of the FRCM jacket was again filled with the mortar, and uniform surface finishing was carried out (Fig. 20d). Bi-directional carbon fibre textiles were used in this project which had a tensile strength and modulus of elasticity of 3400 MPa and 227 GPa, respectively.

5.2 *Strengthening of Cooling Towers, Niederaussem Power Station*

The Niederaussem lignite power station located in Bergheim, Germany was constructed in 1963 with a total output power capacity of 3864 MW. The cooling towers in the power station are always subjected to extreme weather conditions and are continually deteriorating. Due to various causes like high wind pressure, high-temperature stresses and flue gas discharge, a natural cooling tower was damaged. More information about the cooling tower and damage mechanism can be found at Altmeyer et al. (2012). During the inspection, a detailed assessment of the tower was carried out to determine its present load-carrying capacity. From the calculations from the rational methods, it was found that the tower which was built as per the old code did not meet the revised code recommendations. Hence, it was proposed to strengthen the cooling tower for safeguarding its operation. PBO based FRCM was preferred over FRP due to the presence of inorganic cementitious matrix, which can provide better long term durability without degradation under extreme environmental conditions and high-temperature stresses. The strengthening procedure adopted is shown in Fig. 21.

5.3 *Strengthening of RC Columns in Hot Industrial Area, India*

Due to the extreme temperatures more than 100 °C, severe corrosion and micro-cracking were found in the RC frame, which supports a 25 years old coke oven structure. The requirement from the client was to enhance the overall capacity of RC columns with two restrictions such as (a) no change in the section of the column, and (b) proposed strengthening system should work at extreme temperatures more than 100 °C. Carbon FRP was the first choice made for enhancing the load-carrying capacity of RC columns. However, CFRP was not chosen as the proposed strengthening system should be stable under sustained high-temperature exposure.

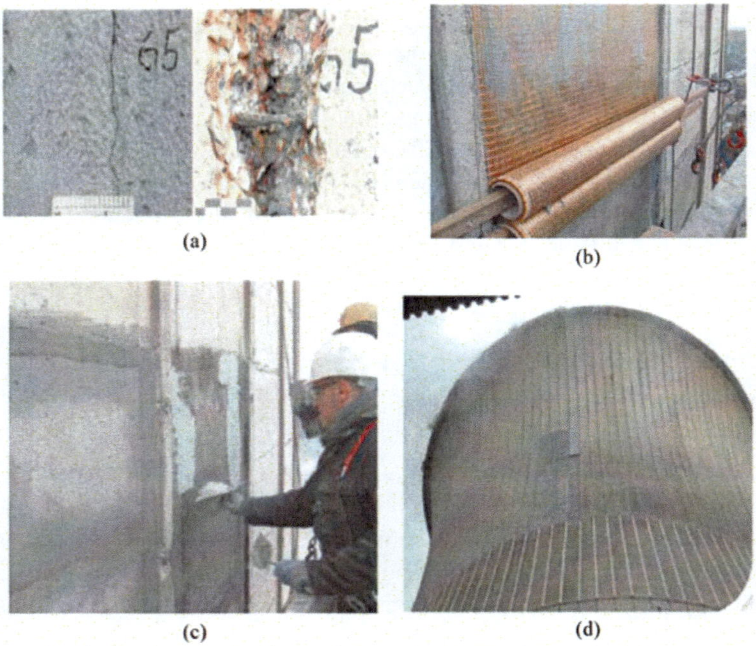

Fig. 21 Strengthening of cooling tower, Germany. **a** Damage extent; **b** strengthening using PBO-FRCM; **c** surface finishing and **d** final view (*Picture Courtesy* Ruredil SPA)

Sanrachana Solutions Ltd., which carried out the renovation work for this work, proposed a combination of shotcrete and SRM hot-wrap FRCM jacketing. SRM hot-wrap comprises of bi-directional carbon fibres with a high tensile strength and modulus of elasticity ($f_t = 3200$ MPa and $E_f = 230$ GPa). Moreover, the mortar used (SRM-CM-MORTAR-250) is a cementitious based hydraulic binder that adheres well with the carbon fabric mesh when used for strengthening concrete or masonry structures. The detailed procedure for the strengthening of RC column using FRCM is shown in Fig. 22. In total, 172 RC columns were strengthened using the proposed FRCM technique. FRCM was successfully applied after conducting a pilot testing program to check both strength enhancement and high temperatures exposure in an accelerated manner. FRCM system was applied on site after pilot tests ensured the desired performance.

5.4 *Romanesque Church of San Roque, Spain*

Romanesque Church of San Roque, Spain belonged to the ancient medieval period and was made of stonework, masonry and timber elements. During a seismic event, the outer walls of the church underwent out of plane separation (Fig. 23a). More-

Fig. 22 Strengthening of corrosion damaged RC column in hot industrial area using FRCM. **a** Site view; **b** extent of damage in column; **c** SRM hot-wrap FRCM jacketing and **d** FRCM strengthened members (*Picture Courtesy* Sanrachana Solutions Ltd.)

over, several locations of the main vault had large cracks and required strengthening (Fig. 23b). Though the church was stable under live loads, few localized zones were severely cracked. FRCM system was applied to the damaged vault to prevent the formation of new cracks and prohibit the growth of the existing longitudinal crack. First, the damaged vault was shored up. Before strengthening, all the dirt and debris were removed. As shown in Fig. 23b, basalt fibre anchors were installed, followed by the installation of basalt fabric. The cementitious mortar was used to cover the fabric and provide surface finishing.

6 Opportunities and Scope for Further Work

The experimental data available until now shows that the use of FRCM will be extensively possible for the strengthening of masonry and reinforced concrete members under different load combinations. To make the design engineers utilize the FRCM based strengthening system for the potential projects, still, some test data is essential along with the availability of proper design guidelines. Recently, some design

Fig. 23 Strengthening of Romanesque Church. **a** Damage due to out of plane separation and **b** FRCM strengthening and Basalt fibres anchors (*Picture Courtesy* Fyfe Europe SA)

specifications and guidelines have been made available by the technical organization for the design of FRCM based strengthening solution (ACI 2013; Triantafillou 2007; Triantafillou 2016). In specific, the use of FRCM may become possible for projects where special design requirements are to be met, i.e., high temperature and durability aspects of the retrofitted structures. A classic example where FRCM system was adopted to strengthen the cooling tower in Niederaussem Lignite power station, Germany, where the cooling tower was predominantly subjected to high temperatures.

Several previous research works have studied the effectiveness of FRCM based strengthening solution for RC and masonry members under different load combinations. However, a lot of research and innovative solutions are required to make it more implementable in the field. One such innovative system is the use of hybrid strengthening system which combines two strengthening techniques such as near-surface mounting and externally bonded FRCM jackets. Some preliminary works on hybrid strengthening systems on the behaviour of RC members under combined compression and bending has been investigated (Bournas and Triantafillou 2009, 2013). However, the effect of hybrid strengthening under shear and torsion loads are still not explored. RC bridge columns with irregular three-dimensional (3D) bridge configurations can undergo axial, bending, and shear forces during earthquake events. The addition of torsion is more likely in skewed or horizontally curved bridges, bridges with unequal spans or column heights, and bridges with outrigger bents. Torsional loadings can significantly affect the flow of internal forces and the deformation capacity of RC columns. Moreover, the presence of torsional loading increases the possibility of brittle shear–dominated failure, which may result in a fatal catastrophe. Besides, the presence of tensile cracks in the principal compression plane softens the stress–strain behaviour of concrete struts. For designing a strengthening system under such a loading scenario where torsion occurs in combination with compression, bending and shear, the hybrid technique can be an effective solution. The hybrid strengthening system should consist of high ratios of confinement (EB fabric) with some NSM as additional longitudinal reinforcement for achieving better performance.

7 Summary

The advantages of FRCM as a strengthening solution for RC and masonry structures were discussed. The behaviour of FRCM strengthened members under various loading conditions such as flexure, shear, torsion, axial loads were explained. The applicability of FRCM strengthening to improve the plastic hinge behaviour of concrete columns and masonry infills under seismic loads was discussed. FRCM has the potential to contribute to the strengthening projects in the coming decades due to its unique advantages such as long durability and high-temperature resistance. Limited case studies were also presented to highlight the improved performance of FRCM to restore the service level of older structures. Opportunities and challenges in increasing the widespread use of FRCM for strengthening applications were also discussed.

Acknowledgements The authors would like to thank Dr Mangesh Joshi, Sanrachana Solutions Ltd, Thane for the help with the required inputs for preparing the case study on FRCM strengthening of severely damaged RC columns.

References

ACI (American Concrete Institute) (2013) Guide to design and construction of externally bonded fabric-reinforced cementitious matrix (FRCM) systems for repair and strengthening concrete and masonry structures. ACI 549.4R. ACI, Farmington Hills, MI

Akhoundi F, Vasconelos G, Lourenco P, Silva LM, Cunha F, Fangueiro R (2018) In-plane behavior of cavity masonry infills and strengthening with textile reinforced mortar. Eng Struct 156:145–160

Alabdulhady MY, Sneed LH, Carloni C (2017) Torsional behaviour of RC beams strengthened with PBO-FRCM composite—an experimental study. Eng Struct 136:393–405

Alabdulhady MY, Sneed LH (2018) A study of the effect of fiber orientation on the torsional behavior of RC beams strengthened with PBO-FRCM composite. Constr Build Mater 166:839–854

Alabdulhady MY, Sneed LH (2019) Torsional strengthening of reinforced concrete beams with externally bonded composites: a state of the art review. Constr Build Mater 205:148–163

Aljazaeri ZR, Myers JJ (2017) Strengthening of reinforced concrete beams in shear with fabric-reinforced cementitious matrix. J Compos Constr 21(5):04017041

Al-Salloum YA, Elsanadedy HM, Alsayed SH, Iqbal RA (2012) Experimental and numerical study for the shear strengthening of reinforced concrete beams using textile-reinforced mortar. J Compos Constr 16(1):74–90

Altmeyer F, Weigl J, Scharf A (2012) Safety analysis and rehabilitation of the natural draft cooling tower Block E at the power plant of Niederaußem. Beton Stahlbetonbau 107(5):318–327 (in German)

Arboleda D, Carozzi FG, Nanni A, Poggi C (2016) Testing procedures for the uniaxial tensile characterization of fabric-reinforced cementitious matrix composites. J Compos Constr 20(3):04015063

Ascione L, De Felice G, De Santis S (2015) A qualification method for externally bonded fibre reinforced cementitious matrix (FRCM) strengthening systems. Compos Part B 78:497–506

ASTM C1583/C1583M (2013) Standard test method for tensile strength of concrete surfaces and the bond strength or tensile strength of concrete repair and overlay materials by direct tension (pull-off method). ASTM International, West Conshohocken, PA

ASTM D3039, D3039M (1996) Standard test method for tensile properties of polymer matrix composite materials. ASTM International, West Conshohocken, USA

Augenti N, Parisi F, Prota A, Manfredi G (2011) In-plane lateral response of a fullscale masonry subassemblage with and without an inorganic matrix-grid strengthening system. J Compos Constr 15:578–590

Awani O, El-Maaddawy T, El Refai A (2016) Numerical simulation and experimental testing of concrete beams strengthened in shear with fabric-reinforced cementitious matrix. J Compos Constr 20(6):04016056

Azam R, Soudki K (2014) FRCM strengthening of shear-critical RC beams. J Compos Constr 18(5):04014012

Babaeidarabad S, Loreto G, Nanni A (2014) Flexural strengthening of RC beams with an externally bonded fabric-reinforced cementitious matrix. J Compos Constr 18(5):04014009

Blanksvärd T, Täljsten B, Carolin A (2009) Shear strengthening of concrete structures with the use of mineral-based composites. J Compos Constr 13(1):25–34

Bournas D (2016) Strengthening of existing structures: selected case studies. In: Triantafillou TC (ed) Textile fibre composites in civil engineering. Woodhead Publishing, Cambridge, UK, pp 389–411 (Chapter 17)

Bournas DA, Triantafillou TC (2009) Flexural strengthening of RC columns with NSM FRP or stainless steel. ACI Struct J 106(4):495–505

Bournas DA, Triantafillou TC (2011a) Bar buckling in RC columns confined with composite materials. J Compos Constr 15(3):393–403

Bournas DA, Triantafillou TC (2011b) Bond strength of lap spliced bars in concrete confined with composite jackets. J Compos Constr 15(2):156–167

Bournas DA, Triantafillou TC (2013) Biaxial bending of reinforced concrete columns strengthened with externally applied reinforcement in combination with confinement. ACI Struct J 110(2):193–204

Bournas DA, Lontou PV, Papanicolaou CG, Triantafillou TC (2007) Textile-reinforced mortar versus fibre-reinforced polymer confinement in reinforced concrete columns. ACI Struct J 104(6):740

Bournas DA, Triantafillou TC, Zygouris K, Stavropoulos F (2009) Textile-reinforced mortar versus FRP jacketing in seismic retrofitting of RC columns with continuous or lap-spliced deformed bars. J Compos Constr 13(5):360–371

Brückner A, Ortlepp R, Curbach M (2006) Textile reinforced concrete for strengthening in bending and shear. Mater Struct 39(8):741–748

Brückner A, Ortlepp R, Curbach M (2008) Anchoring of shear strengthening for T-beams made of textile reinforced concrete (TRC). Mater Struct 41(2):407–418

Cascardi A, Longo F, Micelli F, Aiello MA (2017) Compression strength of confined columns with fiber reinforced mortar (FRM): new design-oriented models. Constr Build Mater 156:387–401

Chellapandian M, Prakash SS, Sharma A (2017) Strength and ductility of innovative hybrid NSM reinforced and FRP confined short RC columns under axial compression. Compos Struct 176:205–216

Chellapandian M, Jain S, Prakash SS, Sharma A (2019) Effect of cyclic damage on the performance of RC square columns strengthened using hybrid FRP composites under axial compression. Fibers 7(10):90. https://doi.org/10.3390/fib7100090

Chinthapalli HK, Chellapandian M, Agarwal A, Prakash SS (2020) Effectiveness of hybrid fibre-reinforced polymer retrofitting on behaviour of fire damaged RC columns under axial compression. Eng Struct 211:1–11

Colajanni P, De Domenico F, Recupero A, Spinella N (2014) Concrete columns confined with fibre reinforced cementitious mortars: experimentation and modeling. Constr Build Mater 52:375–384

Contamine R, Si Larbi A, Hamelin P (2011) Contribution to direct tensile testing of textile reinforced concrete (TRC) composites. Mater Sci Eng A 528(29):8589–8598

Contamine R, Si Larbi A, Hamelin P (2013) Identifying the contributing mechanisms of textile reinforced concrete (TRC) in the case of shear repairing damaged and reinforced concrete beams. Eng Struct 46:447–458

D'Antino T, Papanicolaou C (2017) Mechanical characterization of textile reinforced inorganic-matrix composites. Compos Part B Eng 127(15):78–91

D'Antino T, Papanicolaou C (2018) Comparison between different tensile test setups for the mechanical characterization of inorganic matrix composites. Constr Build Mater 171(May):140–151

De Santis S, Carozzi FG, De Felice G, Poggi C (2017) Test methods for textile reinforced mortar systems. Compos Part B Eng 127:121–132

Ebead U, Shrestha KC, Afzal MS, El Refai A, Nanni A (2017) Effectiveness of fabric-reinforced cementitious matrix in strengthening reinforced concrete beams. J Compos Constr 21(2):04016084

Elsanadedy HM, Almusallam TH, Alsayed SH, Al-Salloum YA (2013) Flexural strengthening of RC beams using textile reinforced mortar—experimental and numerical study. Compos Struct 97:40–55

Escrig C, Gil L, Bernat-Maso E, Puigvert F (2015) Experimental and analytical study of reinforced concrete beams shear strengthened with different types of textile-reinforced mortar. Constr Build Mater 83:248–260

Faella C, Martinelli E, Camorani G, Aiello MA, Micelli F, Nigro E (2011) Masonry columns confined by composite materials: design formulae. Compos Part B Eng 42:705–716

Harajli M, ElKhatib H, San-Jose JT (2010) Static and cyclic out-of-plane response of masonry walls strengthened using textile-mortar system. J Mater Civ Eng 22:1171–1180

Hartig J, Jesse F, Schicktanz K, Haussler-Combe U (2012) Influence of experimental setups on the apparent uniaxial tensile load-bearing capacity of textile reinforced concrete specimens. Mater Struct 45(3):433–446

Jain S, Chellapandian M, Prakash SS (2017) Emergency repair of severely damaged reinforced concrete column elements under axial compression: an experimental study. Constr Build Mater 155:751–761

Jung K, Hong, Han S, Park J, Kim J (2015) Shear strengthening performance of hybrid FRP-FRCM. Adv Mater Sci Eng 2015:11

Kankeri P, Prakash SS, Pachalla SKS (2018) Experimental and numerical studies on efficiency of hybrid overlay and near surface mounted FRP strengthening of pre-cracked hollow core slabs. Structures 15:1–12

Koutas LN, Bournas DA (2017) Flexural strengthening of two way RC slabs with textile-reinforced mortar: experimental investigation and design equations. J Compos Constr 21(1):04016065

Koutas L, Pitytzogia A, Triantafillou TC, Bousias SN (2014) Strengthening of infilled reinforced concrete frames with TRM: Study on the development and testing of textile-based anchors. J Compos Constr 18(3):A4013015

Koutas L, Bousias SN, Triantafillou TC (2015) Seismic strengthening of masonry-infilled RC frames with TRM: experimental study. J Compos Constr 19(2):04014048

Kuntal VS, Chellapandian M, Prakash SS (2017) Efficient near surface mounted CFRP shear strengthening of high strength prestressed concrete beams—an experimental study. Compos Struct 180:16–28

Loreto G, Leardini L, Arboleda D, Nanni A (2013) Performance of RC slab-type elements strengthened with fabric-reinforced cementitious matrix composites. J Compos Constr 18(3):A4013003

Loreto G, Babaeidarabad S, Leardini L, Nanni A (2015) RC beams shear-strengthened with fabric-reinforced-cementitious-matrix (FRCM) composite. Int J Adv Struct Eng 7(4):341–352

Mezrea IA, Yilmaz M, Ispir E, Binbir IE, Bal AI (2017) External jacketing of unreinforced historical masonry piers with open-grid basalt-reinforced mortar. J Compos Constr 21:04016110

Napoli A, Realfonzo R (2015) Reinforced concrete beams strengthened with SRP/SRG systems: experimental investigation. Constr Build Mater 93:654–677

Ombres L (2011) Flexural analysis of reinforced concrete beams strengthened with a cement based high strength composite material. Compos Struct 94(1):143–155

Ombres L (2014) Concrete confinement with a cement based high strength composite material. Compos Struct 109:294–304

Ombres L (2015) Structural performances of reinforced concrete beams strengthened in shear with a cement based fibre composite material. Compos Struct 122:316–329

Pachalla SKS, Prakash SS (2017) Load resistance and failure modes of FRP strengthened precast hollow core slabs with openings. Mater Struct 50(3):14

Papanicolaou CG, Triantafillou TC (2006) Textile-reinforced cementitious or polymeric materials for strengthening of concrete structures. In: Proceedings of the 2nd fib congress. International Federation for Structural Concrete, Lausanne, Switzerland

Papanicolaou CG, Triantafillou TC, Karlos K, Papathanasiou M (2007) Textile reinforced mortar (TRM) versus FRP as strengthening material of URM walls: in-plane cyclic loading. Mater Struct 40:1081–1097

Papanicolaou CG, Triantafillou M, Papathanasiou KK (2008) Textile reinforced mortar (TRM) versus FRP as strengthening material of URM walls: out-of-plane cyclic loading. Mater Struct 41:143–157

Papanicolaou C, Triantafillou TC, Papantoniou I, Balioukos C (2009) Strengthening of two-way slabs with textile-reinforced mortars (TRM). In: Proceedings of the 11th international fib symposium. International Federation for Structural Concrete, Lausanne, Switzerland

Parisi F, Lignola GP, Augenti N, Prota A, Manfredi G (2011) Nonlinear behavior of a masonry subassemblage before and after strengthening with inorganic matrix-grid composites. J Compos Constr 15:821–832

Peled A (2007) Confinement of damaged and nondamaged structural concrete with FRP and TRC sleeves. J Compos Constr 11(5):514–522

Prota A, Marcari G, Fabbrocino G, Manfredi G, Aldea C (2006) Experimental inplane behavior of tuff masonry strengthened with cementitious matrix-grid composites. J Compos Constr 10:1081–1097

Raoof SM, Bournas DA (2017) Bond between TRM versus FRP composites and concrete at high temperatures. Compos Part B Eng 127:150–165

Raoof SM, Koutas LN, Bournas DA (2017) Textile-reinforced mortar (TRM) versus fibre-reinforced polymers (FRP) in flexural strengthening of RC beams. Constr Build Mater 151:279–291

Schladitz F, Curbach M (2012) Torsion tests on textile-reinforced concrete strengthened specimens. Mater Struct 45(1–2):31–40

Schladitz F, Frenzel M, Ehlig D, Curbach M (2012) Bending load capacity of reinforced concrete slabs strengthened with textile reinforced concrete. Eng Struct 40:317–326

Singh MP, Khaleghi B, Saraf VK, Jain SK, Norris G, Goel R, Murty CVR (2002) Roads and Bridges. Earthq Spectra 18(1):363–379
Sneed LH, Verre S, Carloni C, Ombres L (2016) Flexural behaviour of RC beams strengthened with steel-FRCM composite. Eng Struct 127:686–699
Tetta ZC, Koutas LN, Bournas DA (2015) Textile-reinforced mortar (TRM) versus fibre-reinforced polymers (FRP) in shear strengthening of concrete beams. Compos Part B Eng 77:338–348
Tetta ZC, Koutas LN, Bournas DA (2016) Shear strengthening of full-scale RC T-beams using textile-reinforced mortar and textile based anchors. Compos Part B Eng 95:225–239
Tetta ZC, Koutas LN, Bournas DA (2018) Shear strengthening of concrete members with textile-reinforced mortar (TRM): effect of shear span-to-depth ratio, material and amount of external reinforcement. Compos Part B 137:184–201
Thermou GE, Katakalos K, Manos G (2015) Concrete confinement with steel-reinforced grout jackets. Mater Struct 48(5):1355–1376
Trapko T (2014) Effect of eccentric compression loading on the strains of FRCM confined concrete columns. Constr Build Mater 61:97–105
Triantafillou TC (2007) Textile-reinforced mortars (TRM) versus fibre reinforced polymers (FRP) as strengthening and seismic retrofitting materials for reinforced concrete and masonry structures. In Proceedings of the international conference on advanced composites in construction. University of Bath, Bath, UK
Triantafillou TC (2016) Textile fibre composites in civil engineering. Woodhead publishing series in civil and structural engineering, Elsevier Publications
Triantafillou TC, Papanicolaou CG (2006) Shear strengthening of reinforced concrete members with textile reinforced mortar (TRM) jackets. Mater Struct 39(1):93–103
Triantafillou TC, Papanicolaou CG, Zisimopoulos P, Laourdekis T (2006) Concrete confinement with textile reinforced mortar (TRM) jackets. ACI Struct J 103(1):28–37
Trimboli A, Mantegazza G (2004) Special Fiber Reinforced Cementitious Matrix (FRCM) for the shear strengthening of reinforced concrete beams technical report [Online]. Link: http://www.ruredil.com.do/contentManager/productos/files/%5B19%5Darticolosansiro.pdf
Tzoura E, Triantafillou TC (2016) Shear strengthening of reinforced concrete T-beams under cyclic loading with TRM or FRP jackets. Mater Struct 49(1–2):17–28
Zhu D, Peled A, Zaguri E, Mobasher B (2011) Dynamic tensile testing of fabric-cement composites. Constr Build Mater 25(1):385–395
Zou X, Sneed LH (2020) Bond behavior between steel fiber reinforced polymer (SRP) and concrete. Int J Concr Struct Mater 14:46

Material Characterization of Hybrid FRP Bars

Pankaj Munjal and S. B. Singh

Abstract Hybrid fiber-reinforced polymer (FRP) rebar has emerged as one of the most promising and affordable solutions to the brittle failure problems of ordinary FRP rebar in concrete structures. Hybrid FRP comprised of two or more fibers with a single or several matrix which leads to a desirable combination of performance and environmental attributes. In this study, two type of fibers namely carbon fiber and glass fiber embedded in epoxy resin were used for manufacturing of hybrid FRP bars by hand lay-up process. This chapter deals with history of FRP materials, need for hybrid FRP and experimental tests carried out for tensile strength properties of hybrid FRP bars.

1 History of FRP Materials

Incorporation of fiber reinforced polymer (FRP) composite technology into the industrial world was introduced during the late nineties. The first known FRP product is a hull boat manufactured during mid-1930 which was the part of a manufacturing project using a fiber glass fabric and polyester resin (ACI 2007). During and after World War II, the FRP composite products were exploited in aerospace industries for making of aircrafts. The production and use of carbon fiber began in late 1950 with pursuit of improved protective material for rockets (Bacon and Tang 1964). At the same time, Aramid fibers were being produced and appeared first under the trade name of Nomex by DuPont (Choudhury 2012). The FRP industries finally matured in late 1970 when polymer productions increased in comparison to that of

P. Munjal (✉)
Pan-United Concrete Pte Ltd, Singapore, Singapore
e-mail: pankaj710munjal@gmail.com

P. Munjal · S. B. Singh
Birla Institute of Technology & Science, Pilani, Rajasthan 333031, India

S. B. Singh
e-mail: sbsingh@pilani.bits-pilani.ac.in

S. B. Singh et al. (eds.), *Emerging Trends of Advanced Composite Materials in Structural Applications*, Composites Science and Technology,
https://doi.org/10.1007/978-981-16-1688-4_10

steel making and FRP the universal material that it is today. In 1984, the use of FRP composites for strengthening of reinforced concrete (RC) structures was first investigated at the Swiss Federal laboratory where tests on RC beams strengthened with CFRP plates were performed (Meier et al. 1993). In recent years, the applications of FRP composite in constructions have seen rapidly growing around the world in terms of both research activities and practical purpose. Indeed, many have evoked FRP an excellent composite as a new generation of constructions materials following concrete and steel.

1.1 FRP Composite

FRP composites are defined as a polymer matrix, either thermoset or thermoplastic, that is reinforced with a fiber to provide a discernible reinforcing function in one or more directions. FRP composites are anisotropic (properties vary with the direction), whereas steel or aluminum are isotropic (uniform properties in all directions). The performance of any composite depends on the material of which they are made, and the interaction between the materials (fibers and matrix) (ACI 2007). The mechanical properties of the FRP depend upon the arrangement of the primary load-bearing portion of the reinforcing fibers.

(a) ***Reinforcing fibers***

Fibers are the principally load-bearing component of any FRP products. Fibers are pre-assembled into various forms to make the fabrication of composite products. Discontinuous fibers are generally used to make low-cost composite products such as non-woven mats. Fibers, roving, and filaments yarns are generally representing parallel bundles of continuous filaments. The fibers bundles may be used directly in composite fabrications such as by pultrusion process or filament winding to produce bars, or they may be further converted to other reinforced forms such as prepregs, fabrics, and sheets (Rizkalla et al. 2003). Fibers can be both natural and synthetic. However, synthetic fibers are commercially used. Most commercial fibers used in the application of civil engineering are glass, carbon, and aramid. Typical mechanical properties of fibers are given in Table 1. There are three main sources for the production of carbon fibers, namely, Pitch, Polyacrylonitrile (PAN) and Rayon (Madappa 2011). These are manufactured at a processing temperature of around 1000–1500 °C.

Table 1 Typical properties of fiber materials (Correia 2013)

Property	E-Glass	Carbon	Aramid
Tensile strength (MPa)	2350–4600	2600–3600	2800–4100
Elastic modulus (GPa)	73–88	200–400	70–190
Strain at failure (%)	2.5–4.5	0.6–1.5	2.0–4.0
Density (g/cm^3)	2.6	1.7–1.9	1.4

The fibers are classified as high modulus and low modulus fibers based on the fiber microstructure. The fibers with high modulus and low modulus can be identified with value of modulus around 200 GPa and 50 GPa, respectively.

In comparison to carbon fibers, glass fibers are economical and can have strength in a specific direction according to fiber orientation. The main ingredient of glass fiber are silica (50–70%), aluminum, calcium carbonate, iron. The glass fibers are classified as E glass, S-Glass or C-glass based on their ingredients composition present in their microstructures. As the surface of glass fiber is very active, they are susceptible to damage while handling.

The aramid fibers are a class of strong synthetic fibers. The chemical composition of fibers is poly-para-phenylene-terephthalamide (PPD-T). The aromatic ring in the PPD-T gives high thermal stability to the fibers. These fibers are 50% stronger than E glass. These fibers can be degraded by the action of strong acids and bases, but are unaffected by most of the solvents.

(b) ***Resin matrix***

The term matrix used in the context of FRP reinforcements is analogous to the term concrete in reinforced structures. The role of matrix in the FRP composite is to bind the fibers in the desired locations; transfer the load to the fibers and protect them from environmental damages. Three different types of resins, i.e., Epoxy, Polyester, and Vinyl-esters are most commonly used in the production of FRP composite (ACI 2007). Epoxy resin is most widely used in the composite part, structures, and concrete repair. A major benefit of epoxy resin is its lower shrinkage in comparison to polyester resin. Polyester resins are produced by condensation polymerization of the dicarboxylic acid and difunctional alcohols (glycols). Polyesters are versatile because of their capacity to be modified or tailored during the building of the polymer chain. Polyester resins have been found very efficient in all the segments of composite industries. Vinyl-ester resins were developed to combine the advantages of epoxy resins with those of unsaturated polyester resins. These resins are produced by reacting epoxy resin with acrylic or methacrylic acid. This provides as an unsaturated site, much like that produced in polyester resin when maleic anhydride is used. Typical mechanical properties of resin matrix are given in Table 2.

Table 2 Typical properties of resin matrix (Correia 2013)

Property	Epoxy	Polyester	Vinyl-ester
Tensile strength (MPa)	60–80	20–70	68–82
Elastic modulus (GPa)	2–4	2–3	3.5
Strain at failure (%)	1–8	1–5	3–4
Density (g/cm^3)	1.2–1.3	1.2–1.3	1.12–1.16
Glass transition temperature (°C)	100–270	70–120	102–150

1.2 Structural Applications of FRP Materials

There are many applications of FRP materials in civil engineering field. However, the applications are classified into two broad divisions: (a) Applications of FRP materials for new constructions, (b) Retrofitting and strengthening of structures using FRP materials. Architects have also discovered many applications for which FRP can be used such as siding/cladding, roofing, flooring, and partitions.

(a) ***Application of FRP materials for new constructions***

FRP composite materials have demonstrated great potential in construction industries. FRP bars and tendons are widely used for the construction of all flexural members and more specifically bridge decks and bridge girders. Structures such as bridges built completely with FRP composites have demonstrated exceptional durability and effective resistance to the effect of environmental exposure. Bridge street bridge (ACI 440.4R-04) situated in Southfield, Michigan, USA is one of the famous vehicular bridge made of FRP pre-tensioning and external post-tensioning tendons and FRP bars. The first design guidelines in the world for FRP reinforced concrete building structures were established in Japan in 1993. Guide for designs and constructions of structural concrete reinforced with FRP bars (ACI 440.1R-06), Prestressing concrete structures with FRP tendons (ACI 440.4R-04), guidelines for design of concrete structure using FRP material (JSCE Standard 1997), externally bonded FRP reinforcement for RC structures (FIB 9.3 Task Group 2001), and ISIS Canada design manuals (ISIS Canada Design 2001) published in 2001 are the some of the popular design guidelines available for the applications of FRP materials for new constructions. In India, however, no design guidelines are available for FRP materials and constructions.

(b) ***Retrofitting and strengthening of structures using FRP materials***

One of the most common use of FRP, involves the retrofitting and strengthening of damaged or deteriorating structures such as beams, girder, slab, columns, and frames. The FRP materials have shown great potential in replacing the steel reinforcement as a retrofit material because of their advantages such as lightweight, high tensile strength, non-corrodible characteristics, suitable to curved and rough surface, flexible and easy application, less disturbance to occupants during retrofitting work. FRP composites can be used in different forms of strengthening such as FRP wrapping, FRP laminates, FRP prestressing, and FRP bars as near surface mounted reinforcement. Assessment of the effectiveness of FRP composite as retrofitting or strengthening alternative for masonry structures is required to be carried out experimentally.

In 2010, the American Concrete Institute (ACI) published a new standard ACI 440.7R-10 entitled "Guide for the design and construction of externally bonded fiber-reinforced systems for strengthening unreinforced masonry system." This article provides an overview of the ACI design procedures for in-plane and out-of-plane strengthening of URM wall systems. It also provides two example case studies to further detail the design processes of the new standard in an effort to further disseminate the design approach.

1.3 *Need of Hybrid FRP*

In the past few decades, the application of fibre reinforced polymer (FRP) bars became very popular in the construction engineering because of corrosion issues in the ordinary steel bars. Corrosion of steel in bridges, marine and other structures has been a significant problem which cause deterioration of structures leading to failure of structures. Another superior properties of FRP such as high strength, chemical resistance, and high strength to weight ratios have grown the interest of using FRP in concrete structures as an alternative of steel bars. However, the brittle nature of FRP bars decreases the inherent ductility of FRP reinforced and strengthened structures. Many research studies emphasize the elimination of this inadequacy using innovative materials and methods such as pseudo ductile FRP. The development of hybrid ductile FRP rods and fabrics seem to give innovative and promising solutions for ductility related concerns of FRP reinforced sections.

Some recent investigations on design methodology of these high performance composite materials are given herewith. Somboonsong et al. (1998) developed a design methodology to produce hybrid ductile FRP bars. Such hybrid bars, produced by common pultrusion process, comprises of core yarns (aims high strength and stiffness), surrounded by lay-in, braiding yarns (aims higher ductility) and rib yarns (aims better bonding with concrete). Further, they have studied the mechanical properties of such hybrid ductile bars which showed consistent pseudo-ductile behavior. Yan et al. (1999) and Grace et al. (2005) developed FRP fabrics aiming higher ductility and stiffness. These fabrics essentially consisted of carbon and glass fibers tri-axially braided in three different directions. These fibers were selected with different ultimate strains and mixed in such a way that they failed successively producing a yield plateau. Further, the same experimental study investigated the flexural behavior of reinforced concrete beams strengthened using developed FRP fabrics and concluded that beams strengthened with the ductile fabric exhibited higher ultimate loads and achieved higher ductility indexes than those strengthened with the currently available carbon fiber strengthening systems (Grace et al. 2005).

2 Hybrid FRP Bars

Hybrid FRP consists of two or more fibers with a single or several matrix which leads to a desirable combination of performance and environmental attributes. In this study, two types of fibers namely carbon fiber and glass fiber embedded in epoxy (resin) were used for manufacturing of hybrid FRP bars by hand lay-up process. Epoxy (resin) of grade 'Resin 691' and hardener of grade 'Reactive Polyamide 140' were used for making the paste. Resin was used having viscosity of 7000–10,000 mPa s and hardener of viscosity 6000–7000 mPa s. Mixture of resin and hardener was made with proportion of 85:15 by weight as described by the manufacturer. A special twisting equipment as shown in Fig. 1 was prepared in the workshop to manufacture the hybrid FRP bars by hand layup process. In this equipment, hooks are connected to the handle at the ends of the equipment, and fiber rovings are tensioned between the hooks. Fibers are impregnated with epoxy adhesive. After tensioning the fiber between two ends, hooks are rotated for outward twisting. This way the fibers also get entangled with each other and then get stretched outward causing excess resin to drip out as shown in Fig. 2. This also helps in maintaining uniform bar diameter without

Fig. 1 Self-developed FRP bar manufacturing equipment

Fig. 2 Fabrication of CFRP bar

sagging. After the proper number of rotations, the hooks and handles are fixed, and the entire set is left to cure. The FRP bar is cut from the ends after 72 h. The typical length and diameter of the CFRP bars is 1.5 m and 8 mm, respectively. The volume fraction of fiber and resin were maintained 60–65% and 35–40%, respectively. The hybrid FRP bars were left for air curing at room temperature for 15 days before testing. Apart from hand layup bars, five specimens of pultruded CFRP bars supplied by local manufacturing company were also tested for tensile behavior.

2.1 Tensile Strength of FRP Bars

The tensile tests of FRP bars were carried out in accordance with ASTM D7205/D7205M. For the tensile testing of FRP bars, an anchorage system consisting of steel pipe of diameter 38 mm filled with an expansive grout cement (CICO GROUT-V1) was used to avoid premature failure within the grip zones. The grout cement (CICO GROUT-V1) filled into the steel pipe was cured for four weeks to obtain the compressive strength of approximately 75 MPa. The total length of the specimen was 1300 mm and the gauge length was 380 mm. The schematic diagram of anchorage system is shown in Fig. 3. The tests were conducted using UTM with a capacity of 1000 kN and the specimens were fixed at the both top and the bottom ends in frictional metal grips of the UTM (Fig. 4). The load was applied at displacement

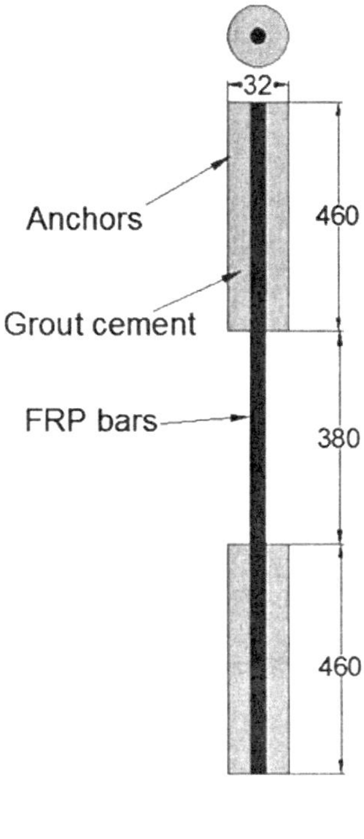

Fig. 3 Anchorage system of FRP bars

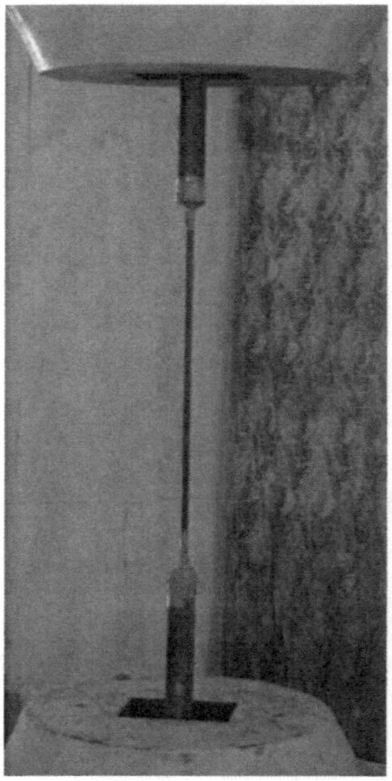

Fig. 4 Tensile testing of FRP bars

control rate of 2 mm/min. A total of 40 samples consisting of 5 specimens for each bar were tested. Table 3 summarizes the average results of tensile testing of FRP bars used in this study. The average tensile stress-strain responses of FRP bars are shown in Figs. 5 and 6. It should be noted that all the reported results represent the average of five tested specimens. The stress-strain behavior of FRP bars approximates that of a

Table 3 Tensile strength properties of FRP bars

S. No.	FRP type	Tensile strength (MPa)	Failure strain (%)	Failure mode
1	Pultruded CFRP	2217	2.23	Pullout
2	Hand lay-up CFRP	1671	2.16	Tensile
3	Hand lay-up GFRP	920	3.50	Tensile
4	Hybrid- 90% glass & 10% carbon fiber	1025	3.10	Combined (tensile and bending)
5	Hybrid—80% glass & 20% carbon fiber	1189	3.01	Combined (tensile and bending)
6	Hybrid—70% glass & 30% carbon fiber	1465	2.90	Combined (tensile and bending)
7	Hybrid—60% glass & 40% carbon fiber	1326	2.85	Combined (tensile and bending)
8	Hybrid—50% glass & 50% carbon fiber	1256	2.79	Combined (tensile and bending)

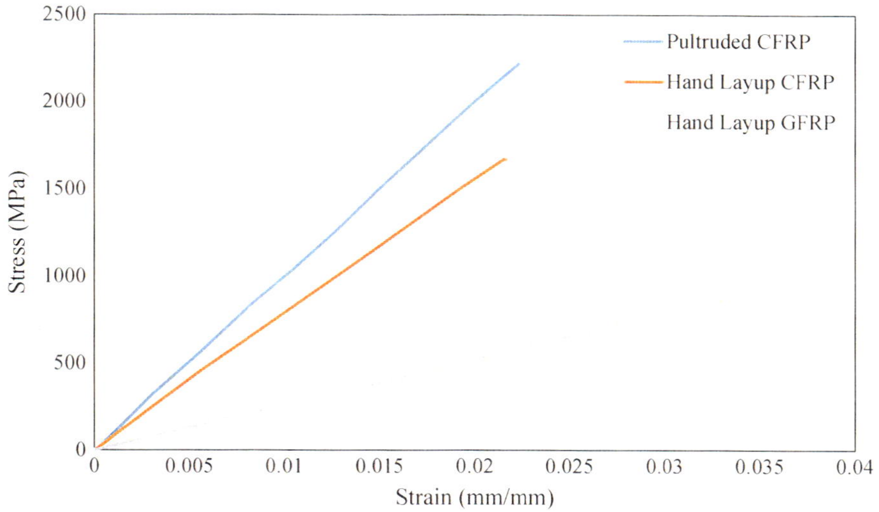

Fig. 5 Tensile stress-strain response of FRP bars

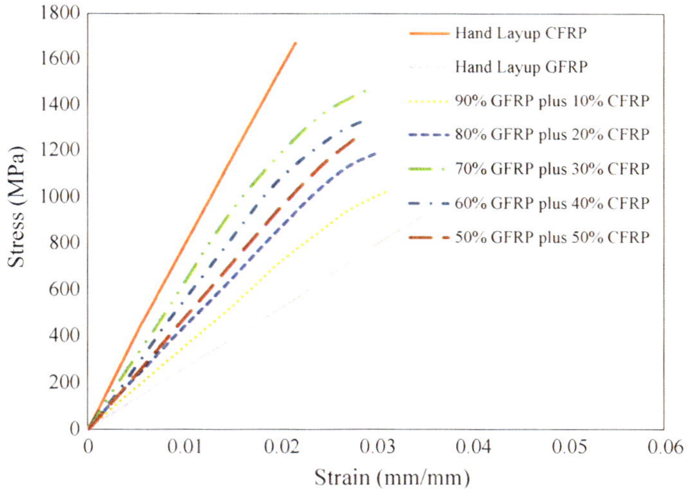

Fig. 6 Tensile stress-strain response of hybrid FRP bars

linear-elastic brittle material and does not have yield point as shown in Figs. 5 and 6. The pultruded CFRP bars have shown the maximum tensile strength in comparison to hand layup CFRP bars. This may be due to product quality is very stable through pultrusion process. It is observed that the hybrid FRP bars containing 70% GFRP and 30% CFRP have the highest strength among all hybrid FRP bars.

Three types of failures (i.e., pullout failure, tensile, and combined tension and bending failure) were observed during the tensile test of FRP bars. The pullout

Fig. 7 Tensile failure of FRP bars

failure was occurred in the case of pultruded bars because of smooth surface of CFRP bars which slips out of the anchor. Most of the hand layup bars failed in tension in which an abrupt failure of fibers occurred and bars completely rupture near to center occurred as shown in Fig. 7. However, some hybrid bars failed in combined tension and bending which is close to the anchor system as shown in Fig. 8.

3 Concluding Remarks

This chapter presents an insight on FRP, need and development of hybrid FRP rebars. The following concluding remarks are made based on the results presented in this chapter.

- New twisting equipment has been self developed for the fabrication of FRP bars aimed at small scale strengthening projects. The quality of FRP bars produced by the equipment is excellent in terms of strength and stiffness with respect to small scale strengthening projects.
- Hybrid FRP bars containing 70% GFRP and 30% CFRP have shown the largest strength and strain capacity among all hybrid FRP bars.
- Three types of failures (i.e. pullout failure, tensile, and combined tension and bending failure) were observed during the tensile test of FRP bars. The pullout

Fig. 8 Combined tension and bending failure of FRP bars

failure occurred in the case of pultruded bars because of smooth surface of CFRP bars which slips out of the anchor.

- Further investigation is required to find the other properties of hybrid FRP bars like as bond test, shear stress test, effects of fiber arrangement in hybrid bars, etc.

Acknowledgements The authors would like to thank to Aditya Birla Group and Department of Science and Technology (DST) (SR/S3/MERC/0051/2012), New Delhi for the financial support.

References

ACI 440.4R-04 (2004) Prestressing concrete structure with FRP tendons. American Concrete Institute, USA, pp 1–35

ACI 440.1R-06 (2006) Design and construction of structural concrete reinforced with FRP bars. American Concrete Institute, USA, pp 1–44

ACI 440R (2007) Report on fiber-reinforced polymer (FRP) reinforcement for concrete structures. American Concrete Institute, USA, pp 1–100

ACI 440.7R-10 (2010) Guide for design and construction of externally bonded FRP systems for strengthening unreinforced masonry structures. American Concrete Institute, USA, pp 1–46

ASTM D7205/D7205M-06 (2011) Standard test method for tensile properties of fiber reinforced polymer matrix composite bars. American Society for Testing Materials, USA

Bacon R, Tang MM (1964) Carbonization of cellulose fibers—I. Low temperature pyrolysis. Carbon 2:211–214

Choudhury SC (2012) Flexural and shear strengthening of RC beams with FRP—an experimental study. M. Tech thesis, NIT Rourkela, India, pp 1–81

Correia JR (2013) The new FRP materials for civil engineering structural applications. In: 57th meeting of the European council of civil engineers (ECCE), Lisbon, Portugal, pp 1–10

FIB 9.3 Task Group (2001) Externally bonded FRP reinforcement for RC structures. International Federation for Structural Concrete, Lausanne, Switzerland, pp 1–130

Grace NF, Wael R, Abdel-Sayed A (2005) Innovative tri-axially braided ductile FRP fabric for strengthening structures. In: 7th international symposium on fiber reinforce polymer for reinforced concrete structures (FRPRCS-7), ACI, Kansas City, MO

ISIS Canada Design Manuals (2001) Strengthening reinforced concrete structures with externally-bonded fibre reinforced polymers. Ontario, Canada, pp 1–215

JSCE (1997) Recommendation for design and construction of concrete structures using continuous fiber reinforcing materials. Concrete engineering series, vol 23. JSCE, Tokyo, pp 1–325

Madappa SVR (2011) Response and micromechanics based design of engineered cementitious composite structure. Ph.D. thesis, BITS Pilani, pp 1–110

Meier U, Deuring M, Meier H, Schwegler G (1993) Fibre-reinforced-plastic (FRP) reinforcement for concrete structures: properties and applications. Elsevier Science, Netherlands, pp 1–20

Rizkalla S, Hassan T, Hassan N (2003) Design recommendations for the use of FRP for reinforcement and strengthening of concrete structures. Prog Struct Mater Eng 5(1):16–28

Somboonsong W, Ko FK, Harris HG (1998) Ductile hybrid fiber reinforced plastic (FRP) rebar for concrete structures: design methodology. ACI Mater J 95(6):655–666

Yan X, Miller B, Nanni A, Bakis CE (1999) Characterization of CFRP rods used as near surface mounted reinforcement. In: 8th international conference on structural faults and repair, pp 1–12

Thermomechanical Elastic–Plastic Stability and Failure Analysis of FGM Plate

Kanishk Sharma and Dinesh Kumar

Abstract A mathematical model based on the first-order shear deformation theory and the von Karman's nonlinear kinematics for buckling, postbuckling and failure analysis of elastic–plastic Functionally Graded Material (FGM) plate under thermo-mechanical is presented. The FGM plate with continuously varying properties along thickness is modeled as a laminate composed of multiple perfectly-bonded layers made of isotropic and homogeneous material having layer-wise constant composition. The thermoelastic properties of FGM are calculated using rule of mixtures and Tamura-Tomota-Ozawa model (TTO model). Whereas, the elastic–plastic material properties are evaluated in accordance with the TTO model, assuming the ceramic phase of FGM to be elastic and the metal phase to be elastic–plastic. Further, the elastic–plastic analysis of FGM is assumed to follow J_2-plasticity with isotropic hardening. Parametric studies are conducted to investigate the effects of plasticity, material inhomogeneity, and thermomechanical loading conditions on the elastic–plastic buckling, postbuckling behavior, and the ultimate load capacity of FGM plate. The postbuckling response of FGM plate is found to be greatly affected by the plasticity consideration. FGM plate with elastic material properties exhibited a continuous increase in the postbuckling strength; whereas, the postbuckling strength of an elastic–plastic FGM plate decreases initially and finally, ultimate failure of the plate occurs.

Keywords Elastic–plastic analysis · J_2-plasticity · Postbuckling · Functionally graded material (FGM) · Nonlinear finite element method

---

K. Sharma
Mechanical Engineering Department, JECRC University, Jaipur 303905, India

D. Kumar (✉)
Mechanical Engineering Department, Malaviya National Institute of Technology, Jaipur 302 017, India
e-mail: vermadinesh2002@gmail.com

S. B. Singh et al. (eds.), *Emerging Trends of Advanced Composite Materials in Structural Applications*, Composites Science and Technology,
https://doi.org/10.1007/978-981-16-1688-4_11

Abbreviations

FEM	Finite element method
FGM	Functionally graded material
FSDT	First order shear deformation theory
TD	Temperature-dependent
TID	Temperature-independent
TTO model	Tamura-Tomota-Ozawa model

List of Symbols

x, y, z	Cartesian coordinate axis
u, v, w	Displacements components in x, y and z directions, respectively
u_0, v_0, w_0	Mid-plane translations corresponding to x, y and z directions, respectively
θ_x	Normal rotation in xz-plane
θ_y	Normal rotation in yz-plane
E	Young's modulus
ν	Poisson's ratio
α	Thermal expansion coefficient
T	Temperature (in Kelvin)
V_m	Volume fraction of metal
V_c	Volume fraction of ceramic
h	Thickness of FGM plate
n	Power law exponent
q	Stress-strain transfer ratio
σ_y	Yield strength of FGM
H	Plastic tangent modulus
$\{\epsilon\}$	Strain tensor
$\{\sigma\}$	Stress tensor
f	Yield surface
κ	Strain hardening parameter
$\{\varepsilon_p\}$	Plastic strain tensor
$\{\Delta\sigma\}$	Incremental stress tensor
$\{\Delta\varepsilon\}$	Incremental strain tensor
$\{\Delta\varepsilon_e\}$	Incremental elastic strain tensor
$\{\Delta\varepsilon_T\}$	Incremental thermal strain tensor
$\{\Delta\varepsilon_{TD}\}$	Incremental strain tensor due to temperature-dependent material properties
$\{\Delta\varepsilon_P\}$	Incremental plastic strain tensor
$[D_{ep}]$	Elasto–plastic constitutive matrix
$\Delta\overline{u}^e$	Incremental displacement vector within element

$\Delta\overline{a}$	Incremental nodal displacement vector
N	Interpolation functions
$[B]$	Strain–displacement matrix
ΔU	Incremental strain energy
ΔW_{ex}	Incremental external work
$\widetilde{\Delta P}$	Incremental tractions at plate surfaces
$\widetilde{\Delta P_e}$	Incremental forces at plate edges
$[K]$	Global stiffness matrix
δ	Deflection at a particular iteration
$\{\Delta R_M\}$	Mechanical incremental load vector
$\{\Delta R_T\}$	Thermal incremental load vector
ψ	Residual force vector
$[K_T]$	Tangent stiffness matrix
λ	Non-dimensionalized in-plane buckling load
$N_{x(or\,y)}$	In-plane compressive load in x-direction or y direction
$\left(\frac{2}{3}\varepsilon_p:\varepsilon_p\right)^{1/2}$	Magnitude of maximum plastic strain

Superscript

e	Element form
T	Transpose matrix
l	Linear
nl	Non-linear

Subscript

c	Ceramic material
m	Metal material
i	Iteration number
max	Maximum

1 Introduction

The advancement in materials has been associated with the man's evolution since ancient time. Yesterday, it was the age of stone, bronze, and iron and today it is the age of advanced materials, such as advanced composites, smart materials and functionally graded materials (FGMs). The most lightweight composite materials with high specific-strength and weight have been used successfully in the aircraft and defense industries and other engineering applications. However, due to strong mismatch of material properties at the interface in the structures made from traditional composites have several disadvantages, like debonding, delamination, plastic deformation and cracking problems, especially at high temperatures and pressures. In order to manufacture thermal barrier materials and also to get rid of the issues associated with the use of conventional laminated composites, a concept of FGMs was proposed in 1984 by the material scientists of Japan (Shiota and Miyamoto 1997). Functionally graded materials (FGMs), the next generation composites, are inhomogeneous material with a smooth and gradual variation in their properties along some specified direction(s), obtained by varying the volume fractions of the constituents (Suresh and Mortensen 1998).

Normally, there are two constituents of FGMs—ceramics and metals. Ceramic provides better thermal, wear and oxidation characteristics, whereas metal imparts high toughness, mechanical strength, and machinability properties to an FGM. Owing to the favorable mechanical and thermal properties, FGM offers a wide range of applications in various engineering fields requiring high temperature resistance combined with good mechnical strength. The promising advantages of using FGMs include decreased thermal stresses (Choules and Kokini 1996), good bonding strength in between joints of dissimilar materials (Howard et al. 1994), and the reduced possibility of catastrophic failure of brittle ceramic materials (Bao and Wang 1995). Further, FGMs are also used in many other applications; for instance, in rocket heat shields, heat engine components, heat exchanger tubes, plasma facings, fusion reactors, nuclear reactor plant, thermo-electric generators, and electrical insulating applications (Mahamood and Akinlabi 2017).

Further, thin-walled elements, such as plates and shells, widely used in many engineering fields, such as civil, aerospace, mechanical, naval, space engineering, and more recently, in micro-engineering, are more vulnerable towards buckling failure due to large deflections, and/or high stresses under in-plane thermo-mechanical loading conditions. Due to the fact that the membrane stiffness of plate-like structures is significantly higher than the bending stiffness that causes these types of structures to absorb a large amount of membrane strain energy with less deformations. However, the deformations are much higher when these structures absorb the same amount of bending energy. If the plate is loaded in such a way that most of its strain energy is contributed by the membrane energy, and under some conditions, such as initial imperfection, eccentric loading, etc., if the stored membrane energy is converted into the equivalent bending energy at some critical load point (called buckling load), then the plate becomes unstable and deforms dramatically in transverse direction producing excessive out-of-plane deflection. This destabilizing phenomenon is well known as buckling. It is well known that for the plate like structures, the buckling does not mean the ultimate failure, and these structures can carry extra load beyond the buckling point which is known as postbuckling strength (Singh and Kumar 1999).

In some applications, these plate-like structural elements are required primarily to resist buckling, and in others, they must carry a load well into the postbuckling range to yield weight savings. Thus, understanding their buckling and postbuckling behavior is needed for the efficient design of these structural elements.

Moreover, it is also noteworthy to mention that the structural failure can occur due to the material failure and/or geometrical instability. Before the material failure, the structure may show inelastic response producing plastic deformation which in turn causes a destabilizing effect on structures under in-plane compression and/or shear loads produced by mechanical and/or thermal loading conditions. Further, due to the safety reasons, machine elements working under the elastic limit, are also designed to carry overloads that can lead to inelastic deformations (Bazant et al. 1993). Therefore, elastic–plastic analysis is required to ensure safe and reliable design of such machine components. Being an important design criterion, the elastic–plastic buckling and postbuckling analysis of homogenous isotropic and composite plates has been reported in various studies (Narayanan and Chow 1984; Shanmugam et al. 1999; El-Sawy, and Martini 2004; Paik 2005; Ghavami and Khedmati 2006; Estefen et al. 2016). In addition, a few studies have also been reported on the inelastic buckling and postbuckling response of FGM structures (Huang and Han 2014; Zhang et al. 2015). The elastic–plastic stability and failure analysis of both imperforated and perforated FGM plates under thermal and/or mechanical loading conditions have been carried out by authors (Sharma and Kumar 2017a, b).

In this chapter, the research findings, along with the complete mathematical formulation, of elastic–plastic buckling, postbuckling and failure analysis of FGM plate under thermomechanical loading conditions, considering the temperature dependent material properties, are reported. The thermoelastic properties (i.e., elastic constants and thermal expansion coefficients) of FGM plate are calculated using theoretical and numerical micromechanics based models—Voigt's model, and Tamura-Tomota-Ozawa model (TTO model). As per the assumption of TTO model, the ceramic phase of FGM is considered to be elastic, whereas the metal phase is assumed to be elastic–plastic. The non-linear FEM formulation for plate analysis is based on the first-order shear deformation theory (FSDT) and von-Kármán's nonlinear kinematics. Further, $J2$-plasticity with isotropic hardening is adopted to perform the elastic–plastic analysis of FGM plate. The incremental solution technique based on Newton–Raphson method is utilized for the solution of nonlinear algebraic equations. Numerical studies are conducted on the buckling, postbuckling and failure responses of elastic–plastic FGM plate, considering the temperature-dependent as well as temperature-independent material properties, under thermomechanical loading condition.

2 Layer-Wise Model of FGM Plate

An FGM plate with continuously varying properties along thickness direction is modeled as a laminate, as shown in Fig. 1, with multiple perfectly-bonded layers of isotropic material having layer-wise constant composition, as used in many of the previous studies (Jin 2002; Shao 2005; Shakeri et al. 2006; Shakeri and Mirzaeifar

Fig. 1 Layer wise model of a continuous FGM plate

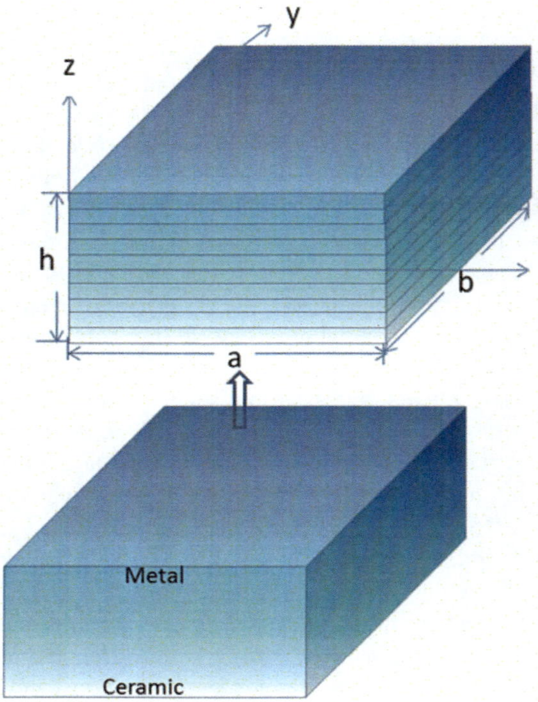

2009; Cinefra et al. 2010; Cinefra and Soave 2011; Yaghoobi et al. 2015). The FGM plate is assumed to be consisting of ceramic and metal phases and the volume fractions of the material constituents are assumed to follow a power law distribution in the thickness direction. In the present study, a coordinate system (x, y, z) is attached, as shown in Fig. 1, in the mid-plane of the plate measuring a, b and h as length, width, and thickness, respectively.

The temperature-dependent material properties Young's modulus E and thermal expansion coefficient α of FGM plate are calculated as follows (Touloukian and Center 1967).

$$P_j(T) = \begin{cases} P_0\left(P_{-1}T^{-1} + 1 + P_1T + P_2T^2 + P_3T^3\right), & for\ T > 0^\circ\,\mathrm{K} \\ P_0 & for\ T \leq 0^\circ\,\mathrm{K} \end{cases} \tag{1}$$

In Eq. (1), $P_j(T)$ represents the temperature-dependent material properties (E or α), and P_0, P_{-1}, P_1, P_2, and P_3 are material specific coefficients given in the Table 1 for both the constituents of FGM: Al_2O_3 (a ceramic phase) and Ni (a metal phase). The effect of temperature on the values of material properties of Al_2O_3 and Ni is depicted in Fig. 2.

Table 1 Temperature-dependent coefficients for thermo-elastic properties of Al_2O_3 and Ni (Reddy and Chin 1998)

Property to be evaluated	Material	Coefficients				
		P_0	P_{-1}	P_1	P_2	P_3
E (in Pa)	Al_2O_3	349.55e9	0.0	−3.853e−4	4.027e−7	−1.673e−10
	Ni	223.95e9	0.0	−2.794e−4	−3.998e−9	0.0
α (in 1/K)	Al_2O_3	6.8260e−6	0.0	1.838e−4	0.0	0.0
	Ni	9.9209e−6	0.0	8.705e−4	0.0	0.0

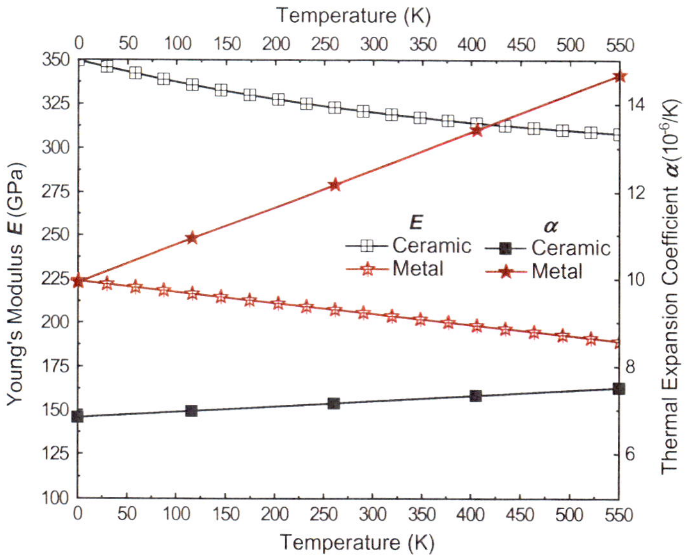

Fig. 2 Variation of Young's modulus (E) and thermal expansion coefficient (α) with temperature

2.1 *Effective Thermoelastic Material Properties of FGM Plate*

The volume fractions of ceramic and metallic constituents of FGM plate are varied in the thickness direction according to the power law, as given below:

$$V_m(z) = \left(\frac{z}{h} + \frac{1}{2}\right)^n; \quad V_c(z) = 1 - V_m(z). \tag{2}$$

Here, subscripts c and m represent the ceramic and the metallic constituents, respectively, and n denotes power law exponent that determines the material gradation profile across the thickness coordinate z, varying from $\frac{h}{2}$ to $-h/2$.

In the present study, the effective elastic material properties of two-phase FGM are calculated using TTO model (i.e., also called modified rule-of-mixtures) (Giannakopoulos et al. 1995). The TTO model has been used extensively in the literature (Giannakopoulos et al. 1995; Jin et al. 2003; Gunes et al. 2011) for accurately predicting the thermoelastic constants of FGM. In TTO model, the effective Young's modulus of two-phase materials, like FGM, is given in terms of Young's moduli (i.e., E_c and E_m) and volume fractions (i.e., V_c and V_m) of the constituting phases (i.e., ceramic and metallic phases), and it is expressed as under (Tamura et al. 1973):

$$E(z,T) = \left\{ V_m(z)E_m(T)\frac{q + E_c(T)}{q + E_m(T)} + (1 - V_m(z))E_c(T) \right\} \times \left\{ V_m(z)\frac{q + E_c(T)}{q + E_m(T)} + (1 - V_m(z)) \right\}^{-1}, \quad (3)$$

where, q represents the stress transfer parameter and it depends upon the properties of constituent materials as well as on the microstructure interaction in the FGM material. The value of q ranges from 0 to ∞; $q \to \infty$ represents the case when the constituent materials deform identically in the loading direction (i.e., Voigt model), while $q = 0$ corresponds to the case wherein the constituent materials experience the same stress level (i.e., Reuss model). Due to the complicated microstructure of FGM, the constituent elements in FGM neither experience equal deformation nor equal stress. Generally, a nonzero finite value of q is assumed to approximately reflect the actual effects of micro-structural interaction in FGM. For instance, for Ni-Al_2O_3 (Giannakopoulos et al. 1995) and TiB-Ti (Jin et al. 2003) FGMs, the value of q is assumed to be 4.5 GPa and for FGM containing Al and SiC phases (Bhattacharyya et al. 2007; Gunes et al. 2011), it is taken as 91.6 GPa.

In the present study, the Poisson's ratio is assumed to be constant (equal to the average value of of Poisson's ratio of metal and ceramic phases) i.e., $\nu = 0.31$, along the thickness of the FGM plate is used. Equation (3) is used to calculate the Young's modulus of FGM plate at a particular value of thickness coordinate. The thermal expansion coefficient at a particular thickness coordinate of FGM plate is calculated using the simple rule-of-mixtures, as expressed below:

$$\alpha(z,T) = \alpha_c(T)V_c(z) + \alpha_m(T)V_m(z). \quad (4)$$

2.2 *Effective Elastic–Plastic Material Properties of FGM Plate*

The yield strength and plastic tangent modulus of FGM are calculated by the same TTO model, as described in the previous Sect. 2.1. According to the TTO model, the overall failure, after elastic–plastic response, of an in-homogenous material, having

Table 2 Temperature-dependent coefficients to evaluate yield strength and tangent modulus of Ni (Williamson et al. 1995)

Property to be evaluated	Coefficients				
	P_0	P_{-1}	P_1	P_2	P_3
σ_{ym} (in Pa)	2.81e6	0.0	516.68e3	−8.79e2	−3.56e −1
H_m (in Pa)	91.75e7	0.0	930.64e4	−15.88e3	75.72e −1

brittle and ductile phases, is governed by the ductile constituent (Jin et al. 2003). This assumption of TTO model is applicable for FGMs (containing ceramic—a brittle phase, and metal—a ductile phase). This is because of the reason that the ductility and good shear strength induced in the FGM by the metal phase relax the stress concentration induced around the inherited cracks and flaws of ceramics through the plastic deformation and hence, eliminate the possibility of brittle failure of FGM (Bandyopadhyay et al. 2000; Soh et al. 2000).

Based on the aforementioned assumption, the overall yield strength and tangent modulus of the FGM plate are calculated using q (stress transfer parameter), σ_{ym} (yield strength of metal) and H_m (tangent modulus of metal), as follows (Giannakopoulos et al. 1995):

$$\sigma_y(z,T) = \sigma_{ym}(T)\left[V_m(z) + \left(\frac{q + E_m(T)}{q + E_c(T)}\right)\frac{E_c(T)}{E_m(T)}(1 - V_m(z))\right], \tag{5}$$

$$H(z,T) = \left\{V_m(z)H_m(T)\frac{q + E_c(T)}{q + H_m(T)} + (1 - V_m(z))E_c(T)\right\} \times \left\{V_m(z)\frac{q + E_c(T)}{q + H_m(T)} + (1 - V_m(z))\right\}^{-1}. \tag{6}$$

In the present study, the metallic phase is assumed to follow bilinear hardening behavior, and the elastic–plastic behavior of FGM is also predicted under the same assumption (Giannakopoulos et al. 1995; Williamson et al. 1995). The values of coefficients to calculate the temperature-dependent yield strength σ_{ym} and tangent modulus H_m of Ni (i.e., using Eq. 1) are given in Table 2 (Williamson et al. 1995).

3 FGM Plate Formulation

3.1 Displacement Field

In the present study, the displacement field is based on first-order shear deformation theory (FSDT), wherein the displacement field, as expressed in Eq. (7), is written in terms of the mid-plane ($z = 0$) translations u_0, v_0, w_0 and the independent normal rotations θ_x and θ_y in the xz- and yz-planes, respectively.

$$u(x,y,z) = u_0(x,y) + z\theta_x(x,y)$$
$$v(x,y,z) = v_0(x,y) + z\theta_y(x,y)$$
$$w(x,y,z) = w_0(x,y) \tag{7}$$

3.2 Strain–Displacement Relationship

Incorporating the von-Kármán's assumptions—derivatives of u and v with respect to x, y, and z are small—and noting that w is independent of z, the strain components for moderately large deformations can be written in the following forms (Reddy 2004):

$$\begin{aligned}
\varepsilon_x &= \frac{\partial u_0}{\partial x} + z\left(\frac{\partial \theta_x}{\partial x}\right) + \frac{1}{2}\left(\frac{\partial w_0}{\partial x}\right)^2, \\
\varepsilon_y &= \frac{\partial v_0}{\partial y} + z\left(\frac{\partial \theta_y}{\partial y}\right) + \frac{1}{2}\left(\frac{\partial w_0}{\partial y}\right)^2, \\
\gamma_{xy} &= \frac{\partial u_0}{\partial y} + \frac{\partial v_0}{\partial x} + z\left(\frac{\partial \theta_x}{\partial y} + \frac{\partial \theta_y}{\partial x}\right) + \left(\frac{\partial w_0}{\partial x}\frac{\partial w_0}{\partial y}\right), \\
\gamma_{xz} &= \frac{\partial w_0}{\partial x} + \theta_x, \\
\gamma_{yz} &= \frac{\partial w_0}{\partial y} + \theta_y.
\end{aligned} \tag{8}$$

Rewriting Eq. (8) into matrix form as

$$\{\epsilon\} = \begin{Bmatrix} \epsilon_x \\ \epsilon_y \\ \gamma_{xy} \\ \gamma_{xz} \\ \gamma_{yz} \end{Bmatrix} = \{\epsilon\}^l + \{\epsilon\}^{nl}, \quad \{\epsilon\}^l = \begin{Bmatrix} \epsilon_p^0 \\ 0 \end{Bmatrix} + \begin{Bmatrix} z\epsilon_b^0 \\ \epsilon_s^0 \end{Bmatrix}, \tag{9}$$

$$\{\epsilon\}^{nl} = \begin{Bmatrix} \epsilon_p^{NL} \\ 0 \end{Bmatrix}, \tag{10}$$

wherein, the linear in-plane strain ϵ_p^0, the bending strain ϵ_b^0, shear strain ϵ_s^0, and the nonlinear in-plane strain terms ϵ_p^{NL} are written as follows:

$$\{\epsilon_p^0\} = \begin{Bmatrix} \frac{\partial u_0}{\partial x} \\ \frac{\partial v_0}{\partial y} \\ \frac{\partial u_0}{\partial y} + \frac{\partial v_0}{\partial x} \end{Bmatrix}, \quad \{\epsilon_b^0\} = \begin{Bmatrix} \frac{\partial \theta_x}{\partial x} \\ \frac{\partial \theta_y}{\partial y} \\ \frac{\partial \theta_x}{\partial y} + \frac{\partial \theta_y}{\partial x} \end{Bmatrix} \tag{11}$$

$$\{\epsilon_s^0\} = \begin{Bmatrix} \frac{\partial w_0}{\partial x} + \theta_x \\ \frac{\partial w_0}{\partial y} + \theta_y \end{Bmatrix}, \quad \text{and} \quad \{\epsilon_p^{NL}\} = \begin{Bmatrix} \frac{1}{2}\left(\frac{\partial w_0}{\partial x}\right)^2 \\ \frac{1}{2}\left(\frac{\partial w_0}{\partial y}\right)^2 \\ \left(\frac{\partial w_0}{\partial x}\frac{\partial w_0}{\partial y}\right) \end{Bmatrix}. \tag{12}$$

3.3 Constitutive Relations

The elastic and elastic–plastic stress–strain relationship for FGM plate with temperature dependent material properties under thermomechanical loading conditions are discussed in the following paragraphs:

3.3.1 Elastic Constitutive Relations

Based on the generalized Hooke's law, the elastic stress–strain relations are given by Reddy (2003) and Sharma and Kumar (2018):

$$\begin{Bmatrix} \sigma_x \\ \sigma_y \\ \tau_{xy} \\ \tau_{yz} \\ \tau_{xz} \end{Bmatrix} = \begin{bmatrix} Q_{11} & Q_{12} & 0 & 0 & 0 \\ Q_{12} & Q_{22} & 0 & 0 & 0 \\ 0 & 0 & Q_{44} & 0 & 0 \\ 0 & 0 & 0 & k_1^2 Q_{55} & 0 \\ 0 & 0 & 0 & 0 & k_2^2 Q_{66} \end{bmatrix} \begin{Bmatrix} \epsilon_x \\ \epsilon_y \\ \gamma_{xy} \\ \gamma_{xz} \\ \gamma_{yz} \end{Bmatrix} \tag{13}$$

or, in the matrix form, we can write:

$$\{\sigma\} = [D]\{\varepsilon\}, \tag{14}$$

Here, k_1^2 and k_2^2 are the shear correction factors, and Q_{ij} are the stiffness matrix components through the thickness of FGM plate and they are functions of material properties, as mentioned below:

$$Q_{11} = \frac{E(z,T)}{1-\nu^2}, \quad Q_{12} = Q_{13} = Q_{23} = \nu Q_{11}, \quad Q_{22} = Q_{33} = Q_{11},$$
$$Q_{44} = Q_{55} = Q_{66} = \frac{E(z,T)}{(1+\nu)}, \tag{15}$$

where, $E(z, T)$ is Young's modulus that varies across the thickness of FGM plate at any temperature T and is calculated using Eq. (3), and ν is the Poisson's ratio that is assumed to be constant through the thickness of FGM plate.

3.3.2 Thermo-Elasto-Plastic Constitutive Relations

The elastic–plastic analysis is carried out under the assumption of von-Mises yielding criterion, and a uniform expansion is assumed to be followed by the yield surface in the stress space with increasing plastic deformation (Hill 1998). The yield function can be expressed as:

$$f = \sqrt{3J_2} - \sigma_{Yield}, \tag{16}$$

where,

$$J_2 = \frac{1}{6}\Big[\big(\sigma_x - \sigma_y\big)^2 + \big(\sigma_y - \sigma_z\big)^2 + (\sigma_x - \sigma_z)^2 + 6\big(\tau_{xy}^2 + \tau_{yz}^2 + \tau_{xz}^2\big)\Big]. \tag{17}$$

Under the strain hardening effect, the initial yield surface expands with plastic deformation, and the equation of yield surface for thermo-elasto-plastic deformation can be written as:

$$f = f(\sigma, \kappa, T), \tag{18}$$

wherein, κ and T are the strain hardening parameter and temperature, respectively.

Using chain rule to differentiate f:

$$df = \left(\frac{\partial f}{\partial \sigma}\right)^T d\sigma + \frac{\partial f}{\partial \kappa} d\kappa + \frac{\partial f}{\partial T} dT. \tag{19}$$

As hardening parameter (κ) is a function of plastic strain (ε_p), Eq. (19) can be rewritten as:

$$df = \left(\frac{\partial f}{\partial \sigma}\right)^T d\sigma + \frac{\partial f}{\partial \kappa}\left(\frac{\partial \kappa}{\partial \varepsilon_p}\right) d\varepsilon_p + \frac{\partial f}{\partial T} dT. \tag{20}$$

The equilibrium conditions under small incremental plastic deformations are maintained only if the plastic strain energy is put equal to zero; hence, we have:

$$df = \left(\frac{\partial f}{\partial \sigma}\right)^T d\sigma + \frac{\partial f}{\partial \kappa}\left(\frac{\partial \kappa}{\partial \varepsilon_p}\right) d\varepsilon_p + \frac{\partial f}{\partial T} dT = 0. \tag{21}$$

Now the total incremental strain includes the incremental parts of elastic strain ($\Delta\varepsilon_e$), thermal strain ($\Delta\varepsilon_T$), strain due to temperature-dependent material properties ($\Delta\varepsilon_{TD}$), and the plastic strain $\left(\Delta\varepsilon_p\right)$, i.e.,

$$\Delta\varepsilon = \Delta\varepsilon_e + \Delta\varepsilon_T + \Delta\varepsilon_{TD} + \Delta\varepsilon_p. \tag{22}$$

Using Hooke's law, to calculate the total incremental stress ($\Delta\sigma$)

$$\Delta\sigma = [D]\{\Delta\varepsilon - (\Delta\varepsilon_T + \Delta\varepsilon_{TD} + \Delta\varepsilon_p)\}. \tag{23}$$

Putting the value of total incremental stress ($d\sigma$) into Eq. (21), we get

$$df = \left(\frac{\partial f}{\partial \sigma}\right)^T [D]\{\Delta\varepsilon - (\Delta\varepsilon_T + \Delta\varepsilon_{TD} + \Delta\varepsilon_p)\} + \frac{\partial f}{\partial \kappa}\left(\frac{\partial \kappa}{\partial \varepsilon_p}\right)\Delta\varepsilon_p + \frac{\partial f}{\partial T}\Delta T = 0. \tag{24}$$

In the associative flow rule of plasticity, the plastic potential function is taken same as the yield function (Chakrabarty 2012) by which the incremental plastic strain can be written as:

$$\Delta\varepsilon_p = d\lambda\frac{\partial f}{\partial \sigma}. \tag{25}$$

The thermal strain ($\Delta\varepsilon_T$) and the strain due to temperature-dependent material properties ($\Delta\varepsilon_{TD}$) can be written as:

$$\Delta\varepsilon_T = \alpha\Delta T, \quad and \quad \Delta\varepsilon_{TD} = \frac{\partial[D]^{-1}}{\partial T}\sigma\Delta T. \tag{26}$$

Let,

$$\left(\frac{\partial f}{\partial \sigma}\right)^T [D]\left(\frac{\partial f}{\partial \sigma}\right) - \frac{\partial f}{\partial k}\left(\frac{\partial k}{\partial \varepsilon_p}\right)^T\left(\frac{\partial f}{\partial \sigma}\right) = \eta^i. \tag{27}$$

Substituting Eqs. (25)–(27) into Eq. (24), to obtain the flow variable $d\lambda$ as:

$$d\lambda = \left[\frac{\left(\frac{\partial f}{\partial \sigma}\right)^T [D]\left\{\Delta\varepsilon - \left(\alpha\Delta T + \frac{\partial[D]^{-1}}{\partial T}\sigma\Delta T\right)\right\} + \frac{\partial f}{\partial T}\Delta T}{\eta^i}\right] \tag{28}$$

The expression for incremental stress $\Delta\sigma$ can be evaluated inserting Eqs. (24)–(28) into Eq. (23) as:

$$\Delta\sigma = [D]\left[(\Delta\varepsilon) - \left(\alpha dT + \frac{\partial[D]^{-1}}{\partial T}\sigma dT\right)\right] - [D]\frac{\partial f}{\partial \sigma}\left[\frac{\left(\frac{\partial f}{\partial \sigma}\right)^T [D]\left\{\Delta\varepsilon - \left(\alpha dT + \frac{\partial[D]^{-1}}{\partial T}\sigma dT\right)\right\} + \frac{\partial f}{\partial T}dT}{\eta^i}\right] \tag{29}$$

Rearranging Eq. (29) to obtain:

$$\Delta\sigma = \left[[D] - \frac{1}{\eta^i}[D]\frac{\partial f}{\partial \sigma}\left(\frac{\partial f}{\partial \sigma}\right)^T [D]\right]\left[(\Delta\varepsilon) - \left(\alpha\Delta T + \frac{\partial [D]^{-1}}{\partial T}\sigma\Delta T\right)\right]$$
$$- \frac{1}{\eta^i}[D]\frac{\partial f}{\partial \sigma}\frac{\partial f}{\partial T}\Delta T. \quad (30)$$

The term $\left[[D] - \frac{1}{\eta^i}[D]\frac{\partial f}{\partial \sigma}\left(\frac{\partial f}{\partial \sigma}\right)^T [D]\right]$ in Eq. (30) is the elasto-plastic constitutive matrix $\left[D_{ep}\right]^i$, that establishes the thermo-elasto-plastic constitutive relationship as follows:

$$\Delta\sigma = \left[D_{ep}\right]^i\left\{\Delta\varepsilon'\right\} \quad (31)$$

$$\Delta\varepsilon' = (\Delta\varepsilon) - \left(\Delta\varepsilon^t\right) \quad (32)$$

$$\Delta\varepsilon^t = \alpha\Delta T + \frac{\partial [D]^{-1}}{\partial T}\sigma\Delta T + \frac{1}{\eta^i}[D]\left[D_{ep}\right]^{-1}\frac{\partial f}{\partial \sigma}\frac{\partial f}{\partial T}\Delta T \quad (33)$$

3.4 Non-linear FEM Formulation for Thermo-Elastic–Plastic Analysis

The procedure to derive the finite element equations for thermo-elastic–plastic analysis of FGM plate is presented in this section. The incremental displacement vector $(\Delta\overline{u}^e)$ within an element is interpolated as:

$$\Delta\overline{u}^e = N\Delta\overline{a}, \quad (34)$$

where, $\Delta\overline{a}$ is the incremental nodal displacement vector and N is the interpolation functions. The incremental strain tensor $(\Delta\varepsilon)$ is expressed in the terms of incremental nodal displacement vector $(\Delta\overline{a})$ and strain–displacement matrix (B), as

$$\Delta\varepsilon = B\Delta\overline{a}, \quad (35)$$

the incremental strain energy (ΔU) may be written as

$$\Delta U = \frac{1}{2}\int_V \Delta\varepsilon'^T \Delta\sigma dV, \quad (36)$$

using Eqs. (31), (32) and (35), to rewrite Eq. (36) in the following form:

$$\Delta U = \frac{1}{2}\Delta\overline{a}^T \int\limits_V B^T\left[D_{ep}\right]^i B\overline{\Delta a} dV - \Delta\overline{a}^T \int\limits_V B^T\left[D_{ep}\right]^i \Delta\varepsilon^t dV$$
$$+ \frac{1}{2}\int\limits_V \Delta\varepsilon^t\left[D_{ep}\right]^i \Delta\varepsilon^t dV, \quad (37)$$

The incremental external work (ΔW_{ex}) due to incremental tractions at plate surfaces (i.e., $z = \pm h/2$) and plate edges can be written as:

$$\Delta W_{ex} = \Delta a^T \int\limits_A N^T \widetilde{\Delta P}\, dA + \Delta a^T \int\limits_s N^T \widetilde{\Delta P_e}\, ds, \quad (38)$$

where,

$$\widetilde{\Delta P} = \left[\Delta P_x \; \Delta P_y \; \Delta P_z \; \overline{\Delta M}_x \; \overline{\Delta M}_y\right]^T, \quad (39)$$

and, the edge forces are given by

$$\widetilde{\Delta P_e} = \left[\overline{\Delta P}_x^e \; \overline{\Delta P}_y^e \; \Delta \overline{P}_z^e \; \overline{\Delta M}_x^e \; \overline{\Delta M}_y^e\right]^T$$
$$= \int\limits_{h/2}^{h/2}\left[\left(\Delta P_x^e \; \Delta P_y^e \; \Delta P_z^e \; \Delta M_x^e \; \Delta M_y^e\right)dz\right]^T \quad (40)$$

The associated equilibrium equation is obtained by applying the principle of variation over the functional ($\Delta\Pi = \Delta U - \Delta W_{ex}$) with respect to the incremental displacement (Δa), as:

$$\frac{\partial(\Delta\Pi)}{\partial(\Delta\overline{a})} = \frac{\partial(\Delta U - \Delta W_{ex})}{\partial(\Delta\overline{a})} = 0 \quad (41)$$

Using Eqs. (37) and (38) into Eq. (41) to get the equilibrium equation:

$$\left(\int\limits_V B^T\left[D_{ep}\right]^i B dV\right)\Delta\overline{a} - \int\limits_V B^T\left[D_{ep}\right]^i \Delta\varepsilon^t dV$$
$$- \int\limits_A N^T \widetilde{\Delta P}\, dA - \int\limits_s N^T \widetilde{\Delta P_e}\, ds = 0. \quad (42)$$

Value of $\Delta\varepsilon^t$ is substituted from Eq. (33) to rewrite Eq. (42) in the following manner:

$$\left(\int_V B^T\left[D_{ep}\right]^i B dV\right)\Delta\overline{a} - \int_V B^T\left[D_{ep}\right]^i$$
$$\times\left\{\alpha\Delta T + \frac{\partial[D]^{-1}}{\partial T}\sigma\Delta T + \frac{1}{\eta^i}[D]\left[D_{ep}\right]^{-1}\frac{\partial f}{\partial\sigma}\frac{\partial f}{\partial T}\Delta T\right\}dV$$
$$-\int_A N^T\widehat{\Delta P}\, dA - \int_s N^T\widehat{\Delta P_e}\, ds = 0. \tag{43}$$

Finally, the set of nonlinear algebraic equations is obtained in the following marix form:

$$[K]\Delta\overline{a} = \Delta R, \tag{44}$$

where, $[K]$ is the structural stiffness matrix given by:

$$[K] = \int_V B^T\left[D_{ep}\right]^i B dV, \tag{45}$$

In Eq. (44), ΔR represents combined thermomechanical incremental load vector due to mechanical loading (ΔR_M) and thermal loading (ΔR_T), i.e.,

$$\Delta R = \Delta R_M + \Delta R_T, \quad \text{having} \tag{46}$$

$$\Delta R_M = \int_A N^T\widehat{\Delta P}\, dA + \int_s N^T\widehat{\Delta P_e}\, ds, \tag{47}$$

and

$$\Delta R_T = \int_V B^T\left[D_{ep}\right]^i\left(\alpha\Delta T + \frac{\partial[D]^{-1}}{\partial T}\sigma\Delta T + \frac{1}{\eta^i}[D]\left\{[D_{ep}]^i\right\}^{-1}\frac{\partial f}{\partial\sigma}\frac{\partial f}{\partial T}\Delta T\right)dV \tag{48}$$

The solution for nonlinear algebraic equations posted in Eq. (44) is obtained using Newton–Raphson iterative solution technique. Newton–Raphson solution technique is based on Taylor's series expansion, and it is convenient to rewrite Eq. (44) as:

$$\psi\{\delta\} = [K(\delta)]\{\delta\} - \Delta R, \tag{49}$$

where, ψ is the residual force and $[K(\delta)]$ is structural stiffness matrix, which is a function of unknown deflection δ. If for an initial estimate of the displacement vector $\{\delta\}_i$ (i.e., for i^{th} iteration) we obtain residual forces $\psi\{\delta\}_i \neq 0$, then an improved solution $\{\delta\}_{i+1}$ is obtained by equating the linearized Taylor's series expansion of

$\psi\{\delta\}_{i+1}$ in the neighborhood of $\{\delta\}_i$ to zero, as below:

$$\psi\{\delta\}_{i+1} \cong \psi\{\delta\}_i + K_T \Delta\{\delta\}_i = 0, \tag{50}$$

where, $\Delta\{\delta\}_i$ is the incremental displacement vector, and K_T is the tangent stiffness matrix evaluated at $\{\delta\}_i$ and is given by:

$$K_T = \left[\frac{\partial\psi\{\delta\}_i}{\partial[\delta]}\right]. \tag{51}$$

The improved solution is then obtained as:

$$\{\delta\}_{i+1} = \{\delta\}_i + \Delta\{\delta\}_i. \tag{52}$$

To improve on the numerical stability and convergence of the solution, the load is applied in small increments. The iterative solution is continued until the following convergence criterion is satisfied.

$$\delta_{i+1} - \delta_i \le \beta, \tag{53}$$

where,β is sufficiently small number, i.e., 0.005%.

3.5 *Ultimate Load Capacity of FGM Plate*

An FEM based strategy that is followed by many authors in the literature (Narayanan and Chow 1984; Shanmugam et al. 1999; El-Sawy and Martini 2004; Paik 2005; Ghavami and Khedmati 2006; Estefen et al. 2016) to predict the ultimate load capacity of elastic–plastic homogeneous isotropic plates is extended to predict ultimate load capacity of elastic–plastic FGM plate. To get the ultimate load carrying capacity of FGM plate, the plate is subjected to the in-plane compressive load in an incremental manner that initially produces recoverable elastic stress and strain. However, a large value of load causes yielding in FGM plate which is followed by recoverable elastic strains as well as irrecoverable plastic strains. As the load further increases, beyond a particular value of load the proportion of plastic strain reaches to an extent where the plate completely loss of its stiffness resulting in the complete collapse and hence, ultimate failure of the FGM plate; the corresponding load is termed as ultimate load capacity. It is noteworthy that in the present elastic–plastic analysis the load is applied in small incremental load steps to ensure that the analysis closely follows the actual load-response curve of FGM plate.

3.6 Computer Implementation

A general purpose finite element package ANSYS® (Ansys Inc 2014) [by encrypting a macro in APDL (i.e., ANSYS Parametric Design Language)] was utilized to implement the formulation and perform thermomechanical elastic–plastic buckling, postbuckling and failure analysis of FGM plate, with temperature-dependent material properties.

The layerwise elastic–plastic FGM plate is modeled using eight-noded shell element, wherein the thermo-elastic–plastic analysis of FGM plate is carried out in two stages. In the first stage the linear buckling analysis is conducted to obtain the critical buckling load and corresponding mode shape of the FGM plate. The linear analysis is followed by nonlinear incremental procedure in which the load is applied gradually in small increments and the corresponding deformation is obtained by updating the temperature-dependent material properties at each load step. It is evident that some sort of imperfection must be introduced into the perfect geometry of the plate to trace its postbuckling response, without any numerical issues, using incremental finite element analysis (Barbero 2013). In the present analysis, the bifurcation buckling in the FGM plate is initiated by superimposing the imperfection equal to *b*/1000 (where, *b* represents width of the plate) and of the form corresponding to the first buckling mode shape.

4 Elastic–Plastic Analysis of FGM Plate

The aim of this section is to present the findings, reported by the authors (Sharma and Kumar, 2017b), obtained by the application of the developed computational model to investigate the thermo-elastic–plastic respose of Ni/Al_2O_3 FGM plate under in-plane compressive loading conditions..

The thermoelastic material properties as well as elastic–plastic strength parameters of FGM plate are taken as non-linear function of temperature, as defined by Eq. (1). Temperature-dependent coefficients for calculating the thermo-elastic properties of Al_2O_3 and Ni are given in Table 1, whereas the elastic–plastic strength coefficients to evaluate yield strength and tangent modulus of metallic phase (i.e., Ni) are given in Table 2. The thermoelastic and elastic–plastic material properties of FGM plate are calculated using rule of mixtures and TTO model, respectively, as discussed in Sect. 2.

Numerical studies are conducted to examine the effects of material, geometrical, and loading parameters on the thermo-elastic–plastic response of FGM plate, and the results are presented in the following form as:

Non-dimensional applied load, $\Lambda = \frac{N_{x\,(\mathrm{or}\,y)}\,\mathrm{b}^2}{E_{\mathrm{m}}\mathrm{h}^3}$;
Non-dimensionalized maximum transverse deflection, $\overline{w} = \frac{w_{\max}}{\mathrm{h}}$;
Magnitude of maximum plastic strain, $(\varepsilon_p)_{eq} = \left(\frac{2}{3}\varepsilon_p\colon\varepsilon_p\right)^{1/2}$.

where, E_c is the Young's modulus of ceramic; h represents the thickness of FGM plate; b is the width of plate; $N_{x(or\,y)}$ is the in-plane compressive load in x- direction (or y-direction) per unit edge length, applied at $x = a$ (or $y = b$); $w_{\max}$ is the maximum transverse deflection, and $\varepsilon_p : \varepsilon_p$ is the scalar product of the plastic strain tensor corresponding to failure point (Fig. 3).

The postbuckling and failure response of both elastic and elastic–plastic simply-supported FGM (for power law exponent, $n = 1$) square plate (of side 1 m) with $b/h = 100$ having temperature-dependent (TD) material properties under various thermomechanical loading conditions are shown in Fig. 4. The corresponding values of yield and failure loads are also marked in the Fig. 4 for the elastic–plastic case. It is to mention here that the yielding point on a particular load–deflection curve corresponds to the minimum value of the load at which elastic to plastic transition of the metallic phase present anywhere in the FGM takes place, whereas the failure point corresponds to the load at which the FGM plate completely losses its stiffness due to excessive plasticity.

It can be observed from the results that irrespective of thermal load, the post-buckling response of FGM plate is significantly affected by the consideration of plasticity and the elastic–plastic FGM plate depicts decrease in the postbuckling strength followed by the ultimate failure.

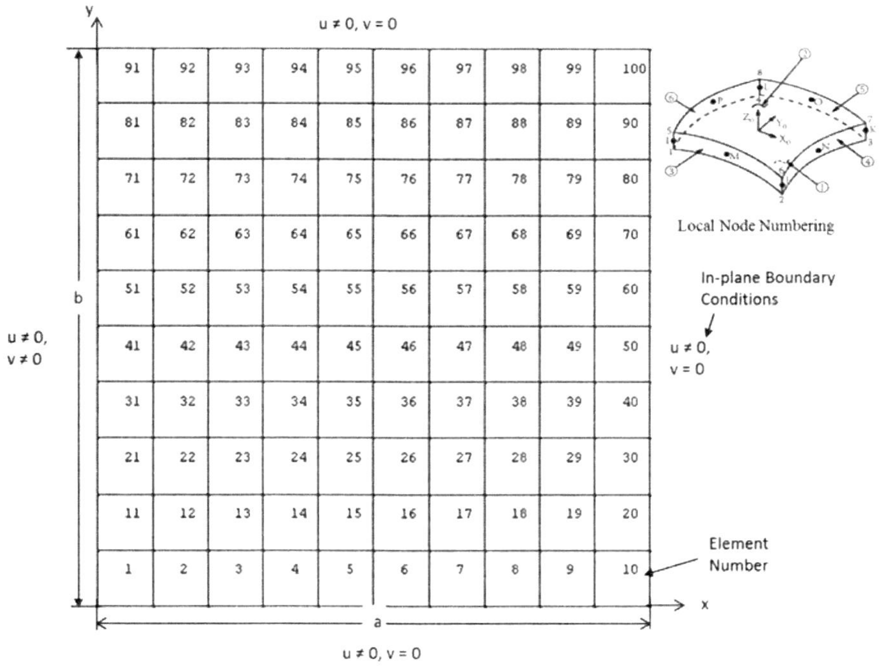

Fig. 3 In-plane boundary conditions for a simply-supported (i.e., $w = 0$ on all edges) FGM plate

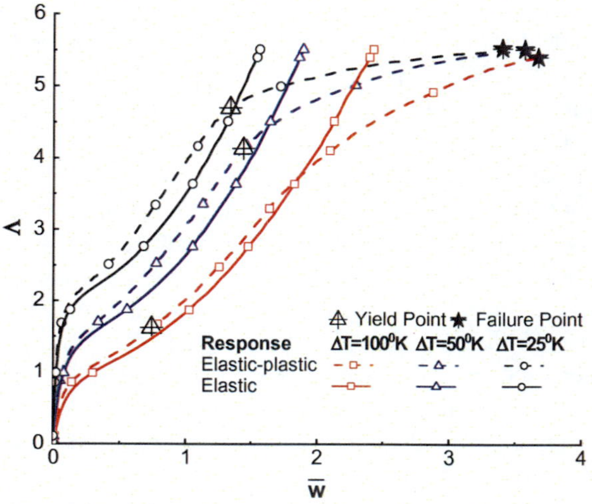

Fig. 4 Buckling and postbuckling response for elastic and elastic–plastic FGM plate under different thermomechanical loadings

Morover, the effect of material property variation and temperature rise on the buckling and postbuckling behaviors of an elastic–plastic FGM plate subjected to uniaxial mechanical compressive loading combined with different thermal load values (i.e., $\Delta T = 25,\ 50$ and $100\,^{\circ}$C) is also examined and the corresponding postbuckling paths are shown in Fig. 4. It is to mention here that different material gradation profiles across the thickness of FGM plate are obtained by changing the value of power law exponent n (i.e., 1, 2 and 3) in Eq. (2).

As depicted in Fig. 5, the postbuckling strength of FGM plate (at a specific value of maximum transverse deflection) decreases considerably with increasing temperature rise. It is also noteworthy that the effect of temperature rise is to increase the rate of

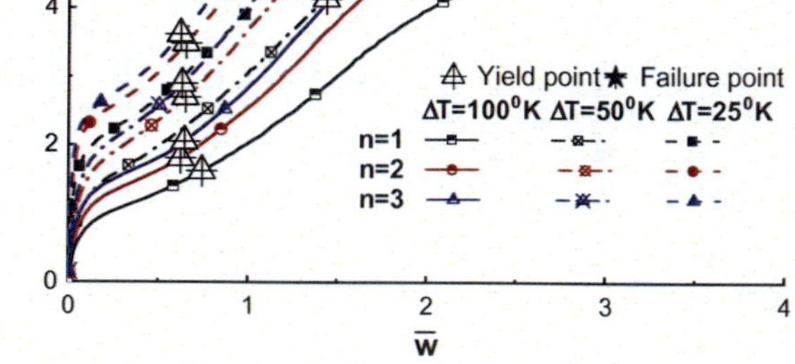

Fig. 5 Nonlinear buckling, postbuckling and failure reasponse of elastic–plastic FGM plate with different value of power law exponent (n) under thermomechanical loading

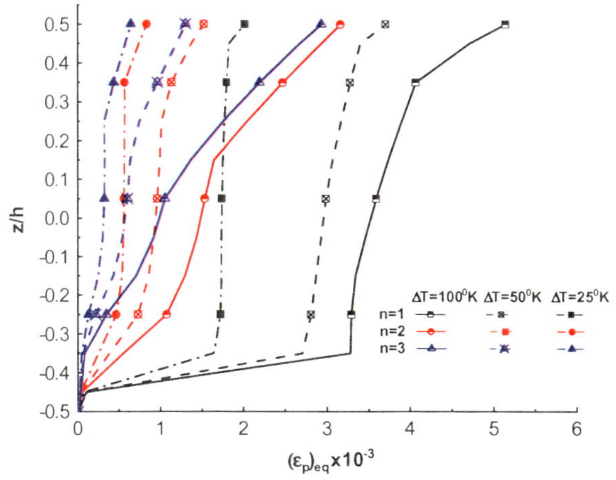

Fig. 6 Varition in the equivalent plastic strain across the thickness of FGM plate with different power law exponents (i.e., n) under thermomechancial loading

plastic flow (as observed subsequently in Fig. 6) which leads to ultimate failure of FGM plate at a relatively lower value of mechanical load. Figure 4 also shows that irrespective of temperature rise, the ultimate load carrying capacity of FGM plate increases with the increase in value of power law index that can be attributed to the higher proportion of ceramic (i.e., Al_2O_3), possessing higher resistance towards buckling failure.

Furthermore, the effect of material gradation profile on the equivalent plastic strain across the thickness of FGM plate at a particular value (taken equal to the failure load for FGM with $n = 1$) of thermomechanical loading is shown in Fig. 6. The flow of plasticity in FGM plate for a particular value of n, is found to be more dominating at the upper region (with higher metal proportion) in comparison to the lower part (with higher ceramic proportion).

Moreover, at a particular value of thickness, the plasticity effect is more pronounced in the case of FGM plate with $n = 1$. It is again demonstrated in Fig. 6 that the effect of higher temperature is to provoke more plasticity effects in the FGM plate.

The postbuckling paths for elastic–plastic FGM plate ($n = 1$) subjected to uni-axial and biaxial loading with different load ratios (i.e., N_y/N_x) under uniform temperature rise (i.e., ΔT) are depicted in Fig. 7. It can be observed from Fig. 7 that irrespective of the value of temperature rise, the maximum buckling load, postbuckling strength (for a particular value of deflection) and failure load are obtained for FGM plate under uni-axial compression (i.e., for $N_y/N_x = 0$), and with the addition of compressive load in y-direction, the values of buckling load, postbuckling strength (for a particular value of deflection) and failure load decrease.

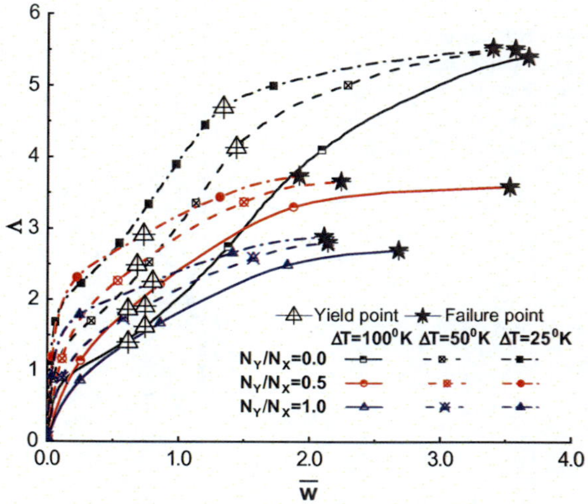

Fig. 7 Effect of loading on elastic–plastic buckling and postbuckling behavior of FGM (for $n = 1$) plate under thermomechanical loading (Sharma and Kumar 2017b)

5 Conclusions

A mathematical formulation for elastic–plastic buckling and postbuckling analysis of FGM plate with temperature-dependent material properties is presented. A layer-wise modeling approach of FGM plate is used, wherein the graded thermoelastic material properties of FGM plate are assumed to be governed by the rule of mixtures and TTO model, whereas the elastic–plastic material characteristics (i.e., yielding strength and tangent modulus) are varied as per the TTO model.

The governing equations for elastic–plastic FGM plate are developed using variational principle. Subsequently, the nonlinear finite element formulation based on the first-order shear deformation theory and the von Karman's nonlinear kinematics is presented. The incremental solution algorithm based on Newton–Raphson method is used to solve the resulting nonlinear algebraic equations. A computer implementation, through ANSYS®, of the formulation is also presented to carry out the analysis. Thereafter, a study on elastic–plastic analysis of FGM plate under in-plane compression caused by the mechanical loading combined with uniform and constant temperature rise is conducted. The effects of plasticity, material inhomogenity, and thermo-mechanical loading on elastic–plastic buckling, postbuckling behavior and ultimate load capacity of FGM plate are analyzed under the framework of J_2 deformation theory associated with the isotropic hardening flow rule.

Based on the study, the following important conclusions are drawn:

- Postbuckling response of FGM plate is found to be greatly affected by the plasticity consideration. FGM plate with elastic material properties exhibits a continous increase in the postbuckling strength; whereas, the postbuckling strength of an elastic–plastic FGM plate decreases initially and finally, ultimate failure occurs.

- Along the thickness of FGM plate, the effect of plastic flow is observed to be more dominating in the metal rich region of FGM plate than the ceramic rich region.
- Effect of rise in temperature difference (i.e., ΔT) is to develop more plastic strain, irrespective of the value of material gradation index n and the mechanical loading condition.
- Buckling load and postbuckling strength of elastic–plastic FGM plate are found to be significantly affected by its material gradation profile; an FGM plate with higher ceramic proportion depicted higher buckling load and postbuckling strength.
- Effect of bi-axial loading is to reduce the buckling load, postbuckling strength (for a particular value of deflection), and failure loads, for all values of temperature difference.

References

ANSYS Inc (2014) ANSYS mechanical theory reference: release 15.0. Canonsburg PA, USA

Arbocz J, Conferenza italiana dei superiori maggiori (1987) Buckling and post-buckling: four lectures in experimental, numerical, and theoretical solid mechanics based on talks given at the CISM-meeting, held in Udine, Italy, 29 Sept–3 Oct 1985

Bandyopadhyay A et al (2000) Mechanical properties of interconnected phase alumina-Al composites. In: Pro SFF, pp 24–31

Bao G, Wang L (1995) Multiple cracking in functionally graded ceramic/metal coatings. Int J Solids Struct 32(19):2853–2871

Barbero EJ (2013) Finite element analysis of composite materials using ANSYS®. CRC Press

Bazant ZP, Cedolin L, Hutchinson JW (1993) Stability of structures: elastic, inelastic, fracture, and damage theories. J Appl Mech. https://doi.org/10.1115/1.2900839

Bhattacharyya M, Kapuria S, Kumar AN (2007) On the stress to strain transfer ratio and elastic deflection behavior for Al/SiC functionally graded material. Mech Adv Mater Struct 14(4):295–302. https://doi.org/10.1080/15376490600817917

Chakrabarty J (2012) Theory of plasticity. Elsevier

Choules BD, Kokini K (1996) Architecture of functionally graded ceramic coatings against surface thermal fracture. J Eng Mater Tech Am Soc Mech Eng 118(4):522–528

Cinefra M, Soave M (2011) Accurate vibration analysis of multilayered plates made of functionally graded materials. Mech Adv Mater Struct 18(1):3–13. https://doi.org/10.1080/15376494.2010.519204

Cinefra M et al (2010) Thermo-mechanical analysis of functionally graded shells. J Therm Stresses 33(10):942–963. https://doi.org/10.1080/01495739.2010.482379

El-Sawy KM, Nazmy AS, Martini MI (2004) Elasto-plastic buckling of perforated plates under uniaxial compression. Thin Wall Struct 42:1083–1101. https://doi.org/10.1016/j.tws.2004.03.002

Estefen SF, Chujutalli JH, Soares CG (2016) Influence of geometric imperfections on the ultimate strength of the double bottom of a Suezmax tanker. Eng Struct 127:287–303. https://doi.org/10.1016/j.engstruct.2016.08.036

Ghavami K, Khedmati MR (2006) Numerical and experimental investigations on the compression behaviour of stiffened plates. J Constr Steel Res. https://doi.org/10.1016/j.jcsr.2006.06.026

Giannakopoulos AE et al (1995) Elastoplastic analysis of thermal cycling: layered materials with compositional gradients. Acta Metall Mater 43(4):1335–1354. https://doi.org/10.1016/0956-7151(94)00360-T

Gunes R et al (2011) The elasto-plastic impact analysis of functionally graded circular plates under low-velocities. Compos Struct 93(2):860–869. https://doi.org/10.1016/j.compstruct.2010.07.008

Hill R (1998) The mathematical theory of plasticity. Oxford University Press
Howard SJ, Tsui YC, Clyne TW (1994) The effect of residual stresses on the debonding of coatings—I. A model for delamination at a bimaterial interface. Acta Metall Mater 42(8):2823–2836
Huang H, Han Q (2014) Elastoplastic buckling of axial loaded functionally graded material cylindrical shells. Compos Struct 117:135–142. https://doi.org/10.1016/j.compstruct.2014.06.018
Jin Z-H (2002) An asymptotic solution of temperature field in a strip a functionally graded material. Int Commun Heat Mass Transfer 29(7):887–895. https://doi.org/10.1016/S0735-1933(02)00409-8
Jin ZH, Paulino GH, Dodds RH (2003) Cohesive fracture modeling of elastic–plastic crack growth in functionally graded materials. Eng Fract Mech 70(14):1885–1912. https://doi.org/10.1016/S0013-7944(03)00130-9
Mahamood RM, Akinlabi ET (2017) Types of functionally graded materials and their areas of application. In: Functionally graded materials. Springer, pp 9–21
Narayanan R, Chow FY (1984) Ultimate capacity of uniaxially compressed perforated plates. Thin-Wall Struct 2(3):241–264. https://doi.org/10.1016/0263-8231(84)90021-1
Paik JK (2005) Ultimate strength of dented steel plates under edge shear loads. Thin-Wall Struct 43(9):1475–1492. https://doi.org/10.1016/j.tws.2005.03.011
Reddy JN (2003) Mechanics of laminated composite plates and shells: theory and analysis, p 840. https://doi.org/10.1007/978-1-4471-0095-9.
Reddy JN (2004) An introduction to nonlinear finite element analysis. Oxford University Press
Reddy JN, Chin CD (1998) Thermomechanical analysis of functionally graded cylinders and plates. J Therm Stress 21(6):593–626. https://doi.org/10.1080/01495739808956165
Shakeri M, Mirzaeifar R (2009) Static and dynamic analysis of thick functionally graded plates with piezoelectric layers using layerwise finite element model. Mech Adv Mater Struct 16(8):561–575. https://doi.org/10.1080/15376490802625514
Shakeri M, Akhlaghi M, Hoseini SM (2006) Vibration and radial wave propagation velocity in functionally graded thick hollow cylinder. Compos Struct 76(1–2):174–181. https://doi.org/10.1016/j.compstruct.2006.06.022
Shanmugam NE, Thevendran V, Tan YH (1999) Design formula for axially compressed perforated plates, vol 34, pp 1–20
Shao ZS (2005) Mechanical and thermal stresses of a functionally graded circular hollow cylinder with finite length. Int J Press Vessels Pip 82(3):155–163. https://doi.org/10.1016/j.ijpvp.2004.09.007
Sharma K, Kumar D (2017a) Elastoplastic stability and failure analysis of FGM plate with temperature dependent material properties under thermomechanical loading. Latin Am J Solids Struct. https://doi.org/10.1590/1679-78253747
Sharma K, Kumar D (2017b) Elastoplastic analysis of FGM plate with a central cutout of various shapes under thermomechanical loading. J Therm Stress. https://doi.org/10.1080/01495739.2017.1323566
Sharma K, Kumar D (2018) Nonlinear stability analysis of a perforated FGM plate under thermal load. Mech Adv Mater Struct 25(2):100–114. https://doi.org/10.1080/15376494.2016.1255817
Shiota I, Miyamoto Y (1997) Functionally graded materials 1996. Elsevier
Singh SB, Kumar A (1999) Postbuckling response and strength of laminates under combined in-plane loads. Compos Sci Technol 59(5):727–736. https://doi.org/10.1016/S0266-3538(98)00125-0
Soh AK, Bian LC, Chakrabarty J (2000) Elastic/plastic buckling of a composite flat plate subjected to uniform edge compression. Thin-Wall Struct 38(3):247–265. https://doi.org/10.1016/S0263-8231(00)00038-0
Suresh S, Mortensen A (1998) Fundamentals of functionally graded materials. The Institute of Materials

Tamura I, Tomato Y, Ozawa H (1973) Strength and ductility of Fe–Ni–C alloys composed of austenite and martensite with various strength. In: Proceedings of the third conference on strength of metals and alloys, vol 1(4). Institute of Metals, Cambridge, pp 611–616

Touloukian YS, Center TPR (1967) Thermophysical properties of high temperature solid materials. vol 1. Elements.-Pt. 1. Macmillan

Williamson RL, Rabin BH, Byerly GE (1995) FEM study of the effects of interlayers and creep in reducing residual stresses and strains in ceramic-metal joints. Compos Eng 5(7):851–863. https://doi.org/10.1016/0961-9526(95)00035-L

Yaghoobi H et al (2015) Thermal buckling analysis of piezoelectric functionally graded plates with temperature-dependent properties. Mech Adv Mater Struct 22(10):864–875. https://doi.org/10.1080/15376494.2013.864436

Zhang Y, Huang H, Han Q (2015) Buckling of elastoplastic functionally graded cylindrical shells under combined compression and pressure. Compos Part B Eng 69:120–126. https://doi.org/10.1016/j.compositesb.2014.09.024

Buckling of Laminated Composite Plate with Imperfections Subjected to In-Plane Pulse Loads

Vasanth Keshav, Shuvendu Narayan Patel, and Rajesh Kumar

Abstract In this article, the stability of a laminated composite plate when subjected to in-plane compressive pulse load is investigated in the finite element method framework. Convergence and validation studies are carried out using the current mathematical formulation and compared with the results from the existing literatures. The effects of loading duration, imperfection and ply orientation on the dynamic buckling behavior of the plate with irregular imperfection are studied in detail and the results are reported. It is observed that the plate having irregular imperfection of the order of 20% of the plate thickness has a lower non-linear dynamic buckling load than the plate with 15% irregular imperfection.

Keywords Dynamic buckling · Laminated composite plate · In-plane loads

1 Introduction

Composite materials have wide applications in aerospace, biotechnology, marine, automobile and civil engineering industries where high strength and light weight design are essential. Analyzing a complete structure is time consuming and costly. Thus, parts of the complete structures are analyzed separately. Laminated composite plates are part of a complex structure. During the service life of the structure, the plate may be subjected to loads in the axial direction from its neighboring components which act for a short duration. Due to these loads, the plate can become unstable and render the entire structure vulnerable to failure. The instability in the plate can be caused due to static loads and dynamic loads. In case of dynamic loads, the loss of stability caused due to in-plane pulse loads is called dynamic buckling (Jansen 2005).

V. Keshav (✉) · S. N. Patel · R. Kumar
Department of Civil Engineering, BITS Pilani, Pilani Campus, Pilani, Rajasthan 333031, India
e-mail: keshav.vasanth@gmail.com

S. N. Patel
e-mail: shuvendu@pilani.bits-pilani.ac.in

S. B. Singh et al. (eds.), *Emerging Trends of Advanced Composite Materials in Structural Applications*, Composites Science and Technology,
https://doi.org/10.1007/978-981-16-1688-4_12

The instability in isotropic plates due to static loads is studied by many authors. In this regard, Budiansky (1948) investigated the static buckling behavior of isotopic long plates simply-supported by rigid posts at the inner side of the plate using analytical method. Libove et al. (1949) calculated the static buckling load of plates with simply supported edges and of uniform thickness, subjected to non-uniform compressive stresses using Rayleigh–Ritz method. Gerard and Becker (1957) reported the theories and experimental data pertaining to static buckling of plates used in aircrafts. The authors presented charts for non-dimensional static buckling load for plates with various boundary conditions. Bauer and Reiss (1965) investigated the non-linear static buckling of plate using numerical method and the reported that for loads lower than the static linear buckling load, buckled states are not observed. Smith and Sridharan (1978) investigated the static buckling behavior of plates supported at the ends using rigid diaphragms using finite strip method and reported better computational economy when compared to the results obtained from finite element method.

Apart from these, Dawe and Roufaeil (1982) presented the results of static buckling analysis of isotropic and transversely isotopic plates using numerical method with the incorporation of transverse shear effects of the plate. Manickarajah et al. (1998) conducted optimization technique for plates subjected to in-plane loads and optimized the thickness of the plate for a given volume so as to have the maximum static buckling load. Audoly et al. (2002) studied the static buckling behavior of long plates supported by two clamped edges and subjected to biaxial loading using numerical method. Ruocco et al. (2011) investigated the effect of boundary conditions on the static buckling behavior of a plate using analytical method by assuming a combination of trigonometric and polynomial displacements to solve the buckling problem. Ádány et al. (2018) reported the results of the static buckling of plates using finite element method based on constrained finite strip method. The authors reported that the proposed method required rectangular discretization of the plate and reported the results for various boundary and loading conditions.

Moreover, the static buckling behaviour of laminated composite plates has also been investigated by many authors. Jones (1973) presented exact solutions for buckling and vibration of simply supported cross-ply plates using classical plate theory. The author reported that plates with large number of layers do not guarantee higher static buckling loads. An exhaustive literature survey on the static buckling behavior and post-buckling behavior of plates was reported by Leissa (1987). The static buckling analysis of composite plate subjected to axial and shear loads was performed by Narita and Leissa (1990) using numerical methods. Hu et al. (2003) studied the static buckling behaviour of plate subjected to parabolic loads in the longitudinal and transverse directions using Rayleigh–Ritz method. The authors reported that the static buckling load for plates subjected to parabolic loads was lower than the static buckling loads when subjected to uniform loads. Shufrin et al. (2008) investigated the static buckling behaviour of plates subjected to a combination of shear, compressive and tensile loads using semi-analytical method based on classical plate theory. Sayyad and Ghugal (2014) presented the results of the static buckling analysis of isotropic and composite plates using analytical method based on trigonometric shear and normal deformation theory.

Furthermore, the instability in isotropic and composite plates due to periodic loads has also been studied by a few authors. Chen and Yang (1990) used finite element method and reported the results of parametric instability analysis of angle-ply composite plates. Ganapathi et al. (2000) discussed a plate element in the framework of finite element method to analyze the non-linear parametric instability of isotropic plates. Chattopadhyay and Radu (2000) studied the parametric instability of isotropic plates using finite element method based on higher-order shear deformation theory and compared the results with other plate theories. Sahu and Datta (2007) reported an exhaustive review of studies between the years 1987 and 2005 on the parametric instability of isotropic plates and composite plates. Ramachandra and Panda (2012) investigated the parametric instability of composite plate subjected to non-uniform in-plane loads by means of finite element method and reported the results of the effect of static loads factor on instability region. Sahoo and Singh (2015) studied the parametric instability of composite plates subjected to static and periodic loads using finite element method based on inverse hyperbolic zigzag theory. Yusof and Rasid (2016) reported the results of parametric instability analysis on angle-ply composite plate using finite element method. Darabi and Ganesan (2018) studied the non-linear parametric instability of cross-ply laminated plates using numerical method.

In addition, the instability of plates due to in-plane pulse loads has also been investigated by a few authors. The dynamic buckling behavior of shells was considered by Hutchinson and Budiansky (1966), Budiansky and Hutchinson (1966) and Budiansky (1967). The authors formulated criteria for determining the dynamic buckling load of imperfection sensitive structures. The dynamic buckling behavior of plates subjected to fluid–solid slamming was investigated experimentally by Cui et al. (1999). Experimental results were compared with finite element results (Cui et al. 2001) and the authors defined the dynamic buckling load as the load at which the transverse load increases rapidly. Petry and Fahlbusch (2000) presented a criterion for calculating the dynamic buckling load of isotropic plates based on the stress in the plate. Kowal-Michalska (2010) investigated the dynamic buckling behavior of plates subjected to sinusoidal and rectangular pulse loads using finite element method and reported that Volmir's criterion, in many cases, gives the results which are very close to the results obtained from Budiansky-Hutchinson criterion. Azarboni et al. (2015) reported the results of dynamic buckling analysis of isotropic plates subjected to dynamic and quasi-static loads for various boundary conditions and loading functions (rectangular, sinusoidal, exponential and damping) using numerical methods. The dynamic buckling analysis of isotropic plates with elastically restrained edges was performed by Yang and Wang (2017) using numerical method. Mondal and Ramachandra (2020) investigated the dynamic buckling of composite plates with imperfections in the plates using finite element method. Rostamijavanani (2020) studied the dynamic buckling behaviour of composite cylindrical panels using finite element method considering Budiansky-Roth criterion for dynamic buckling load. Apart from these, a few studies are available on the investigation of dynamic buckling behaviour of various structures subjected to pulse loads (Yang et al. 2020; Box et al. 2020; Sellitto et al. 2020).

From the literature survey it is evident that the instability of isotropic and composite plates subjected to static and periodic loads has been investigated thoroughly. Some studies are also available for the dynamic buckling behavior of composite plates, but these studies consider imperfection only in the form of the first buckling mode of the plate. In this investigation, an irregular-shaped imperfection comprising of a combination of first four buckling modes is considered for the plate. Imperfection is considered as a certain percentage of the thickness of the plate. The effects of loading duration, ply orientation and the imperfection on the stability of the plate subjected to in-plane pulse loads is evaluated.

2 Theory

The non-linear dynamic equilibrium equation solved by ABAQUS/explicit is given in Eq. (1).

$$[M]\{\ddot{u}\} + [K(\{u\})]\{u\} = \{F(t)\} \tag{1}$$

where,

$$[M] = \int N^T \rho N dx\, dy\, dz$$

$$[K] = \int B^T D B dx\, dy\, dz$$

$$\{F\} = \int N^T \overline{b} dV + \int N^T \tau dS + \int N^T \overline{p} dA$$

where, '$[M]$' and '$[K]$' are the mass and the stiffness matrices. The stiffness matrix is a function of displacements. This is due to incorporation of geometric non-linearity in the panel. '$[F(t)]$' is the load vector. The nodal acceleration vector, velocity vector and the displacement vector are represented as '$\{\ddot{u}\}$', '$\{\dot{u}\}$' and '$\{u\}$', respectively. 'τ', 'ρ', '$\overline{p}$' and '$\overline{b}$' are the boundary traction over the surface, mass density, pressure acting on the surface, and body force per unit volume, respectively. 'B' is the matrix connecting displacements and strains; and, 'N' is the shape function matrix.

In ABAQUS, S4R element is used to model the plate. S4R is a four-noded element with four corner nodes and one integration point at the center (Fig. 1). The non-linear kinematic relationships due to large deformations are considered. The calculations are done at local coordinates and the results are utilized in the subsequent steps (Abaqus 2013). In ABAQUS/Explicit, the time integration is done using central difference method.

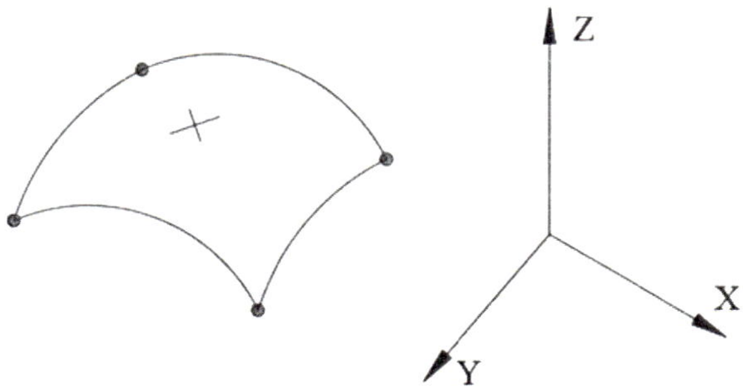

Fig. 1 S4R element used in the current study

3 Results and Discussion

In this section, first the results of the convergence and validation studies are presented and then, the results of the present investigation are reported.

3.1 Convergence and Validation Studies

The results of the convergence and validation studies are presented in this section. For this, the static buckling load of a composite plate, shock spectrum of an aluminum plate and the dynamic response of an orthotropic plate are considered.

3.1.1 Static Buckling of a Composite Plate

The static buckling load of a composite plate is calculated, and the results are compared with the results reported by Narita and Leissa (1990). The geometric and material properties used are: $a/b = 1$, $a/h = 100$, $E_{11} = 138$ GPa, $E_{22} = 8.96$ GPa, $G_{12} = G_{13} = G_{23} = 7.1$ GPa and $\nu_{12} = 0.3$. The stacking sequence considered is $(\theta/-\theta/\theta/-\theta/\theta)$ where $\theta = 0°, 15°, 30°, 45°, 60°, 75°, 90°$. The geometry of the plate is shown in Fig. 2, wherein, 'b' is the length of the loaded edge and 'h' is the thickness of the plate. The pre-buckling boundary conditions are shown in Fig. 3 and the buckling boundary condition are shown in Fig. 4. In Figs. 3 and 4, 'u', 'v' and 'w' are the displacements along X, Y and Z axes, respectively. Furthermore, 'θ_X', 'θ_Y' and 'θ_Z' are the rotations about X, Y and Z axes, respectively. The non-dimensional static buckling load is calculated using Eq. 2. The non-dimensional static buckling load for various mesh sizes is shown in Fig. 5. The results of the present investigation are compared with the finite element results of Narita and Leissa (1990) in Table 1.

$$\frac{N_{XX}a^2}{D_0}$$

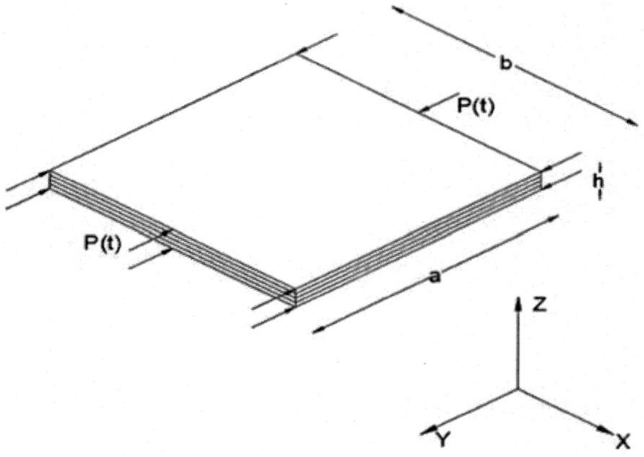

Fig. 2 Geometry of composite plate

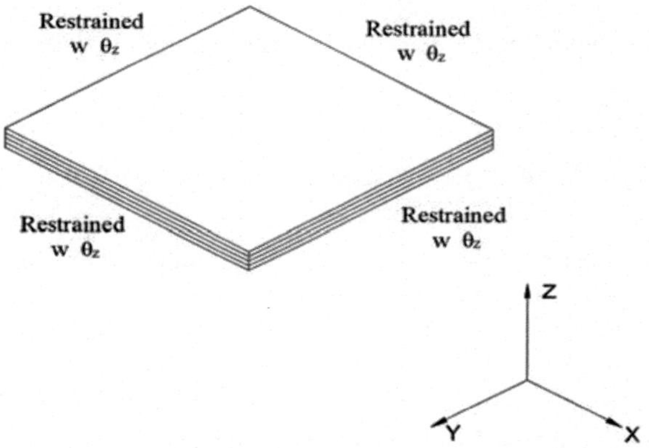

Fig. 3 Pre-buckling boundary conditions

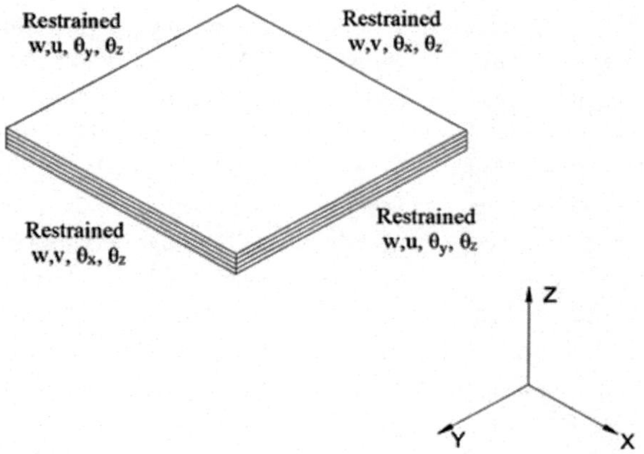

Fig. 4 Buckling boundary conditions

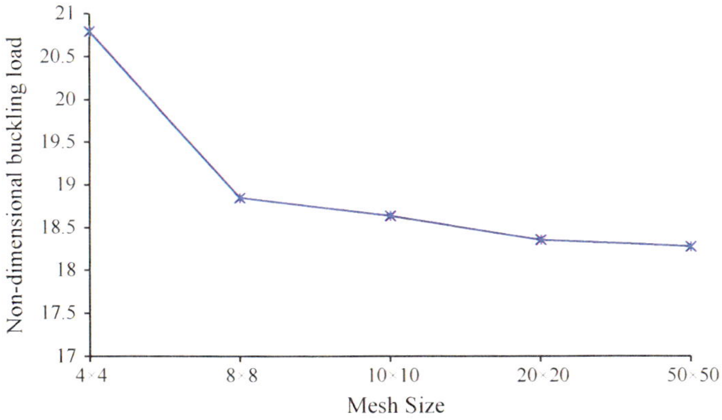

Fig. 5 Non-dimensional static buckling load for various mesh sizes

Table 1 Non-dimensional static buckling load for the plate with stacking sequence (θ/−θ/θ/−θ/θ)

Θ (°)	Narita and Leissa (1990)	Present
0	12.91	12.94
15	14.21	14.18
30	17.11	16.92
45	18.66	18.36
60	16.49	16.35
75	9.94	9.91
90	7.43	7.45

where,

$$D_0 = \frac{E_{11}h^3}{12(1 - \upsilon_{12}\upsilon_{21})}$$

$$\upsilon_{21} = \left(\frac{E_{22}}{E_{11}}\right)\upsilon_{12} \tag{2}$$

From Fig. 5 it is observed that the mesh size of 20 × 20 is converged; and, from Table1 it is observed that the results of the present study match well with the results reported by Narita and Leissa (1990).

3.1.2 Shock Spectrum of an Aluminium Plate

The shock spectrum of an aluminium plate is presented in this section. A square plate is subjected to dynamic load. The maximum deflection of the plate is observed for various loading durations. The geometric properties and the material properties of

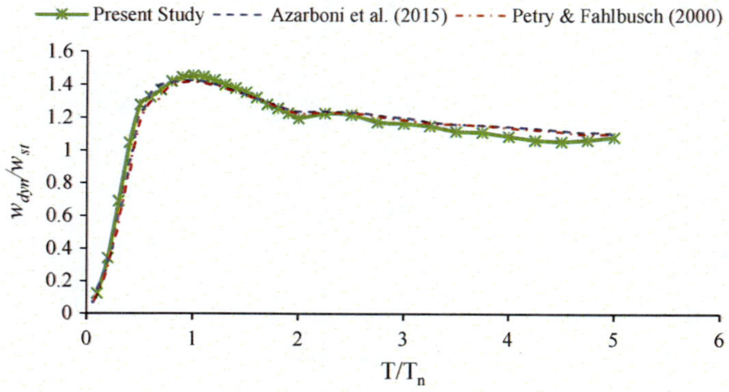

Fig. 6 Shock spectrum of unstiffened plate subjected to load $N_{dyn} = 3N_{st}$

the plate as described in Keshav et al. (2019) are used. The results of present study are presented along with the FEM/numerical results from the references (Petry and Fahlbusch 2000; Azarboni et al. 2015) in Fig. 6. In Fig. 6, 'T_n' is the first natural period of the plate.

It is observed from Fig. 6 that the results of the present study agree well with the results from the literature. The peak response can be observed in the neighborhood the $T/T_n = 1$.

3.1.3 Validation of Dynamic Response of an Orthotropic Plate

In this section, the dynamic response of an orthotropic plate is validated. It is subjected to rectangular pulse load (Fig. 7). The geometric properties and the material properties use are same as presented in Kowal-Michalska and Mania (2008). Imperfection in the plate is in the form of the first buckling mode shape. The boundary conditions for pre-buckling and buckling stage are same as shown in Fig. 3b. Figure 8 depicts the

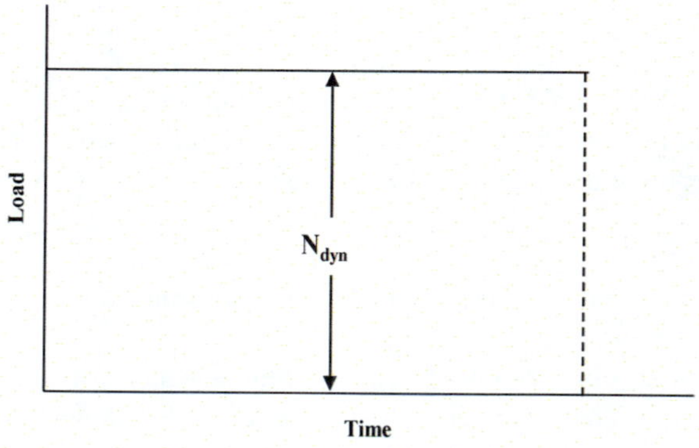

Fig. 7 Rectangular pulse loading function. N_{dyn} is the maximum amplitude of the dynamic load applied

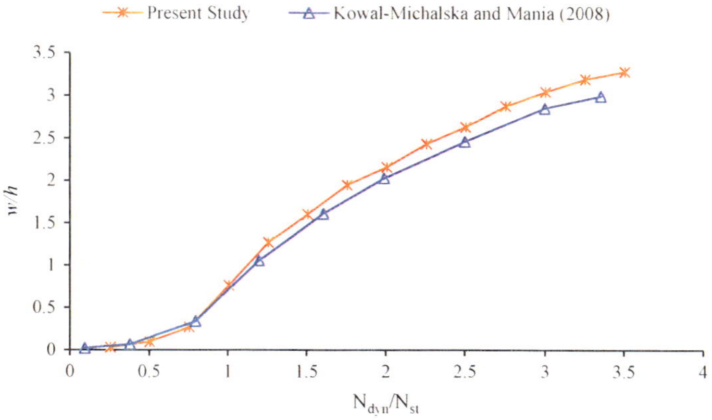

Fig. 8 Non-dimensional load versus non-dimensional displacement for orthotropic plate subjected to rectangular pulse load

plot of N_{dyn}/N_{st} versus w/h for the plate subjected to rectangular pulse loading. The results are compared with the FEM results of Kowal-Michalska and Mania (2008) and a good agreement in the results can be observed in Fig. 8.

From the convergence and validation studies, it can be seen that the results of the present study match well with the results from the literature. Based on the results of convergence study, mesh size of 20 × 20 is considered in the subsequent part of this study; Further, the same boundary conditions for pre-buckling and buckling stages, as shown, respectively, in Figs. 3 and 4, are considered for evaluating the static buckling load of the plate.

3.2 *Dynamic Buckling Studies*

The stability of a laminated composite plate considered in the convergence and validation study (Sect. 3.1.1), is considered when the plate is subjected to an in-plane pulse load. The geometry of the plate is shown in Fig. 2 and the boundary conditions are shown in Fig. 4. The plate is subjected to rectangular pulse load as shown in Fig. 7; wherein, N_{dyn} is the maximum amplitude of the dynamic load the plate is subjected to. The dynamic buckling load is calculated in accordance with Vol'mir criterion: the in-plane dynamic load at which the maximum displacement is equal to the thickness of the plate (Vol'mir 1974). The results obtained from this criterion and the results from Budiansky-Hutchinson criterion vary by 10% (Kubiak 2013).

For the analysis, first the static buckling load of the plate is calculated. Next, the first natural period of the plate is calculated. The imperfection is applied in the plate as a combination of the first four static buckling modes of the plate. This results in an irregular shape of the plate, which is more realistic than considering only the first

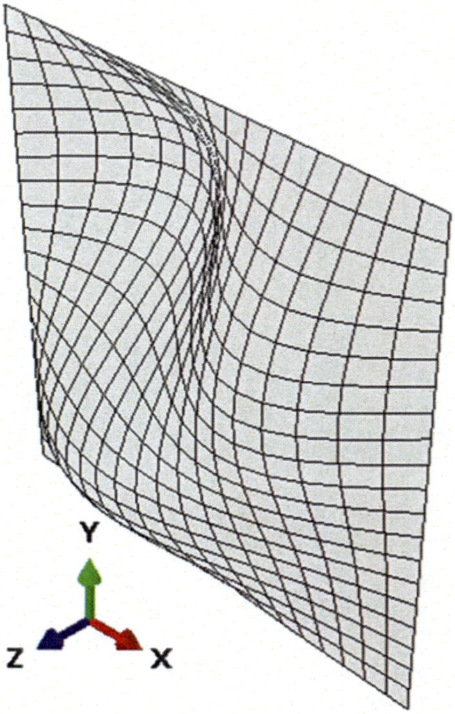

Fig. 9 Shape of the plate with 0.1 h imperfection in the form of first four buckling modes. Scale factor = 100

buckling mode of the plate. The resulting pate with imperfection is shown in Fig. 9 with the scale factor of 100 and the imperfection equal to 10% of the thickness of the plate. This means that the imperfection value of each of the mode shape considered is 10% of the thickness of the plate. For various magnitudes of dynamic loads, the maximum transverse displacements are evaluated and the dynamic load at which the maximum displacement reaches a value equal to the thickness of the plate is called the dynamic buckling load.

In the succeeding sections, the effect of loading duration, imperfection and the ply orientation on the stability of the plate under in-plane pulse load is investigated.

3.2.1 Effect of Loading Duration

The influence of the loading duration on the stability of laminated composite plate subjected to in-plane pulse load is studied in this section. The plate with $a/b = 1$, $a/h = 100$ and stacking sequence $(\theta/-\theta/\theta/-\theta/\theta)$; $\theta = 45°$ is considered. The loading durations considered are $0.4T_n$, $0.2T_n$, T_n, $2.5T_n$ and $5T_n$. The non-dimensional load versus non-dimensional displacement plots are shown in Fig. 10 for the plate subjected to rectangular pulse load for various loading durations.

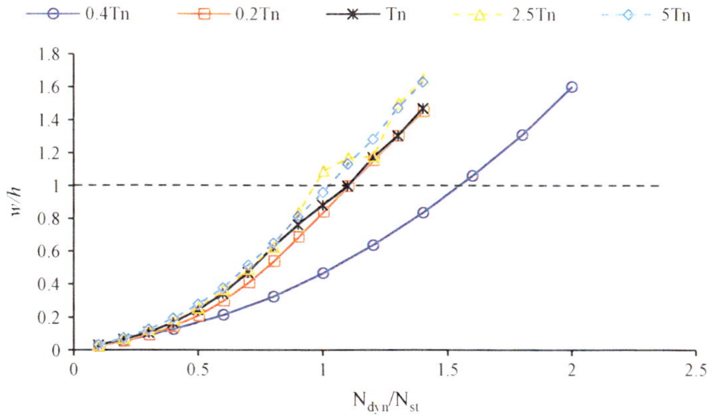

Fig. 10 Non-dimensional load versus non-dimensional displacement for the plate with 10% imperfection subjected to rectangular in-plane pulse load for various loading durations

It is observed from Fig. 10 that the loading duration near the first natural period of the plate is critical. The plate will have the lowest dynamic buckling load when subjected to a load till its first natural period. The dynamic buckling load of the plate will be the highest when the duration of loading is very short ($0.4T_n$ in Fig. 10). The same conclusion was drawn from Fig. 6 as well.

3.2.2 Effect of Imperfection

The influence of imperfection on the stability of the composite plate subjected to in-plane pulse load is checked in this section. The plate with $a/b = 1$, $a/h = 100$ and stacking sequence ($\theta/-\theta/\theta/-\theta/\theta$) with $\theta = 45°$ is considered. The imperfections equal to 5, 10 15 and 20% of the plate thickness in the form of first four buckling modes of the plate are considered. The non-dimensional load versus non-dimensional displacement plots are shown in Fig. 11 for the plate subjected to rectangular pulse load for various imperfection values.

It is observed from Fig. 11 that the plate with 15% imperfection has the lowest dynamic buckling which is slightly lower than the static buckling load of the plate. The highest dynamic buckling load is of the plate with 5% imperfection; however, the value ($1.16N_{st}$) is also close to the static buckling load of the plate.

3.2.3 Effect of Stacking Sequence of the Plate

The influence of stacking sequence on the stability of the composite plate subjected to in-plane pulse load is investigated in this section. The plate with $a/b = 1$, $a/h = 100$ and imperfection $= 0.1\ h$ is considered. The stacking sequence ($\theta/-\theta/\ \theta/-\theta/\theta$) with $\theta = 0°, 15°, 30°, 45°, 60°, 75°$ and $90°$ is considered. The non-dimensional

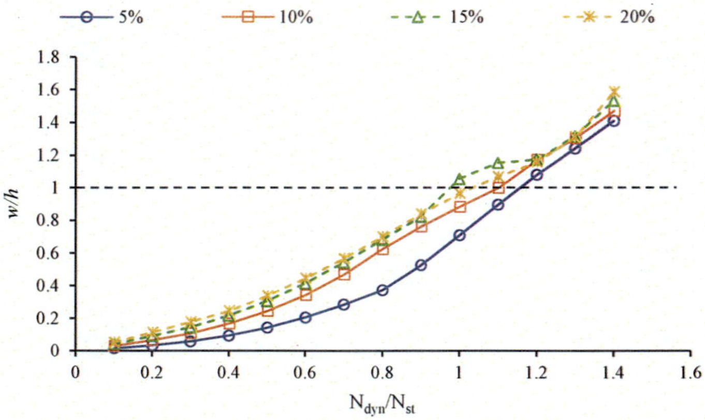

Fig. 11 Non-dimensional load versus non-dimensional displacement for the plate subjected to rectangular in-plane pulse load for various imperfection values

load versus non-dimensional displacement plots are shown in Fig. 12 for the plate subjected to rectangular pulse load for various stacking sequences.

It is observed form Fig. 12 that the dynamic buckling load for the plate with stacking sequence ($\theta/-\theta/\theta/-\theta/\theta$) with $\theta = 0°$ is the lowest and it is found to be lower than the static buckling load of the plate. The dynamic buckling load for the plate with stacking sequence ($\theta/-\theta/\theta/-\theta/\theta$) with $\theta = 15°$ is equal to the static buckling load of the plate. The dynamic buckling loads for the rest of the stacking sequences are slightly higher than the static buckling load of the plate.

Figures 13, 14 and 15 show the plots of non-dimensional load versus non-

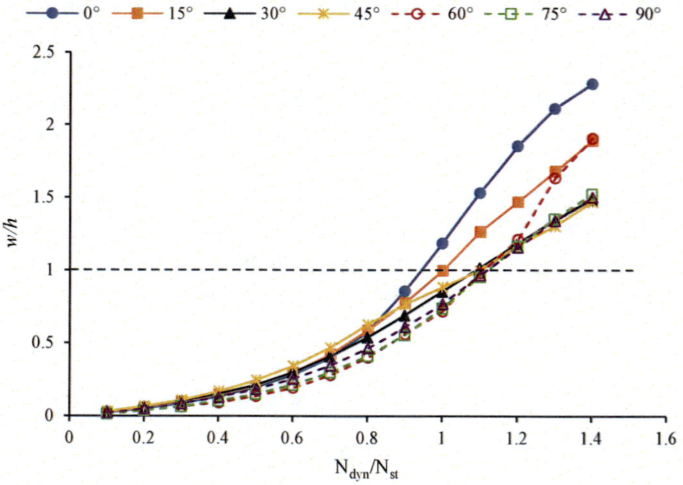

Fig. 12 Non-dimensional load versus non-dimensional displacement for the plate with 10% imperfection subjected to rectangular in-plane pulse load for various stacking sequences ($\theta/-\theta/\theta/-\theta/\theta$)

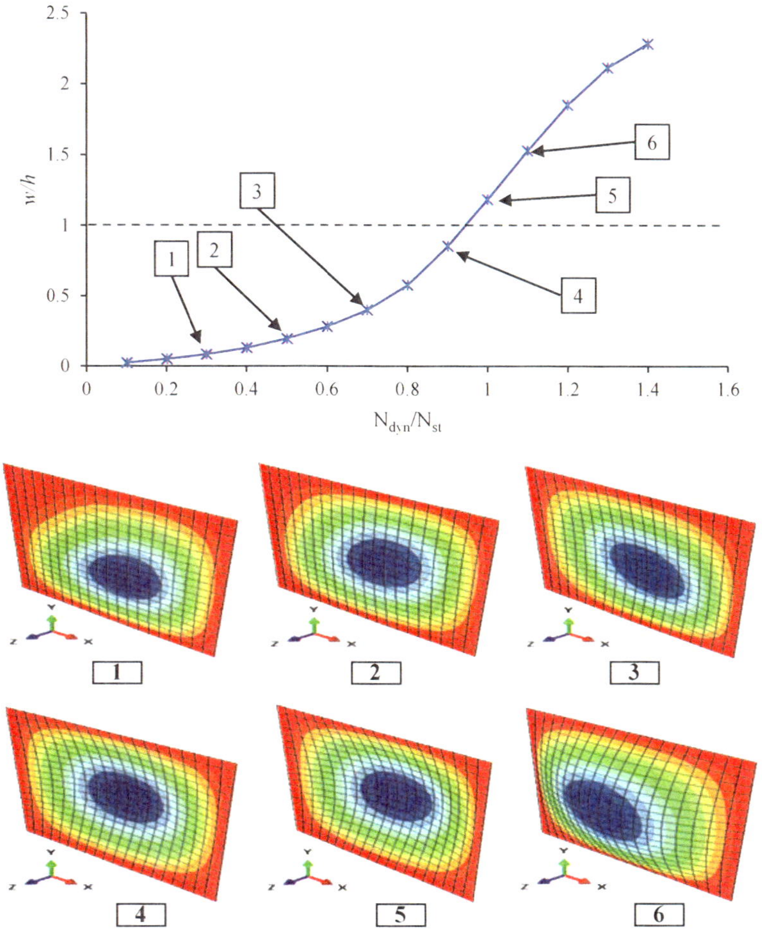

Fig. 13 Non-dimensional load versus non-dimensional displacement for the plate with stacking sequence (θ/−θ/θ/−θ/θ); θ = 0° and 10% imperfection subjected to rectangular in-plane pulse load along with deformed shapes at various magnitude of loads

dimensional displacement of the plate for stacking sequence (θ/−θ/θ/−θ/θ) with θ = 0°, 45° and 90° respectively, having imperfection = 0.1 h. In these figures, the deformed shapes of the plate at various magnitude of loads are also shown. The scale factor for the deformed shapes is 10 for all the cases. It can be observed that the in the case of plate with θ = 45°, the deformed shapes at $N_{dyn} = 0.8N_{st}$ and $N_{dyn} = N_{st}$ are different from other cases. This is because, with the change in the stacking sequence, the laminated composite plates deform in a difference manner even though the same imperfection profile is given in all the cases (Fig. 9).

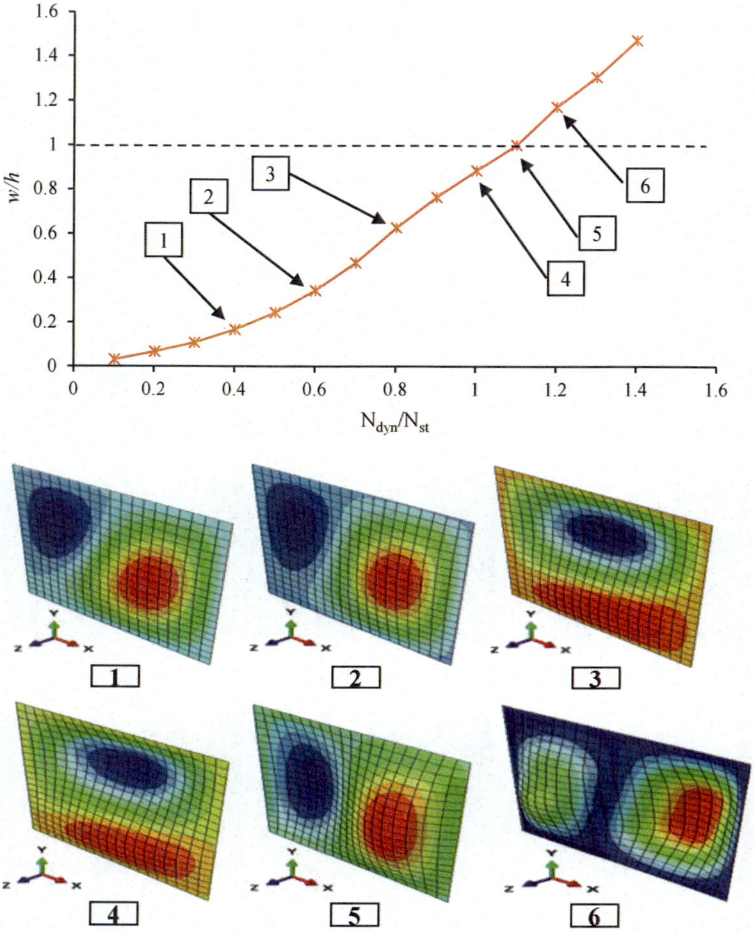

Fig. 14 Non-dimensional load versus non-dimensional displacement for the plate with stacking sequence (θ/−θ/θ/−θ/θ); θ = 45° and 10% imperfection subjected to rectangular in-plane pulse load along with deformed shapes at various magnitude of loads

4 Conclusions

The stability of the laminated composite plate with irregular imperfection profile subjected to in-plane pulse load is investigated. It is seen that the dynamic buckling load of the plate is lower than its static buckling load when the plate has an irregular imperfection profile. Laminated composite plate when subjected to a pulse load till its first natural period has a lower dynamic buckling load than when subjected to a pulse duration which is very short. This observation is in line with the fact that the duration of loading near the first natural period of the structure is critical in a structure (example: earthquake loading). It is expected that the plate with a higher imperfection

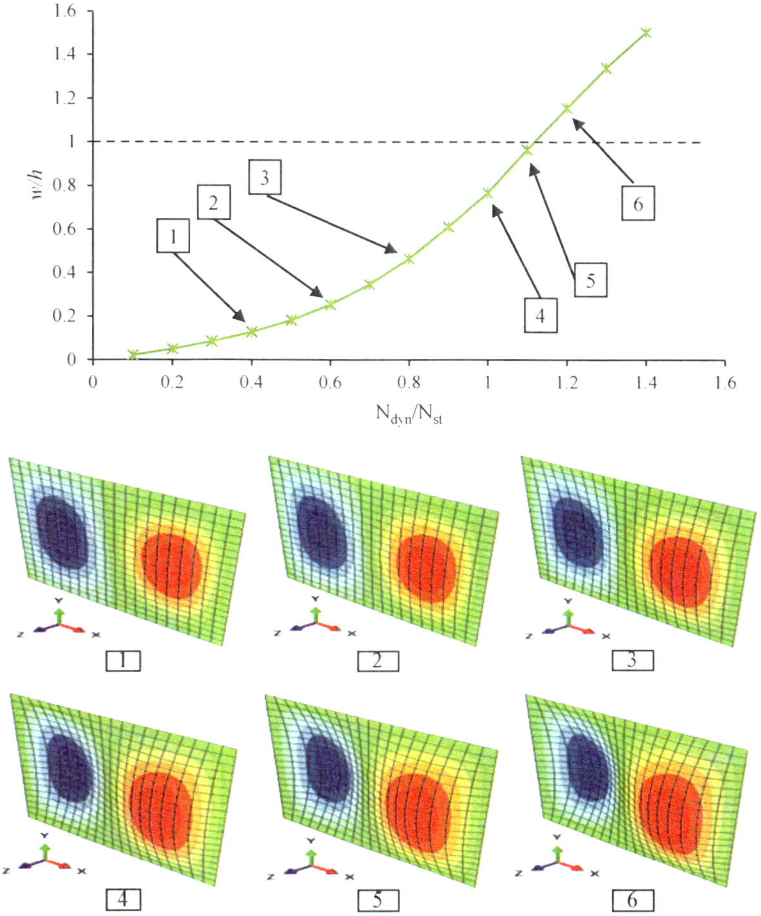

Fig. 15 Non-dimensional load versus non-dimensional displacement for the plate with stacking sequence (θ/−θ/θ/−θ/θ); θ = 90° and 10% imperfection subjected to rectangular in-plane pulse load along with deformed shapes at various magnitude of loads

value will have a lower dynamic buckling load. But it is not the case when irregular imperfections are present in the plate. It is seen that the dynamic buckling load of the plate with 15% imperfection is lower than the dynamic buckling load of the plate with 20% imperfection. The panel with ply orientation (θ/−θ/θ/−θ/θ); θ = 0°, has dynamic buckling load lower than its static buckling load and the panel with θ = 45° and θ = 90°, has dynamic buckling loads higher than their corresponding static buckling loads. With the change in the ply orientation, the plate deforms in a different manner in all three cases even though the same imperfection is present in the plate.

References

Abaqus (2013) ABAQUS documentation. Dassault Systèmes, Providence, RI

Ádány S, Visy D, Nagy R (2018) Constrained shell finite element method, part 2: application to linear buckling analysis of thin-walled members. Thin-Wall Struct 128:56–70. https://doi.org/10.1016/j.tws.2017.01.022

Audoly B, Roman B, Pocheau A (2002) Secondary buckling patterns of a thin plate under in-plane compression. Eur Phys J B Condens Matter Complex Syst 27(1):7–10. https://doi.org/10.1140/epjb/e20020124

Azarboni HR, Darvizeh M, Darvizeh A, Ansari R (2015) Nonlinear dynamic buckling of imperfect rectangular plates with different boundary conditions subjected to various pulse functions using the Galerkin method. Thin Wall Struct 94:577–584. https://doi.org/10.1016/j.tws.2015.04.002

Bauer L, Reiss EL (1965) Nonlinear buckling of rectangular plates. J Soc Ind Appl Math 13(3):603–626. https://doi.org/10.1137/0113039

Box F, Kodio O, O'Kiely D, Cantelli V, Goriely A, Vella D (2020) Dynamic buckling of an elastic ring in a soap film. Phys Rev Lett 124(19):198003. https://journals.aps.org/prl/abstract/10.1103/PhysRevLett.124.198003

Budiansky B (1948) Compressive buckling of flat rectangular plates supported by rigid posts. NACA research memorandum no. L8130b

Budiansky B (1967) Dynamic buckling of elastic structures: criteria and estimates. In: Dynamic stability of structures. Pergamon, pp 83–106. https://doi.org/10.1016/B978-1-4831-9821-7.50010-7

Budiansky B, Hutchinson JW (1966) A survey of some buckling problems. AIAA J 4(9):1505–1510. https://doi.org/10.2514/3.3727

Chattopadhyay A, Radu AG (2000) Dynamic instability of composite laminates using a higher order theory. Comput Struct 77(5):453–460. https://doi.org/10.1016/S0045-7949(00)00005-5

Chen LW, Yang JY (1990) Dynamic stability of laminated composite plates by the finite element method. Comput Struct 36(5):845–851. https://doi.org/10.1016/0045-7949(90)90155-U

Cui S, Cheong HK, Hao H (1999) Experimental study of dynamic buckling of plates under fluid–solid slamming. Int J Impact Eng 22(7):675–691. https://doi.org/10.1016/S0734-743X(99)00022-6

Cui S, Hao H, Cheong HK (2001) Numerical analysis of dynamic buckling of rectangular plates subjected to intermediate-velocity impact. Int J Impact Eng 25(2):147–167. https://doi.org/10.1016/S0734-743X(00)00035-X

Darabi M, Ganesan R (2018) Nonlinear dynamic instability analysis of laminated composite thin plates subjected to periodic in-plane loads. Nonlinear Dyn 91(1):187–215. https://doi.org/10.1007/s11071-017-3863-9

Dawe DJ, Roufaeil OL (1982) Buckling of rectangular Mindlin plates. Comput Struct 15(4):461–471. https://doi.org/10.1016/0045-7949(82)90081-5

Ganapathi M, Patel BP, Boisse P, Touratier M (2000) Non-linear dynamic stability characteristics of elastic plates subjected to periodic in-plane load. Int J Non-Linear Mech 35(3):467–480. https://doi.org/10.1016/S0020-7462(99)00034-7

Gerard G, Becker H (1957) Handbook of structural stability: part I, buckling of flat plates. NACA Technical (no. 3781)

Hu H, Badir A, Abatan A (2003) Buckling behavior of a graphite/epoxy composite plate under parabolic variation of axial loads. Int J Mech Sci 45(6–7):1135–1147. https://doi.org/10.1016/j.ijmecsci.2003.08.003

Hutchinson JW, Budiansky B (1966) Dynamic buckling estimates. AIAA J 4(3):525–530. https://doi.org/10.2514/3.3468

Jansen EL (2005) Dynamic stability problems of anisotropic cylindrical shells via a simplified analysis. Nonlinear Dyn 39(4):349–367. https://doi.org/10.1007/s11071-005-4343-1

Jones RM (1973) Buckling and vibration of unsymmetrically laminated cross-ply rectangular plates. AIAA J 11(12):1626–1632. https://doi.org/10.2514/3.50660

Keshav V, Patel SN, Kumar R (2019) Stability and failure study of suddenly loaded laminated composite cylindrical panel. Int J Appl Mech 11(10):1950093. https://doi.org/10.1142/S1758825119500935

Kowal-Michalska K, Mania RJ (2008) Some aspects of dynamic buckling of plates under in-plane pulse loading. Mech Mech Eng 12(2):135–146

Kowal-Michalska K (2010) About some important parameters in dynamic buckling analysis of plated structures subjected to pulse loading. Mech Mech Eng 14(2):269–279

Kubiak T (2013) Static and dynamic buckling of thin−walled plate structures. Springer, Switzerland, 2013. https://doi.org/10.1007/978-3-319-00654-3

Leissa AW (1987) A review of laminated composite plate buckling. ASME Appl Mech Rev 40(5):575–591. https://doi.org/10.1115/1.3149534

Libove C, Ferdman S, Reusch JJ (1949) Elastic buckling of a simply supported plate under a compressive stress that varies linearly in the direction of loading. NACA technical note no. 1891

Manickarajah D, Xie YM, Steven GP (1998) An evolutionary method for optimization of plate buckling resistance. Finite Elem Anal Des 29(3–4):205–230. https://doi.org/10.1016/S0168-874X(98)00012-2

Mondal S, Ramachandra LS (2020) Nonlinear dynamic pulse buckling of imperfect laminated composite plate with delamination. Int J Solids Struct. https://doi.org/10.1016/j.ijsolstr.2020.04.010

Narita Y, Leissa AW (1990) Buckling studies for simply supported symmetrically laminated rectangular plates. Int J Mech Sci 32(11):909–924. https://doi.org/10.1016/0020-7403(90)90063-O

Petry D, Fahlbusch G (2000) Dynamic buckling of thin isotropic plates subjected to in-plane impact. Thin-Wall Struct 38(3):267–283. https://doi.org/10.1016/S0263-8231(00)00037-9

Ramachandra LS, Panda SK (2012) Dynamic instability of composite plates subjected to non-uniform in-plane loads. J Sound Vib 331(1):53–65. https://doi.org/10.1016/j.jsv.2011.08.010

Rostamijavanani A (2020) Dynamic buckling of cylindrical composite panels under axial compressions and lateral external pressures. J Fail Anal Prev 1–10. https://doi.org/10.1007/s11668-020-01032-3

Ruocco E, Minutolo V, Ciaramella S (2011) A generalized analytical approach for the buckling analysis of thin rectangular plates with arbitrary boundary conditions. Int J Struct Stab Dyn 11(01):1–21. https://doi.org/10.1142/S0219455411003963

Sahoo R, Singh BN (2015) Dynamic instability of laminated composite and sandwich plates using a new inverse hyperbolic zigzag theory. J Aerosp Eng 28(4):04014109. https://doi.org/10.1061/(ASCE)AS.1943-5525.0000440

Sahu SK, Datta PK (2007) Research advances in the dynamic stability behavior of plates and shells: 1987–2005—part I: conservative systems. Appl Mech Rev 60(2):65–75. https://doi.org/10.1115/1.2515580

Sayyad AS, Ghugal YM (2014) On the buckling of isotropic, transversely isotropic and laminated composite rectangular plates. Int J Struct Stab Dyn 14(07):1450020. https://doi.org/10.1142/S0219455414500205

Sellitto A, Di Caprio F, Guida M, Saputo S, Riccio A (2020) Dynamic pulse buckling of composite stanchions in the sub-cargo floor area of a civil regional aircraft. Materials 13(16):3594. https://doi.org/10.3390/ma13163594

Shufrin I, Rabinovitch O, Eisenberger M (2008) Buckling of laminated plates with general boundary conditions under combined compression, tension, and shear—a semi-analytical solution. Thin-Wall Struct 46(7–9):925–938. https://doi.org/10.1016/j.tws.2008.01.040

Smith TG, Sridharan S (1978) A finite strip method for the buckling of plate structures under arbitrary loading. Int J Mech Sci 20(10):685–693. https://doi.org/10.1016/0020-7403(78)90055-3

Vol′mir AS (1974) The nonlinear dynamics of plates and shells. Foreign technology division wright-Patterson AFB Ohio FTD-HC-23-851-74

Yang B, Wang D (2017) Buckling strength of rectangular plates with elastically restrained edges subjected to in-plane impact loading. Proc Inst Mech Eng Part C J Mech Eng Sci 231(20):3743–3752. https://doi.org/10.1177/0954406216652171

Yang Z, Liu A, Yang J, Fu J, Yang B (2020) Dynamic buckling of functionally graded graphene nanoplatelets reinforced composite shallow arches under a step central point load. J Sound Vib 465:115019. https://doi.org/10.1016/j.jsv.2019.115019

Yusof Z, Rasid ZA (2016) Numerical modelling of parametric instability problem for composite plate using finite element method. In: AIP conference proceedings, vol 1750, no 1, p 030044. AIP Publishing LLC. https://doi.org/10.1063/1.4954580

Parametric Instability Analysis of Functionally Graded CNT-Reinforced Composite (FG-CNTRC) Plate Subjected to Different Types of Non-uniform In-Plane Loading

Vishal Singh, Rajesh Kumar, and Shuvendu Narayan Patel

Abstract Carbon nanotube has attracted many researchers from last two decades due to its exceptional mechanical and multiuse properties. In this article, a semi-analytical study is performed to determine the dynamic instability of a Functionally Graded Carbon Nanotube Reinforced Composite (FG-CNTRC) plate exposed to uniform and various non-uniform in-plane loadings. The efficient mechanical properties for the plate are estimated using rule of mixture where CNTs are distributed aligned and distributed across the plates' thickness such as Uniformly distributed (UD) and Functionally Graded (FG-X and FG-O). Here, The FG-CNTRC plate is modeled by means of higher order shear deformation theory (HSDT) and the stress distributions (σ_{xx}, σ_{yy}, τ_{xy}) within the plate because of non-uniform loadings are calculated using Airy's stress method. Then, the Hamilton's principle is applied to obtain the governing partial differential equations of the FG-CNTRC plate, and which is later solved with the help of Galerkin's method to convert it to ordinary (Mathieu type) differential equations. Next, these Mathieu type equations are solved employing Bolotin's method to trace the instability boundaries corresponding to period 2T. At last, the consequence of different parameters like volume fraction of CNT, types of non-uniform loading, static load factor, types of CNTs distribution on instability of the FG-CNTRC plate are examined.

Keywords FG-CNTRC · Galerkin's method · Bolotin's method · Non-uniform loading

1 Introduction

Carbon nanotubes (CNT), because of its versatile nature in different applications has drawn the attention of many investigators after its discovery by scientist Iijima (1991) and which is due to its very effective thermal, electrical, and mechanical properties (Ciecierska et al. 2013). CNTs when mixed with polymer epoxy is found to be

V. Singh · R. Kumar (✉) · S. N. Patel
Department of Civil Engineering, Birla Institute of Technology and Science, Pilani 333031, India
e-mail: rajesh.kr@pilani.bits-pilani.ac.in

S. B. Singh et al. (eds.), *Emerging Trends of Advanced Composite Materials in Structural Applications*, Composites Science and Technology,
https://doi.org/10.1007/978-981-16-1688-4_13

increasing the strength and stiffness of the matrix (Gojny et al. 2004; Liew et al. 2015) and also increases the strength-weight and stiffness-weight ratios of the plate (Macías et al. 2017). Mehrabadi et al. (2012) and Arani et al. (2011) reported in their study that the increase in CNTs volume fraction within the matrix, increases the buckling load carrying capacity of plate. In this viewpoint, Kiani (2017) has investigated an FG-CNTRC plate loaded with parabolic loading for the analysis of buckling load, where the plate has been modeled using FSDT to approximate the kinematics of plate and solved using Ritz method and Airy stress function formulation. On similar field, Malekzadeh and Dehbozorgi (2016) works on FG-CNTRC skew plate to check its low velocity impact behavior employing first order shear deformation theory (FSDT) and solved via. finite element method (FEM) combining Newmark and Newton–Raphson methods.

It is very well understood from the above investigation that how addition of CNTs increase in strength and stiffness of the plate and which ultimately increases the buckling load carrying capacity of the plate. But estimating the buckling load and strength increment for composite plates is not only the parameters for the design of a composite plate, the effected due to dynamic loading and dynamic instability plays a important role in the design of aerospace related applications. Thus, the knowledge for dynamic stability of plate is required for proper design of composite plates. In this context CNT has some beneficial impact on the plate's dynamics stability. Rafiee et al. (2014) has investigated a piezoelectric FG-CNTRC plate with imperfection to study the nonlinear dynamic instability using FSDT and von-kármán geometric nonlinearity and solve employing Galenkin's method. Again, a temperature-dependent FG-CNTR visco-plate was analyzed for its dynamic instability behavior by Kolahchi et al. (2016) using FSDT and solved via. Generalized differential quadrature method (GDQM). In the same context, Thanh et al. (2017) has investigated nonlinear dynamic response of FG-CNTRC plate with temperature-dependent material properties using Reddy's higher order shear deformation theory (HSDT) and solved using Airy stress function, Galerkin method and fourth-order Runge–Kutta method. A dynamic instability analysis of CNTs reinforced sandwich plate under uniform in-plane loading was investigated by Sankar et al. (2016) after modeling the plate with both HSDT and FSDT, which was later solved using shear flexible QUAD-8 serendipity element. Lee (2018) performed the dynamic instability of the CNT reinforced fiber composite skew plate with delamination based on HSDT using FEM. In an investigation by Pathi and Vasudevan (2019), rotating CNT reinforced composite plate has been modeled using FSDT for the dynamic instability in the presence uniform in-plane periodic loading.

From the above literature survey, it can be predicted that study on dynamic instability of FG-CNTRC has been performed by many researchers but those were mainly oriented to uniform in-plane loading or thermal loading conditions but dynamic instability of FG-CNTRC exposed to non-uniform loading is not yet report as per the authors knowledge. The aim of this study is to investigate the dynamic instability region and behavior of UD-CNTRC and FG-CNTRC plate exposed to non-uniform in-plane loading where CNTs are aligned and distributed throughout the thickness of the plate. Mainly, the consequence of different parameters like volume fraction of CNT, types of non-uniform loading, types of CNTs distribution, static load factor on dynamic instability of the UD-CNTRC and FG-CNTRC plate (FG-O and FG-X) are examined and the results obtained from the current work will help in appropriate design of UD-CNTRC and FG-CNTRC plate against dynamic instability under non-uniform loading conditions.

2 Formulation

In the current study, a functionally graded carbon nanotube reinforced composite (FG-CNTRC) plate is semi-analytically analyzed for dynamic instability using the formulation, as stated in this section. The constituents used in each lamina are SWCNT (chiral indices $(n_0, m_0) = (10, 10)$) and polymer matrix (epoxy resin). The effective mechanical properties of the lamina (CNT embedded matrix) are obtained using rule of mixture technique as given in next subsection. Figure 1 shows the pictorial representation of the FG-CNTRC plate and CNTs distribution across the thickness of plate.

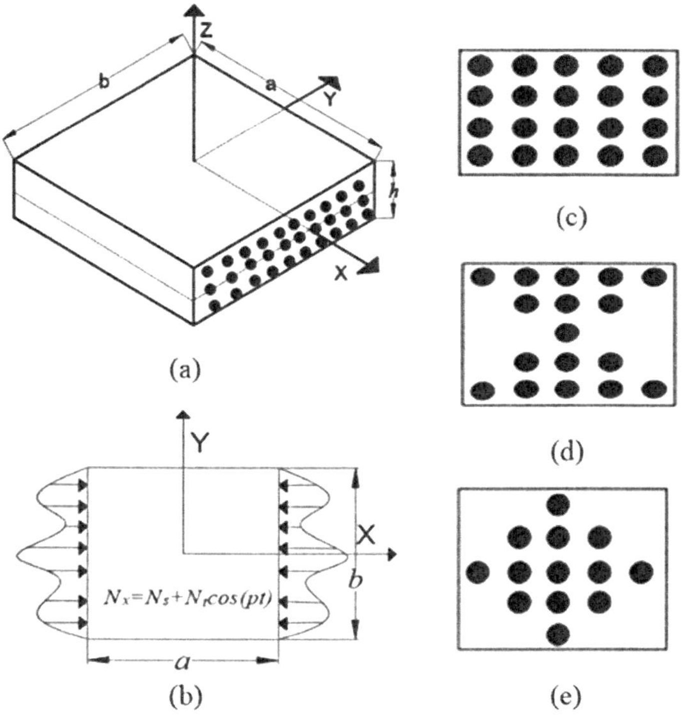

Fig. 1 Schematic view of **a** UD-CNTRC plate, **b** non-uniform loading **c** UD-CNTRC, **d** FG-X CNTRC and **e** FG-O CNTRC

Table 1 Various cases of CNTs distribution as a function thickness (Kiani 2017)

Distribution of CNTs in CNTRC plate	V_{CNT}
UD	$V_{\_cnt}$
FG-O	$2V_{\_cnt}\left(1-2\frac{\lvert z\rvert}{h}\right)$
FG-X	$4V_{\_cnt}\frac{\lvert z\rvert}{h}$

2.1 *The Rule of Mixture*

The effective mechanical properties of the FG-CNTRC plate are obtained using the rule of mixtures. However, the scale dependent properties of nanocomposite media are accounted using efficiency parameters which is mentioned in result and discussion section. According to the rule, the effective mechanical properties are estimated as:

$$E_{11} = \eta_1 V_{CNT} E_{11}^{CNT} + V_m E_m \tag{1}$$

$$\frac{\eta_2}{E_{22}} = \frac{V_{CNT}}{E_{22}^{CNT}} + \frac{V_m}{E_m} \tag{2}$$

$$\frac{\eta_3}{G_{12}} = \frac{V_{CNT}}{G_{12}^{CNT}} + \frac{V_m}{G_m} \tag{3}$$

$$V_{CNT} + V_m = 1 \tag{4}$$

$$\mu_{12} = V_{\_cnt}\mu_{11}^{CNT} + V_m \mu_m \tag{5}$$

In Eqs. (1)–(5), η_1, η_2, and η_3 are the efficiency parameters. The efficiency parameters are considered to equate the values acquired of shear modulus and Young's modulus of the current modified rule of mixtures with that of the results acquired according to the molecular dynamics simulations (Shen 2011). Besides, E_{11}^{CNT}, E_{22}^{CNT}, and G_{12}^{CNT} are the elastic moduli and shear modulus of SWCNTs, respectively. Moreover, E_m and G_m are the properties of isotropic matrix. Also, V_{CNT} and V_m denotes the volume fraction of CNTs and matrix, respectively (Kiani 2017) (Table 1).

The volume fraction of all these cases is equal to $V_{\_cnt}$ although the distribution of the CNTs are different in all cases.

2.2 *Kinematics of CNTRC Plate*

In the current study, the CNTRC plate is modeled considering HSDT developed by Reddy (1985). The displacement fields according to this theory, such that the transverse shear strains at both the top and bottom surfaces are zero, for rectangular plate is as stated below:

$$u = u^0 + z\varphi_x - \frac{4z^3}{3h^2}\left(\varphi_x + w^0_{,x}\right) \tag{6}$$

$$v = v^0 + z\varphi_y - \frac{4z^3}{3h^2}\left(\varphi_y + w^0_{,y}\right) \tag{7}$$

$$w = w^0 \tag{8}$$

In the above equations, u, v, and w indicate the displacements of a material point (x, y) which is at a distance 'z' away from the neutral surface of the plate in the three principal directions. Similarly, u^0, v^0 and w^0 denote the displacements of the point on the neutral surface. φ_x and φ_y represent the rotation of the cross-section perpendicular to the x-axis and y-axis, respectively. The suffix $()_{,x}$ and $()_{,y}$ symbolize the differentiation with respect to x and y respectively. To simplify Eqs. (6) and (7), $(\varphi_x + w^0_{,x})$ is represented by ϕ^0_x and $(\varphi_y + w^0_{,y})$ is represented by ϕ^0_y. On further simplifying these equations (Eqs. 6 and 7) as per Soldatos (1991),

$$u = u^0 - zw^0_{,x} + f(z)\phi^0_x \tag{9}$$

$$v = v^0 - zw^0_{,y} + f(z)\phi^0_y \tag{10}$$

$$w = w^0 \tag{11}$$

where $f(z) = z\left(1 - \frac{4z^2}{3h^2}\right)$. At 'z' distance ahead of the neutral surface of the plate, the strain–displacement equations can be written as:

$$\varepsilon_{xx} = \varepsilon^0_{xx} - zw^0_{,xx} + f(z)\phi^0_{x,x} \tag{12}$$

$$\varepsilon_{yy} = \varepsilon^0_{yy} - zw^0_{,yy} + f(z)\phi^0_{y,y} \tag{13}$$

$$\gamma_{xy} = \gamma^0_{xy} - 2zw^0_{,xy} + f(z)\phi^0_{x,y} + f(z)\phi^0_{y,x} \tag{14}$$

$$\gamma_{xz} = u_{,z} + w_{,x} = f'(z)\phi^0_x \tag{15}$$

$$\gamma_{yz} = v_{,z} + w_{,y} = f'(z)\phi^0_y \tag{16}$$

where, ε^0_{xx}, ε^0_{yy} and γ^0_{xy} are the strains at the neutral surface of the plate as defined in Eqs. (17)–(19).

$$\varepsilon_{xx}^{0} = u_{,x}^{0} \tag{17}$$

$$\varepsilon_{yy}^{0} = v_{,y}^{0} \tag{18}$$

$$\gamma_{xy}^{0} = u_{,x}^{0} + v_{,y}^{0} \tag{19}$$

According to constitutive law, the stress and strain of a lamina are related as:

$$\begin{Bmatrix} \sigma_{xx} \\ \sigma_{yy} \\ \tau_{xy} \\ \tau_{yz} \\ \tau_{xz} \end{Bmatrix} = \begin{bmatrix} Q_{11} & Q_{12} & 0 & 0 & 0 \\ Q_{12} & Q_{22} & 0 & 0 & 0 \\ 0 & 0 & Q_{66} & 0 & 0 \\ 0 & 0 & 0 & Q_{44} & 0 \\ 0 & 0 & 0 & 0 & Q_{55} \end{bmatrix} \begin{Bmatrix} \varepsilon_{xx} \\ \varepsilon_{yy} \\ \gamma_{xy} \\ \gamma_{yz} \\ \gamma_{xz} \end{Bmatrix} \tag{20}$$

where, Q_{ij} ($i, j = 1, 2, 6$) refers to the plane stress material stiffness constants which are expressed in terms of engineering constants of CNT embedded matrix as follows:

$$Q_{11} = Q_{22} = E_{hm}/\left(1 - \nu_{hm}^{2}\right), \quad Q_{12} = \nu_{hm} E_{hm}/\left(1 - \nu_{hm}^{2}\right),$$
$$Q_{66} = Q_{44} = Q_{55} = E_{hm}/2(1 + \nu_{hm}).$$

The force, moment, and additional moment resultants in curvature due to additional changes are related to strains as defined in Eqs. (21)–(24).

$$\begin{Bmatrix} N_{xx} \\ N_{yy} \\ N_{xy} \end{Bmatrix} = \begin{bmatrix} A_{11} & A_{12} & A_{16} \\ A_{12} & A_{22} & A_{26} \\ A_{16} & A_{26} & A_{66} \end{bmatrix} \begin{Bmatrix} \varepsilon_{xx}^{0} \\ \varepsilon_{yy}^{0} \\ \gamma_{xy}^{0} \end{Bmatrix} + \begin{bmatrix} B_{11} & B_{12} & B_{16} \\ B_{12} & B_{22} & B_{26} \\ B_{16} & B_{26} & B_{66} \end{bmatrix} \begin{Bmatrix} -w_{,xx}^{0} \\ -w_{,yy}^{0} \\ -2w_{,xy}^{0} \end{Bmatrix}$$
$$+ \begin{bmatrix} C_{11} & C_{12} & C_{16} \\ C_{12} & C_{22} & C_{26} \\ C_{16} & C_{26} & C_{66} \end{bmatrix} \begin{Bmatrix} \phi_{x,x}^{0} \\ \phi_{y,y}^{0} \\ \phi_{x,y}^{0} + \phi_{y,x}^{0} \end{Bmatrix} \tag{21}$$

$$\begin{Bmatrix} M_{xx} \\ M_{yy} \\ M_{xy} \end{Bmatrix} = \begin{bmatrix} B_{11} & B_{12} & B_{16} \\ B_{12} & B_{22} & B_{26} \\ B_{16} & B_{26} & B_{66} \end{bmatrix} \begin{Bmatrix} \varepsilon_{xx}^{0} \\ \varepsilon_{yy}^{0} \\ \gamma_{xy}^{0} \end{Bmatrix} + \begin{bmatrix} D_{11} & D_{12} & D_{16} \\ D_{12} & D_{22} & D_{26} \\ D_{16} & D_{26} & D_{66} \end{bmatrix} \begin{Bmatrix} -w_{,xx}^{0} \\ -w_{,yy}^{0} \\ -2w_{,xy}^{0} \end{Bmatrix}$$
$$+ \begin{bmatrix} E_{11} & E_{12} & E_{16} \\ E_{12} & E_{22} & E_{26} \\ E_{16} & E_{26} & E_{66} \end{bmatrix} \begin{Bmatrix} \phi_{x,x}^{0} \\ \phi_{y,y}^{0} \\ \phi_{x,y}^{0} + \phi_{y,x}^{0} \end{Bmatrix} \tag{22}$$

$$\begin{Bmatrix} M^a_{xx} \\ M^a_{yy} \\ M^a_{xy} \end{Bmatrix} = \begin{bmatrix} C_{11} & C_{12} & C_{16} \\ C_{12} & C_{22} & C_{26} \\ C_{16} & C_{26} & C_{66} \end{bmatrix} \begin{Bmatrix} \varepsilon^0_{xx} \\ \varepsilon^0_{yy} \\ \gamma^0_{xy} \end{Bmatrix} + \begin{bmatrix} E_{11} & E_{12} & E_{16} \\ E_{12} & E_{22} & E_{26} \\ E_{16} & E_{26} & E_{66} \end{bmatrix} \begin{Bmatrix} -w^0_{,xx} \\ -w^0_{,yy} \\ -2w^0_{,xy} \end{Bmatrix}$$

$$+ \begin{bmatrix} F_{11} & F_{12} & F_{16} \\ F_{12} & F_{22} & F_{26} \\ F_{16} & F_{26} & F_{66} \end{bmatrix} \begin{Bmatrix} \phi^0_{x,x} \\ \phi^0_{y,y} \\ \phi^0_{x,y} + \phi^0_{y,x} \end{Bmatrix} \tag{23}$$

$$\begin{Bmatrix} Q^a_{yz} \\ Q^a_{xz} \end{Bmatrix} = \begin{bmatrix} H_{44} & H_{45} \\ H_{45} & H_{55} \end{bmatrix} \begin{Bmatrix} \phi^0_y \\ \phi^0_x \end{Bmatrix} \tag{24}$$

where, N_{xx}, N_{yy} and N_{xy} are the force resultants; M_{xx}, M_{yy} and M_{xy} are the moment resultants; M^a_{xx}, M^a_{yy}, and M^a_{xy} are the additional moment resultants in curvature due to additional change, and Q^a_{yz} and Q^a_{xz} are the transverse shear force resultants. The additional change of curvature is denoted by $\phi^0_{x,x}$, $\phi^0_{y,y}$ and $\phi^0_{x,y} + \phi^0_{y,x}$. The overall CNTRC plate stiffness constants; A_{ij}, B_{ij}, C_{ij}, D_{ij}, E_{ij}, F_{ij}, and H_{ij} are expressed as stated in Eqs. (25)–(27).

$$\left(A_{ij}, B_{ij}, D_{ij}\right) = \int_{-h/2}^{h/2} Q_{ij}\left(1, z, z^2\right)dz \quad (i, j) = (1, 2, 6) \tag{25}$$

$$\left(C_{ij}, E_{ij}, F_{ij}\right) = \int_{-h/2}^{h/2} Q_{ij}(1, z, f(z))f(z)dz \quad (i, j) = (1, 2, 6) \tag{26}$$

$$\left(H_{ij}\right) = \int_{-h/2}^{h/2} Q_{ij} f'(z) f'(z)dz \quad (i, j) = (4, 5) \tag{27}$$

2.3 In-Plane Elasticity Problem

It is well known that due to uniform in-plane loading at the edge of the plate, the developed pre-buckling stress distribution is also uniform and uni-axial and matches with the applied loading. However, when the non-uniform loading is applied, these three stresses components (σ_{ij}, $(i, j = x, y)$) are developed within the plate. This needs to be evaluated for estimating the internal stress resultants ($(n_{ij}$, $(i, j = x, y))$ due to various non-uniform loadings which are applied in-plane at the edge of the plate, to develop the governing equation of motions of the FG-CNTRC plate. The explicit analytical expressions for the pre-buckling stresses (σ_{ij}, $(i, j = x, y)$) within the FG-CNTRC plate under non-uniform in-plane mechanical loadings are developed by solving in-plane elasticity problem using Airy's approach. Furthermore, equilibrium equation for in-plane stress in terms of Airy's stress function (ϕ) for FG-CNTRC plate is estimated using strain-compatibility conditions and is given as,

$$a_{22}\frac{\partial^4\phi}{\partial x^4} + (2a_{12} + a_{66})\frac{\partial^4\phi}{\partial x^2\partial y^2} + a_{11}\frac{\partial^4\phi}{\partial y^4} = 0 \tag{28}$$

and Airy's stress function (ϕ) is described by

$$\eta_{xx} = \frac{\partial^2\phi}{\partial y^2}, \quad \eta_{yy} = \frac{\partial^2\phi}{\partial x^2}, \quad \eta_{xy} = -\frac{\partial^2\phi}{\partial x\partial y} \tag{29}$$

where, $(A)^{-1}$ is the flexibility matrix of the FG-CNTRC plate.

$$(A)^{-1} = \begin{bmatrix} a_{11} & a_{12} & a_{16} \\ a_{12} & a_{22} & a_{26} \\ a_{16} & a_{26} & a_{66} \end{bmatrix} \tag{30}$$

Here, $A = A_{ij}$ (i, j = 1, 2, 6) is the extensional stiffness of the CNTRC plate and obtained using Eq. (25)

Now, Airy's stress function is assumed in the form of series as,

$$\phi(x, y) = \sum_{i=1}^{\infty} r_i(y)\cos(\alpha_i x) + \sum_{j=1}^{\infty} s_j(x)\cos\left(\beta_j y\right) + R_0 y^2 \tag{31}$$

where, $\alpha_i = 2i\pi/a$, $\beta_j = 2j\pi/b$, $r_i(y)$ and $s_j(x)$ are unknown functions in y and x, respectively. Substituting the above expression in the in-plane stress equilibrium Eq. (28) and then the coefficients of $\cos(\alpha_i x)$ and $\cos\left(\beta_j y\right)$ are equated which gives out the results in two ordinary differential equations in $r_i(y)$ and $s_j(x)$ respectively,

$$a_{11}\frac{\partial^4 r_i(y)}{\partial y^4} - (2a_{12} + a_{66})\alpha_i^2\frac{\partial^2 r_i(y)}{\partial y^2} + a_{22}\alpha_i^4 r_i(y) = 0 \tag{32a}$$

$$a_{22}\frac{\partial^4 s_j(x)}{\partial x^4} - (2a_{12} + a_{66})\beta_j^2\frac{\partial^2 s_j(x)}{\partial x^2} + a_{11}\beta_j^4 s_j(x) = 0 \tag{32b}$$

Substituting $r_i(y) = \exp(\overline{\lambda_2} y)$ and $s_j(x) = \exp(\overline{\lambda_1} x)$ in the above equations, roots of the above equation are $\overline{\lambda_2} = \pm\alpha_{i1}, \pm\alpha_{i2}$ and $\overline{\lambda_1} = \pm\beta_{j1}, \pm\beta_{j2}$.

Where, $\alpha_{i1}, \alpha_{i2} = \alpha_i\sqrt{\frac{(2a_{12}+a_{66})\pm\sqrt{(2a_{12}+a_{66})^2-4a_{11}a_{22}}}{a_{11}}}$ and $\beta_{j1}, \beta_{j2} = \beta_j\sqrt{\frac{(2a_{12}+a_{66})\pm\sqrt{(2a_{12}+a_{66})^2-4a_{11}a_{22}}}{a_{22}}}$. Since the functions $r_i(y)$ and $s_j(x)$ are symmetric about y and x axes respectively, we can write

$$r_i(y) = R_{i1}\cos h(\alpha_{i1}y) + R_{i2}\cos h(\alpha_{i2}y) \tag{33}$$

$$s_j(x) = S_{j1}\cos h(\beta_{j1}x) + S_{j2}\cos h(\beta_{j2}x) \tag{34}$$

Substituting the expressions for $r_i(y)$ and $s_j(x)$ in Eq. (31), the expression for Airy's stress function is written as,

$$\phi(x, y) = \sum_{i=1}^{\infty}\{R_{i1}\cos h(\alpha_{i1}y) + R_{i2}\cos h(\alpha_{i2}y)\}\cos(\alpha_i x) + \sum_{j=1}^{\infty}\{S_{j1}\cos h(\beta_{j1}x) + S_{j2}\cos h(\beta_{j2}x)\}\cos(\beta_j y) + R_0 y^2 \tag{35}$$

The in-plane stress resultants are determined by substituting the stress function (Eq. 35) in Eq. (29). Thus,

$$\eta_{xx} = \sum_{i=1}^{\infty}\cos(\alpha_i x)\left(R_{i1}\cos h(\alpha_{i1}y)\alpha_{i1}^2 + R_{i2}\cos h(\alpha_{i2}y)\alpha_{i2}^2\right) - \sum_{j=1}^{\infty}\cos(\beta_j y)\left(S_{j1}\cos h(\beta_{j1}x) + S_{j2}\cos h(\beta_{j2}x)\right)\beta_j^2 + 2R_0 \tag{36}$$

$$\eta_{yy} = -\sum_{i=1}^{\infty}\cos(\alpha_i x)(R_{i1}\cos h(\alpha_{i1}y) + R_{i2}\cos h(\alpha_{i2}y))\alpha_i^2 + \sum_{j=1}^{\infty}\cos(\beta_j y)\left(S_{j1}\cos h(\beta_{j1}x)\beta_{j1}^2 + S_{j2}\cos h(\beta_{j2}x)\beta_{j2}^2\right) \tag{37}$$

$$\eta_{xy} = \sum_{i=1}^{\infty}\sin(\alpha_i x)\alpha_i(R_{i1}\sin h(\alpha_{i1}y)\alpha_{i1} + R_{i2}\sin h(\alpha_{i2}y))\alpha_{i2} + \sum_{j=1}^{\infty}\sin(\beta_j y)\beta_j\left(S_{i1}\sin h(\beta_{j1}x)\beta_{j1} + S_{i2}\cos h(\beta_{j2}x)\beta_{j2}\right) \tag{38}$$

The coefficients, $R_{i1}, R_{i2}, S_{j1}, S_{j2}$ in expressions $\eta_{xx}(x, y)$, $\eta_{yy}(x, y)$ and $\eta_{xy}(x, y)$ are calculated using in-plane stress boundary conditions, which are written as

$$\eta_{xx}\left(\pm\frac{a}{2}, y\right) = R(y),\quad \eta_{xy}\left(\pm\frac{a}{2}, y\right) = 0,\quad \eta_{xy}\left(x, \pm\frac{b}{2}\right) = 0,\quad \eta_{yy}\left(x, \pm\frac{b}{2}\right) = 0 \tag{39}$$

where, $R(y)$ represents different types of in-plane non-uniform mechanical edge load distributions. Satisfying the in-plane stress boundary conditions which results in following set of simultaneous equations in the form of unknown coefficients,

$$R_{i1} = -\left(\frac{\alpha_{i2}}{\alpha_{i1}}\right)\frac{\sin h\frac{\alpha_{i2}b}{2}}{\sin h\frac{\alpha_{i1}b}{2}}R_{i1} \tag{40}$$

$$S_{j1} = -\left(\frac{\beta_{j2}}{\beta_{j1}}\right)\frac{\sin h\frac{\beta_{j2}a}{2}}{\sin h\frac{\beta_{j1}a}{2}}S_{j2} \tag{41}$$

$$\begin{aligned}&\alpha_i^2\left(\cos h\frac{\alpha_{i2}b}{2} - \frac{\alpha_{i2}}{a_{i1}}\cot h\frac{\alpha_{i1}b}{2}\sin h\frac{\alpha_{i2}b}{2}\right)R_{i2}\\ &= -\left(\frac{2}{a}\right)\sum_{j=1}^{\infty}\frac{\beta_{j2}\cos\frac{\beta_j b}{2}}{\sin h\frac{\beta_{j1}a}{2}}\left(\beta_{j1}\sin h\frac{\beta_{j2}a}{2}I_1 - \beta_{j2}\sin h\frac{\beta_{j1}a}{2}I_2\right)S_{j2}\end{aligned} \tag{42}$$

$$\begin{aligned}&\beta_j^2\left(\cos h\frac{\beta_{j2}a}{2} - \frac{\beta_{j2}}{\beta_{j1}}\cot h\frac{\beta_{j1}a}{2}\sin h\frac{\beta_{j2}a}{2}\right)S_{j2} + \left(\frac{2}{b}\right)I_0\\ &= -\left(\frac{2}{b}\right)\sum_{i=1}^{\infty}\frac{\alpha_{i2}\cos\frac{\alpha_i a}{2}}{\sinh\frac{\alpha_{i1}b}{2}}\left(\alpha_{i1}\sin h\frac{\alpha_{i2}b}{2}I_3 - \alpha_{i2}\sin h\frac{\alpha_{i1}b}{2}I_4\right)R_{i2}\end{aligned} \tag{43}$$

$$R_0 = \frac{1}{b}\int_0^{b/2} R(y)dy \tag{44}$$

Here,

$$I_0 = 2\int_0^{b/2} R(y)\cos(\beta_j y)dy, \quad I_1 = 2\int_0^{a/2}\cos h(\beta_{j1}x)\cos(\alpha_i x)dx,$$

$$I_2 = 2\int_0^{a/2}\cos h(\beta_{j2}x)\cos(\alpha_i x)dx, \quad I_3 = 2\int_0^{b/2}\cos h(\alpha_{j1}y)\cos(\beta_i y)dy,$$

$$I_4 = 2\int_0^{b/2}\cos h(\alpha_{j2}y)\cos(\beta_j y)dy.$$

2.4 Governing Equations

Hamilton's principle Eq. (45) is employed to get the equations of motion for the FG-CNTRC plate in terms of forces, moments, additional moments and shear resultants,

$$\delta^{(1)}\left(\int_{t_0}^{t1}(U - W - T)\right) = 0 \tag{45}$$

In the above equation, U implies strain energy, W is the external work done by the applied loads and T represents plate kinetic energy in the time interval t_0 to t_1 whereas $\delta^{(1)}$ denotes the first variation. The partial differential equations of the CNTRC plate exposed to non-uniform in-plane compressive loading (time-dependent) as follows:

$$\hat{N}_{xx,x} + \hat{N}_{xy,y} = \rho_g u^0_{,tt} \tag{46}$$

$$\hat{N}_{xy,x} + \hat{N}_{yy,y} = \rho_g v^0_{,tt} \tag{47}$$

$$M_{xx,xx} + 2M_{xy,xy} + M_{yy,yy} + \left(\hat{N}_{xx}w_{,x} + \hat{N}_{xy}w_{,y}\right)_{,x} + \left(\hat{N}_{xy}w_{,x} + \hat{N}_{yy}w_{,y}\right)_{,y} = \rho_g w^0_{,tt} \tag{48}$$

$$M^a_{xx,x} + M^a_{xy,y} - Q^a_{xz} = \rho_h \phi^0_{x,tt} \tag{49}$$

$$M^a_{xy,x} + M^a_{yy,y} - Q^a_{yz} = \rho_h \phi^0_{y,tt} \tag{50}$$

In the above equations, $\rho_g = \int_{-h/2}^{h/2} \rho_{hm} dz$, $\rho_h = \int_{-h/2}^{h/2} \rho_{hm} z^2 dz$ and $\hat{N}_{ij} = \left[N_{ij} - n_{ij}\right]$, where $i, j = (x, y)$ and n_{ij} are the internal stress resultants due to applied non-uniform in-plane loading, and N_{ij} are the stress resultants due to the large deformation. Therefore, $\hat{N}_{ij}$ are the net stress resultants within the FG-CNTRC plate.

2.5 Galerkin's Method

The Galerkin's method is employed to minimize the error by orthogonalizing it with respect to a set of assumed basis shape function satisfying the prescribed boundary conditions. This method helps in reducing the governing partial differential equations into the set of ordinary differential equations. The displacement fields that satisfy the boundary conditions are expressed as,

$$u^0 = \sum_{m=1}^{M^*} \sum_{n=1}^{N^*} U_{mn}(t)\Theta^1_{mn}(x, y) \tag{51}$$

$$v^0 = \sum_{m=1}^{M^*} \sum_{n=1}^{N^*} V_{mn}(t)\Theta^2_{mn}(x, y) \tag{52}$$

$$w^0 = \sum_{m=1}^{M^*} \sum_{n=1}^{N^*} W_{mn}(t)\Theta^3_{mn}(x, y) \tag{53}$$

$$\phi^0_x = \sum_{m=1}^{M^*} \sum_{n=1}^{N^*} K_{mn}(t)\Theta^4_{mn}(x, y) \tag{54}$$

$$\phi^0_y = \sum_{m=1}^{M^*} \sum_{n=1}^{N^*} L_{mn}(t)\Theta^5_{mn}(x, y) \tag{55}$$

where, $U_{mn}(t)$, $V_{mn}(t)$, $W_{mn}(t)$, $K_{mn}(t)$ and $L_{mn}(t)$ are undetermined coefficients. $\Theta^1_{mn}(x, y)$, $\Theta^2_{mn}(x, y)$, $\Theta^3_{mn}(x, y)$, $\Theta^4_{mn}(x, y)$ and $\Theta^5_{mn}(x, y)$ are assumed basis functions satisfying the boundary conditions of the given problem. m and n are the numbers of modes considered in the approximate displacement fields $\left(u^0, v^0, w^o, \phi^0_x \text{ and } \phi^0_y\right)$ along x and y directions respectively. Here, the total number of terms is $5 \times M^* \times N^*$. Where, M^* and N^* are decided based on the converged solution. The simply supported boundary conditions along all the edges are considered in which only normal in-plane displacement is allowed and in-plane tangential displacements and out of plane displacements are restricted. This may be written as,

$$n_{xx} - N_{xx} = -\hat{N}_{xx}, \quad M^a_{xx} = M_{xx} = v^0 = w^0 = \phi^0_y = 0 \quad \text{at } x = -a/2, a/2$$

and

$$n_{yy} - N_{yy} = -\hat{N}_{yy}, \quad M^a_{yy} = M_{yy} = u^0 = w^0 = \phi^0_x = 0 \quad \text{at } y = -b/2, b/2.$$

The trigonometric basis functions, which satisfy the above boundary conditions at all, the edges of plate can be expressed as,

$$\Theta^1_{mn}(x, y) = \sin\left(\frac{m\pi x}{a}\right)\cos\left(\frac{n\pi y}{b}\right) \tag{56}$$

$$\Theta^2_{mn}(x, y) = \cos\left(\frac{m\pi x}{a}\right)\sin\left(\frac{n\pi y}{b}\right) \tag{57}$$

$$\Theta^3_{mn}(x, y) = \cos\left(\frac{m\pi x}{a}\right)\cos\left(\frac{n\pi y}{b}\right) \tag{58}$$

$$\Theta^4_{mn}(x, y) = \sin\left(\frac{m\pi x}{a}\right)\cos\left(\frac{n\pi y}{b}\right) \tag{59}$$

$$\Theta^5_{mn}(x, y) = \cos\left(\frac{m\pi x}{a}\right)\sin\left(\frac{n\pi y}{b}\right) \tag{60}$$

Galerkin's method implies that, $\iint_A L_i\left(u^o, v^o, w^o, \phi^0_x, \phi^0_y\right)\Theta^i_{mn}(x, y)_j dxdy = 0$ for $i = 1, 2, 3, 4, 5$ and $j = 1, 2, \ldots M^* \times N^*$ where $L_i$is the non-linear partial differential equations and A is the total area of the plate.

2.6 Dynamic Instability

The applied time-dependent non-uniform in-plane load is considered to be of the form ($N_x = N_s + N_t\cos(pt)$), where N_s is the static load component, N_t is the dynamic load component and 'p' denotes the excitation frequency. The dynamic instability behavior of the plate is explained by the following ordinary differential equation (i.e., Mathieu-Hill equation):

$$[M]\{\ddot{\delta}\} + [K - (N_S + N_T\cos(pt))[K_G]]\{\delta\} = \{0\} \tag{61}$$

According to Eq. (61), $[M]$ stands for mass matrix, $[K]$ stands for stiffness matrix and $[K_G]$ stands for the geometric stiffness matrix of the plate. Bolotin's method (Bolotin 1964) is employed to trace the boundaries of instability regions. In the above equations, dynamic and static loading component are varied as $N_T = \eta N_{cr}$ and $N_S = \mu N_{cr}$ such that $(\mu + \eta) \leq 1$, where N_{cr} is the buckling load. The Eq. (61) has periodic solutions on the boundaries, with period $2T$. These solutions are assumed in terms of Fourier series as shown in Eqs. (62) and (63), where a_k and b_k are some arbitrary constants. In the graph of dimensionless excitation frequency (Ω) vs dynamic load factor (η), the region between curves with period $2T$ is known as the principal dynamic instability region. The region between curves of a different time period, is the region of stability.

$$\delta(t) = b_0 + \sum_{k=2,4,6}^{\infty}\left(a_k \sin\frac{kpt}{2} + b_k \cos\frac{kpt}{2}\right) \tag{62}$$

$$\delta(t) = \sum_{k=1,3,5}^{\infty}\left(a_k \sin\frac{kpt}{2} + b_k \cos\frac{kpt}{2}\right) \tag{63}$$

On substituting the above solutions of $\delta(t)$ in Eq. (61) and equating coefficients of identical sine and cosine terms, we get a homogeneous algebraic equation in terms of arbitrary constants a_k and b_k. For the solution to be non-trivial, the determinant of coefficients of a_k and $b_k$must be equal to zero. The equations thus obtained give the boundaries of instability. Equation (64) gives the upper and lower boundaries of the first-order approximation of principal dynamic instability region (period $2T$), while Eq. (65) gives a corrected principal instability region considering second-order approximation.

$$\left| K^* \pm 0.5\beta N_{cr} K_G - 0.25 M p_1^2 \right| = 0 \tag{64}$$

$$\begin{bmatrix} K^* \pm 0.5\beta N_{cr} K_G & -0.5\beta N_{cr} K_G \\ -\beta N_{cr} K_G & K^{\cdot} - 2.25 M p_1^2 \end{bmatrix} - p_2^2 \begin{bmatrix} 0.25M & 0 \\ 0 & 0 \end{bmatrix} = 0 \tag{65}$$

where $K^{*}= [K_L]$-$N_S[K_G]$.

3 Result and Discussion

Following material properties is consider for the present study as per Kiani (2017). The Young's modulus (E_m) is 2.5 GPa, Mass density (ρ_m) is 1150 kg/m^3 and Poisson's ratio (μ_m) is 0.34 for matrix and for armchair single-walled carbon nanotube (SWCNT) with chiral indices ($n_0 = m_0 = 10$) having T(k) = 300, $E_{11}^{CNT} = 5.6466\,\text{Tpa}$, $E_{22}^{CNT} = 7.08\,\text{Tpa}$, $G_{12}^{CNT} = 1.9445\,\text{Tpa}$, $\mu_{12}^{CNT} = 0.175$ $\rho_{\text{CNT}} = 1400$ kg/m^3. The efficiency parameters for three different CNTs volume fraction are: $\eta_1 = 0.137$ and $\eta_2 = 1.022$ for $V_{\_cnt} = 0.12$, $\eta_1 = 0.142$ and $\eta_2 = 1.626$ for $V_{\_cnt} = 0.17$, and $\eta_1 = 0.141$ and $\eta_2 = 1.585$ for $V_{\_cnt} = 0.28$. Here, the efficiency parameter η_3 is considered equal to 0.7 η_2. The shear modulus $G_{13} = G_{12}$, whereas G_{23} is 1.2 times of G_{12}. The properties mentioned here will be used in the study further unless mentioned separately. In this section, various non-uniform in-plane edge loadings such as parabolic, concentrated, partial edge loading are considered along with uniform loading for evaluating the buckling load and boundaries of instability of the FG-CNTRC plate. The various non-uniform in-plane loading functions are given below,

Partial edge loading function is expressed as,

$$N_s = N_t = \bar{N}_0 \frac{b}{d} \left(\frac{d}{b} + \sum_{r=1}^{\infty} \frac{2}{\pi} \frac{1}{r} \sin \frac{r\pi d}{b} \cos \frac{2r\pi y}{b} \right) \tag{66}$$

Here, partial edge loading at the edge of the plate is modeled using a single Fourier series along the y-direction. In which 50 terms (i.e., $r = 1$ to 50) in Fourier series are considered for the converged pre-buckling stresses (σ_{ij},($i, j = x. y$)) within the CNTRC plate.

Parabolic loading function is expressed as,

$$N_s = N_t = \frac{3}{2} \bar{N}_0 \left(1 - 4 \frac{y^2}{b^2} \right) \tag{67}$$

Concentrated loading function is expressed as,

$$N_x = N_t = \frac{\bar{N}_0}{c\sqrt{\pi}} exp\left(-\frac{y^2}{c^2}\right) \tag{68}$$

In this above expression, $c = 1/24$ is chosen based on converged pre-buckling stresses (σ_{ij}, $(i, j = x, y)$) within the CNTRC plate. Here, the loading function for all the different types of loadings is considered in such a way that total load (i.e., area due to the loading distribution at the edge of plate) is same.

3.1 Validation Study

To validate the accuracy and effectiveness of the current semi-analytical model, the obtained results of validation study from the present semi-analytical model along with the published one has been carried out and presented in Table 2 for the buckling load parameter $\left(k_{cr} = \frac{\lambda_{cr} b^2}{\pi^2 D};\ D = \frac{E_1 h^3}{12(1-\mu_{12}^2)}\right)$ of FG-CNTRC plates exposed to parabolic in-plane edge loading $N_x = \bar{N}_0\left(1 - \frac{4y^2}{b^2}\right)$ for three cases of CNT distribution patterns, four aspect ratios with simply supported boundary conditions. The length-to-thickness ratio is chosen as $b/h = 50$, and CNTs volume fraction of set to V_cnt = 0.17. The results are well matched with the published results of Kiani (2017).

Table 3 presents the dimensionless fundamental frequencies parameter ($\Omega_n = \omega_n a^2/h\sqrt{\rho_{ep}/E_{ep}}$) of FG-CNTRC plates under uniform edge loading for three different CNT distribution patterns, three side-to-thickness ratios along with simply supported (SSSS) boundary conditions. The length-to-thickness ratio is chosen as $b/h = 50$, and the volume fraction of CNT ($V_{\_cnt}$) is considered as 0.11, 0.14 and 0.17. The results are very well matched with the published results given by Zhu et al. (2012).

The first and second (corrected) order approximation of principal instability region of a SSSS composite plate ($a/b = 1$, $b/h = 25$, 0/90/90/0, $= V_{\_cnt}0$) having the mechanical properties as $E_1/E_2 = 40$, $E_2 = 6.595\ GPa$, $G_{12}/E_2 = 0.6$, $G_{23}/E_2 = 0.5$,

Table 2 Buckling load coefficient (k_{cr}) of FG-CNTRC plate for $V_{\_cnt} = 0.17$ with $b/h = 50$

Distribution	Method	a/b			
		0.8	1	1.5	2
UD	Kiani (2017)	85.8715	59.1758	32.8725	26.0732
	Present	86.4724	59.1761	32.7528	26.0783
FG-X	Kiani (2017)	122.1227	84.3354	45.5584	34.2909
	Present	123.2382	84.3704	45.3957	34.2520
FG-O	Kiani (2017)	47.4787	33.2042	20.3205	18.3286
	Present	48.0089	33.4528	20.4415	18.4381

Table 3 Comparison of dimensionless fundamental natural frequency ($\Omega_n = \omega_n a^2/h\sqrt{\rho_{ep}/E_{ep}}$) of a simply supported square plate ($a/b = 1$) with different distribution of CNTs, volume fraction of CNTs and edge-thickness ratios

$V_{\_cnt}$			0.11		0.14		0.17	
b/h	Distribution	Method	(1,1)	(1,2)	(1,1)	(1,2)	(1,1)	(1,2)
10	UD	Zhu et al. (2012)	13.532	17.7	14.306	18.362	16.815	22.063
		Present	13.59	17.85	14.391	18.552	16.882	22.257
	FG-X	Zhu et al. (2012)	14.616	18.646	15.368	19.385	18.278	23.541
		Present	14.711	18.803	15.425	19.51	18.21	22.553
	FG-O	Zhu et al. (2012)	11.55	16.265	12.338	17.003	14.282	20.091
		Present	11.395	16.301	12.188	16.902	14.168	20.224
20	UD	Zhu et al. (2012)	17.355	21.511	18.921	22.867	21.456	26.706
		Present	17.336	21.495	18.931	22.885	21.428	26.68
	FG-X	Zhu et al. (2012)	19.939	23.776	21.642	25.359	24.735	29.809
		Present	19.935	23.763	21.621	25.335	24.631	29.665
	FG-O	Zhu et al. (2012)	13.523	18.486	14.784	19.462	16.628	22.739
		Present	13.43	18.423	14.703	19.418	16.552	22.714
50	UD	Zhu et al. (2012)	19.233	23.408	21.354	25.295	23.697	28.987
		Present	19.159	23.286	21.322	25.205	23.613	28.829
	FG-X	Zhu et al. (2012)	22.91	26.66	25.555	29.192	22.416	33.434
		Present	22.904	26.635	25.499	29.065	22.287	33.187
	FG-O	Zhu et al. (2012)	14.302	19.373	15.801	20.563	17.544	23.783
		Present	14.252	19.279	15.772	20.494	17.494	23.697

$\nu_{12} = 0.25$, $G_{13} = G_{12}$, $\nu_{13} = \nu_{12}$ is compared with the principal instability region given by Moorthy et al. (1990) in Fig. 2. It can be seen that the region of instability is in proximity to those of the result given by authors.

3.2 *Influence of CNT Distribution*

The dimensionless buckling load coefficient (k_{cr}) and dimensionless fundamental natural frequency (Ω_n) of the SSSS UD-CNTRC and FG-CNTRC plate for (b/h =

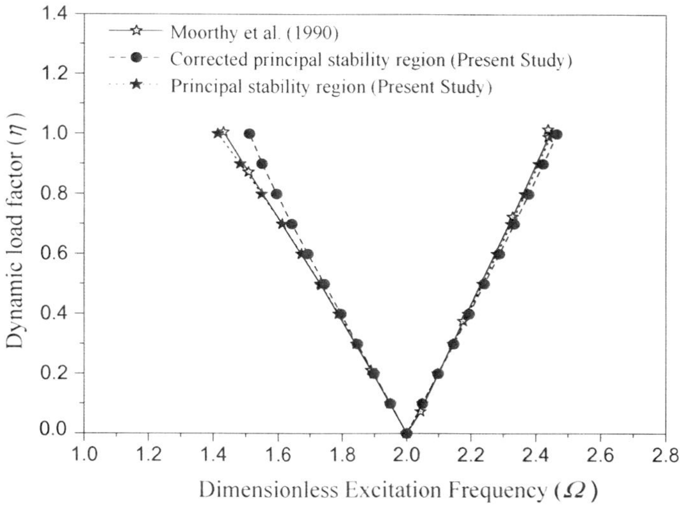

Fig. 2 Validation of principal instability region of SSSS composite plate (a/b = 1, b/h = 25, 0/90/90/0, $V_{\_cnt}$ = 0)

20) under uniform and different types of non-uniform loadings with varying volume fraction of CNT ($V_{\_cnt}$) is given in Table 4. It can be noted from the results that the dimensionless fundamental natural frequency (Ω_n) has no effect of the types of loading applied at the edge of the plate rather it is affected by the volume fraction of CNT and type of CNT distribution over the thickness of the plate. While dimensionless buckling load coefficient (k_{cr}) is affected by both the type of loading applied at the edges of the CNTRC plate and volume fraction of CNT ($V_{\_cnt}$). It is also observed that the value of k_{cr} decreases in sequences as uniform > parabolic > partial edge (d/b = 0.25) > concentrated loading but rises with the increase in CNTs volume fraction, this is because the increase in CNTs volume fraction increases the stiffness of the FG-CNTRC plate while concentration of loading at the edge of the plate, decreases its stiffness. Also, the K_{cr} value of FG-X distribution is higher in comparison to K_{cr} of other two distributions, which shows that FG-X shape distribution increases the stiffness of the plate.

As per Fig. 3, the dynamic instability region (DIR) of a SSSS UD-CNTRC and FG-CNTRC plate (a/b = 1, $V_{\_cnt}$ = 0.17, b/h = 20) under in-plane partial edge (d/b = 0.25) loading has been plotted for different CNT distributions with respect to the K_{cr} of FG-O. The origin of instability of FG-X distribution is having higher frequency compared to UD and FG-O type distribution, which shows that the FG-X has higher stiffness of plate than the other two. At the same time, the width of DIR at dynamic load factor $\lambda_d = 0.5$ in decreasing sequence is given as 8.17 $h\sqrt{(Eep/\rho ep)}$, 6.38$h\sqrt{(Eep/\rho ep)}$, 5.56 $h\sqrt{(Eep/\rho ep)}$ for FG-O, UD and FG-X respectively. This is because UD have uniformly distributed CNTs, FG-O has CNTs distribution as such that the top and the bottom layers have minimum and middle layer has maximum

Table 4 Dimensionless buckling load coefficient (k_{cr}) and dimensionless fundamental natural frequency (Ω_n) for different CNT distribution model of CNTRC plate (*b/h* = 20) subjected to uniform and different types of non-uniform loadings with varying volume fraction ($V_{\_cnt}$)

Distribution	$V_{\_cnt}$	K_{cr}/Ω_n	Uniform	Parabolic	Partial edge (d/b = 0.25)	Concentrated
UD	0.12	K_{cr}	28.103	21.716	15.049	11.834
		Ω_n	15.853	15.853	15.853	15.853
	0.17	K_{cr}	42.348	32.765	22.688	18.249
		Ω_n	19.359	19.359	19.359	19.359
	0.28	K_{cr}	61.765	47.658	33.066	25.178
		Ω_n	23.113	23.113	23.113	23.113
FG-X	0.12	K_{cr}	36.774	28.418	19.692	15.500
		Ω_n	18.133	18.133	18.133	18.133
	0.17	K_{cr}	55.812	43.188	29.904	24.100
		Ω_n	22.223	22.223	22.223	22.223
	0.28	K_{cr}	78.881	60.8847	42.230	32.412
		Ω_n	26.117	26.117	26.117	26.117
FG-O	0.12	K_{cr}	17.171	13.269	9.195	7.237
		Ω_n	12.393	12.393	12.393	12.393
	0.17	K_{cr}	25.650	19.848	13.743	11.076
		Ω_n	15.067	15.067	15.067	15.067
	0.28	K_{cr}	38.080	29.392	20.387	15.647
		Ω_n	18.150	18.150	18.150	18.150

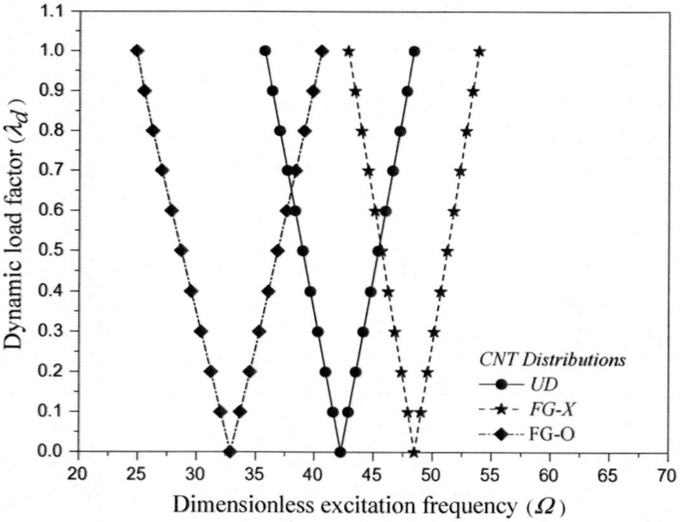

Fig. 3 Effect of CNTs distribution on principal instability zone of a SSSS UD and FG-CNTRC plate (*a/b* = 1, V_cnt = 0.17, *b/h* = 20) under in-plane partial edge (*d/b* = 0.25) loading

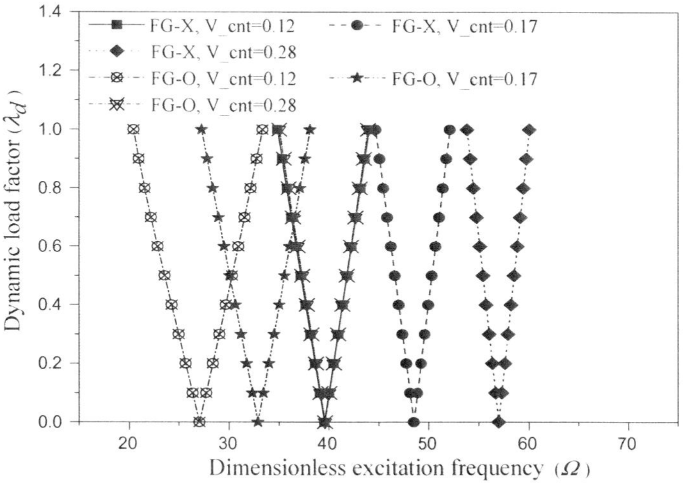

Fig. 4 Effect of type of CNTs distribution and volume fraction on principal instability zone of a SSSS FG-CNTRC plate ($a/b = 1$, $b/h = 20$) under in-plane partial edge ($d/b = 0.25$) loading

volume fraction CNTs, while, FG-X has CNTs distribution as such that the top and bottom layers have maximum and middle layer has minimum volume fraction CNTs.

In case of Fig. 4, dynamic instability region for both the types of CNTs distribution (FG-X and FG-O) are plotted with different volume fraction of CNT ($V_{\_cnt}$) for a SSSS FG-CNTRC plate ($a/b = 1$, $b/h = 20$) under in-plane partial edge (d/b = 0.25) loading with respect to the Kcr of FG-O with $V_{\_cnt} = 0.12$ and it is observed that the DIR of FG-X with $V_{\_cnt} = 0.12$ is overlapped to DIR of FG-O with $V_{\_cnt} = 0.28$, which signifies that the stiffness of CNTRC plate having FG-X with $V_{\_cnt} = 0.12$ is equal to CNTRC plate having FG-O with $V_{\_cnt} = 0.28$.

3.3 *Influence of Various Non-uniform Loadings on Dynamic Instability Region*

As per Fig. 5, it is observed that under the action of various non-uniform in-plane periodic loadings, the width of principal instability zone of a SSSS FG-X CNTRC plate ($a/b = 1$, $b/h = 20$, $V_{\_cnt} = 0.17$) with static load factor (λ_s) = 0 is minimum for uniform loading and maximum for concentrated loading. The width of the dynamic instability region (DIR) at dynamic load factor (λ_d) = 0.5 can be represented as $5.23h\surd(Eep/\rho ep)$, $6.75h\surd(Eep/\rho ep)$, $8.40h\surd Eep/\rho ep)$, $9.73h\surd(Eep/\rho ep)$ and $12.05h\surd(Eep/\rho ep)$ for uniform, parabolic, partial edge ($d = 0.5$), partial edge ($d = 0.25$) and concentrated loadings respectively with respect to the buckling load of concentrated loading.

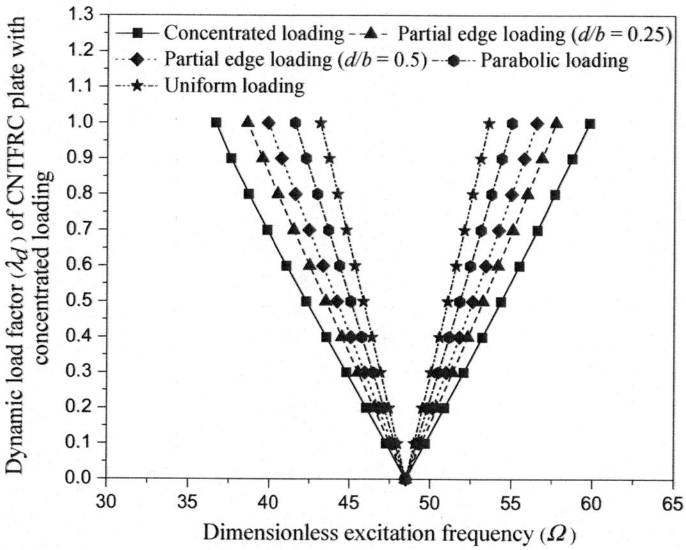

Fig. 5 Effect of different loading conditions on principal instability zone of a SSSS FG-X CNTRC Plate ($a/b = 1$, $b/h = 20$, $\lambda_s = 0$)

This reveals that with increase in concentration of loads at the edge of FG-X CNTRC plate, stiffness of the plate decreases which leads to the increase in the width of instability region. Moreover, the origin of stability region is same for all the cases of loading due to $\lambda_s = 0$. Again, from Fig. 6, it is observed that under the action of various non-uniform periodic loadings, a SSSS FG-X CNTRC plate ($a/b = 1$, $b/h = 20$, $V_{\_cnt} = 0.17$) with static load factor (λ_s) $= 0.4$ shows that the origin of instability region is different for every loading as compared to Fig. 5. The width of the dynamic instability region (DIR) at dynamic load factor (λ_d) $=$ 0.5 can be represented as $5.74h\surd(Eep/\rho ep)$, $7.65h\surd(Eep/\rho ep)$, $9.86h\surd Eep/\rho ep)$, $11.75h\surd(Eep/\rho ep)$ and $15.26h\surd(Eep/\rho ep)$ for uniform, parabolic, partial edge (d = 0.5), partial edge (d = 0.25) and concentrated loadings respectively with respect to the buckling load of concentrated loading.

4 Conclusion

In this article, authors have investigated the dynamic instability region of a simply supported (SSSS) FG-CNTRC plate under uniform and various types of non-uniform in-plane loadings. Here, the effect of different parameters like CNTs volume fraction, types of non-uniform loading, CNTs distribution types (UD, FG-X or FG-O), static load factor on dynamic instability of the CNTRC plate are examined. The remarks from the present investigation are summarized as,

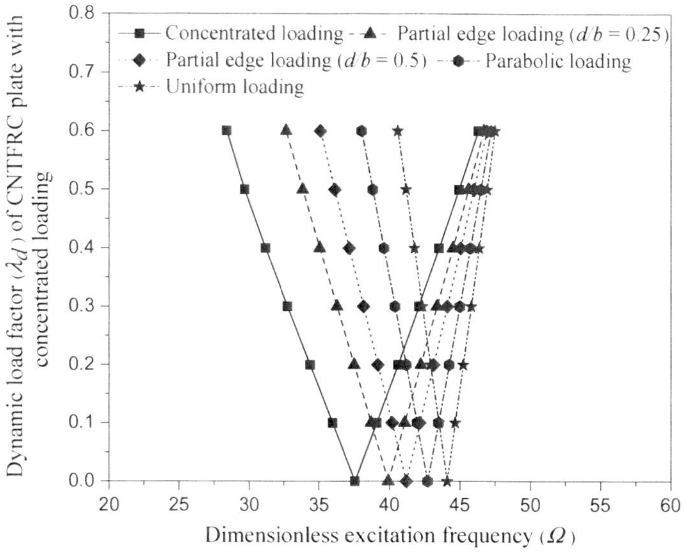

Fig. 6 Effect of different loading conditions on principal instability zone of a simply supported FG-X CNTRC Plate ($a/b = 1$, $b/h = 20$, $\lambda_s = 0.4$)

- With the increase in CNTs volume fraction for any type of CNTs distribution in the FG-CNTRC plate, the stiffness of the plate increases which results in reducing the width of instability region.
- It was observed that the instability region for FG-X with $V_{\_cnt} = 0.12$ is same as the instability region of FG-O with $V_{\_cnt} = 0.28$. Which indicates the affect of CNT distribution on composite plate and minimization in use of volume fraction of CNT.
- Out of various loading conditions, concentrated load has maximum width of instability region and uniform loading shows minimum width of instability.

References

Arani AG, Maghamikia S, Mohammadimehr M, Arefmanesh A (2011) Buckling analysis of laminated composite rectangular plates reinforced by SWCNTs using analytical and finite element methods. J Mech Sci Technol 25(3):809–820

Bolotin VV (1964) The dynamics stabililty of elastic system. Holden day, CA, San Francisco

Ciecierska E, Boczkowska A, Kurzydlowski KJ, Rosca ID, Hoa SV (2013) The effect of carbon nanotubes on epoxy matrix nanocomposites. J Therm Anal Calorim 111:1019–1024

Gojny FH, Wichmann MHG, Köpke FB, Schulte K (2004) Carbon nanotube-reinforced epoxy-composites: enhanced stiffness and fracture toughness at low nanotube content. Compos Sci Technol 64:2363–2371

Iijima S (1991) Helical microtubules of graphitic carbon. Nature 354:56–58

Kiani Y (2017) Buckling of FG-CNT-reinforced composite plates subjected to parabolic loading. Acta Mech 228:1303–1319

Kolahchi R, Safari M, Esmailpour M (2016) Dynamic stability analysis of temperature-dependent functionally graded CNT-reinforced visco-plates resting on orthotropic elastomeric medium. Compos Struct 150:255–265

Lee SY (2018) Dynamic instability assessment of carbon nanotube/fiber/polymer multiscale composite skew plates with delamination based on HSDT. Compos Struct 200:757–770

Liew KM, Lei ZX, Zhang LW (2015) Mechanical analysis of functionally graded carbon nanotube reinforced composites: A review. Compos Struct 120:90–97

Macías EG, Tembleque LR, Triguero RC, Saez A (2017) Eshelby-Mori-Tanaka approach for post-buckling analysis of axially compressed functionally graded CNT/polymer composite cylindrical panels. Compos B 128:208–224

Malekzadeh P, Dehbozorgi M (2016) Low velocity impact analysis of functionally graded carbon nanotubes reinforced composite skew plates. Compos Struct 140:728–748

Mehrabadi SJ, Aragh BS, Khoshkhahesh V, Taherpour A (2012) Mechanical buckling of nanocomposite rectangular plate reinforced by aligned and straight single-walled carbon nanotubes. Compos B Eng 43(4):2031–2040

Moorthy J, Reddy JN, Plaut RH (1990) Parametric instability of laminated composite plates with transverse shear deformation. Int J Solids Struct 26(7):801–811

Pathi JL, Vasudevan R (2019) Numerical investigation of dynamic instability of a rotating CNT reinforced composite plate. AIP Conf Proc 2057:1–10

Rafiee M, He XQ, Liew KM (2014) Non-linear dynamic stability of piezoelectric functionally graded carbon nanotube-reinforced composite plates with initial geometric imperfection. Int J Non-Linear Mech 59:37–51

Reddy JN, Liu CF (1985) A higher-order shear deformation theory of laminated elastic shells. Int J Eng Sci 23(3):319–330

Sankar A, Natarajan S, Ganapathi M (2016) Dynamic instability analysis of sandwich plates with CNT reinforced facesheets. Compos Struct 146:187–200

Shen HS (2011) Postbuckling of nanotube-reinforced composite cylindrical shells in thermal environments, part I: axially-loaded shells. Compos Struct 93:2096–2108

Soldatos KP (1991) A refined laminated plate and shell theory with applications. J Sound Vib 144(1):109–129

Thanh NV, Khoa ND, Tuan ND, Tran P, Duc ND (2017) Nonlinear dynamic response and vibration of functionally graded carbon nanotube-reinforced composite (FG-CNTRC) shear deformable plates with temperature-dependent material properties and surrounded on elastic foundations. J Therm Stresses 40:1257–1274

Zhu P, Lei ZX, Liew KM (2012) Static and free vibration analyses of carbon nanotube-reinforced composite plates using finite element method with first order shear deformation plate theory. Compos Struct 94:1450–1460

Recent Advancements in the Application of Natural Fiber Based Composites in Structural Engineering—A Review

A. S. Mehra and S. B. Singh

Abstract In the last decade, natural fibers have established themselves as a possible alternative to synthetic fibers in composites materials due to properties that range from low cost, low density, biodegradability, carbon positive nature, i.e., absorbing more carbon dioxide than produced, easy processing, etc. This chapter presents an overview of recent developments on the use of natural fiber reinforced polymer composites in structural engineering applications primarily focusing on their use as flexure elements, as flexure and shear strengthening medium for existing reinforced cement concrete elements, and the response of such materials to dynamic loadings. Also, the various mechanical properties of the natural fiber-reinforced composites as obtained from the material characterization of the composite during the above-mentioned investigations, and those reported in the various literatures are discussed.

Keywords Natural fiber composites · FRP · Shear strengthening · Flexure strengthening · Composite beams

1 Introduction

A structural composite material is a material system made of two or more macroscopic scale phases such that the mechanical properties of the resulting material are superior to those of the constituent materials when acting alone. In a fiber-reinforced composite material, one of the phase is that of the fibers and the other is of the resin. In addition, the matrix can also contain fillers, catalysts, pigments, UV stabilizers, fire retarders, etc. The fibers are stronger, stiffer, continuous, or discontinuous constituents and are called reinforcement while the resin is the weaker phase and

A. S. Mehra (✉) · S. B. Singh
Department of Civil Engineering, Birla Institute of Technology & Science, Pilani, Rajasthan 333031, India
e-mail: p20190440@pilani.bits-pilani.ac.in

S. B. Singh
e-mail: sbsingh@pilani.bits-pilani.ac.in

S. B. Singh et al. (eds.), *Emerging Trends of Advanced Composite Materials in Structural Applications*, Composites Science and Technology,
https://doi.org/10.1007/978-981-16-1688-4_14

called as matrix. The fiber reinforcements are the backbone of the composite material and govern its stiffness and strength in the direction of fibers. On the other hand, the matrix protects the sensitive fibers, holds the fibers in place, and transfer the stresses from the composite to the fibers and also between the fibers.

A two-phase composite material can be classified into three categories based on their constituents. Firstly, the particulate composites consisting of randomly shaped and sized particles dispersed in a matrix. Secondly, the discontinuous or short-fiber composites consisting of short fibers, nanotubes, or whiskers as the reinforcing phase. Thirdly, the continuous fiber composites, that are reinforced by long continuous fibers and are found to be the most efficient among the three from the structural strength and stiffness point of view. The continuous fibers can be all parallel in form of rovings (producing a unidirectional composite), can be perpendicular to one another in form of mats or fabrics (producing a bidirectional composite), or can be randomly oriented in form of chopped strand mat (producing a multidirectional composite) (Daniel and Ishai 2007).

In a unidirectional composite laminate (generally employed in structural engineering applications), the fibers are oriented only in one direction, i.e., along the longitudinal direction. The longitudinal tensile strength of such a composite is dominated by the reinforcing fibers and can be expressed as a function of the properties of fibers and the matrix. Whereas the other properties like the longitudinal compressive strength, the transverse tensile strength, the transverse compressive strength, and the in-plane shear strength are dominated by the matrix and are significantly lower than the longitudinal tensile strength. To date, no reliable theoretical formulations are available for relating the matrix-dominated properties as a function of the properties of fibers and the matrix (Bank 2006).

Fiber-reinforced plastics (FRP) have proven themselves to be valuable for use in the construction of new buildings and bridges and the upgrading of existing buildings and bridges for almost over 50 years. Fiber-reinforced polymer composites have various advantages unique to them over the conventional monolithic materials like high specific strength, high specific stiffness, low density, high fatigue resistance, non-corrosive, high wear resistance, thermal and acoustic insulation, magnetic resistance, etc. Besides, such materials are anisotropic and heterogeneous allowing the designer to tailor the properties as per the requirement and produce an optimized design configuration (Daniel and Ishai 2007).

Conventionally glass, carbon, aramid, or a hybrid combination of these are used as reinforcing fibers, and thermosetting epoxies, polyesters, and vinyl esters are used as the resin matrix. But since synthetic fibers and resin are non-renewable, non-biodegradable, their usage leads to serious environmental issues. As in recent years, sustainability and recyclability have become much more important concerns in the construction industry than structural efficiency, the abundant use of such environmentally unfriendly materials in the massive construction sector is being rendered as questionable. A possible solution to this problem could be found in the use of natural fibers produced from plants such as hemp, sisal, flax, and bamboo, etc., and from animal waste like feathers, wool, etc. as reinforcements, and organic resins like those made from soy oil, and polylactic acid (PLA) (PLA is a biodegradable thermoplastic polymer which is generally obtained through bacteria aided fermentation of corn (Morales et al. 2017).), etc., as the matrix, to produce the fiber-reinforced polymer composites called as bio-composites (Bank 2006).

Bio-composites are currently being employed in a number of experimental investigations, but no commercial utilization of them as structural elements is reported anywhere till date, also no significant design guidelines, codes, technical reports, or even manufacturer's manuals are available for them. This chapter compiles and summarizes the recent developments in the use of natural fiber reinforced polymer composites in structural engineering applications.

2 Natural Fibers and Their Characteristics

Composite materials that are made up of all-natural fibers in a synthetic or bio-polymer matrix are called as Bio-composites (Wool and Sun 2005). Examples of some of the natural fibers are flax, hemp, jute, banana, bamboo, sisal, etc. Figure 1 shows the classification of the natural fibers that are or can be employed for reinforcing the polymer matrix of the composite laminates.

Natural fibers have emerged as possible replacements for synthetic fibers conventionally employed as reinforcements in FRP composites due to their high strength to weight ratio, low carbon footprint, non-corrosive nature, combustible, nontoxic, and lower cost of processing (Susheel et al. 2009; Rajesh and Pitchaimani 2016). Natural fibers are renewable, produces fewer pollutants and greenhouse gases during manufacturing, requires less labor and energy for processing, and are bio-degradable (Cevallos et al. 2015; Cevallos and Olivito 2015; Yan and Chouw 2013).

With properties such as low-density, natural fibers can be used for producing composites that are lightweight and have high specific properties (Mohanty et al. 2001b). Table 1 gives the various properties of the natural fibers as reported in the various literature. Studies have shown that natural fibers have the potential to

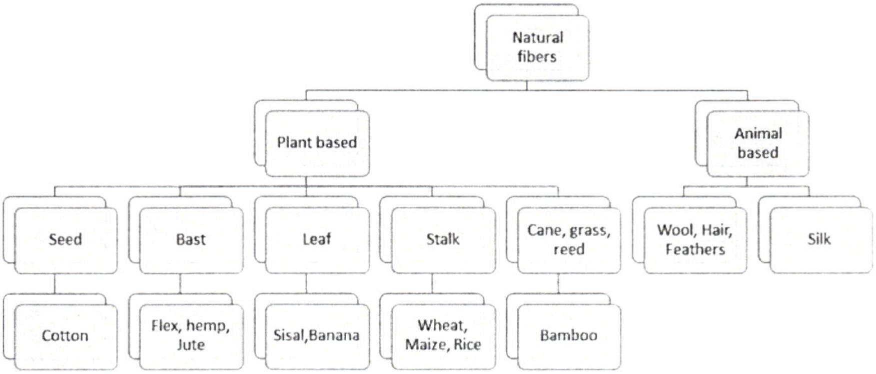

Fig. 1 Classification of natural fibers (Mohanty et al. 2001b; John and Uomas 2008)

Table 1 Typical properties of natural and synthetic fibers (Bank 2006; Daniel and Ishai 2007; John and Uomas 2008; Mallick 2007; Mukhopadhyay 2004; Mohanty et al. 2001b; Singh 2014; Suhad et al. 2015; Tong et al. 2017; Yan et al. 2014b)

Fibers	Density (g/cm^3)	Tensile strength (N/mm^2)	Elastic Modulus (kN/mm^2)	Elongation at failure (%)	Moisture absorption (%)
E-glass	2.50–2.58	1000–3500	70.0–76.0	0.5–4.8	–
Aramid	1.40–1.44	3000–3450	63.0–67.0	1.9–2.8	–
Carbon	1.40–1.80	3790–4000	200.0–240.0	0.3–1.4	–
Flax	1.40–1.50	343–2000	27.6–103.0	1.2–3.3	12
Hemp	1.40–1.50	270–900	23.5–90.0	1.0–3.5	12
Jute	1.30–1.49	302–800	10.0–30.0	1.0–1.8	17
Sisal	1.33–1.50	363–700	9.0–38.0	2.0–7.0	14
Kenaf	1.2–1.40	223–930	14.5–53.0	1.5–2.7	17
Coir	1.15–1.46	95–230	2.8–6.0	15.0–51.4	13
Cotton	1.51–1.60	287–800	5.5–12.6	3.0–10.0	08–25
Ramie	1.00–1.55	400–1000	24.5–128.0	1.2–4.0	12–17
Bamboo	0.60–1.10	140–800	11.0–32.0	2.5–3.7	–

produce composites having properties similar to those produced from glass fibers (Wallenberger and Weston 2003). Natural fibers reinforced composites can be used for enhancing the structural response of the existing reinforced cement concrete elements without causing harm to the environment (Yan et al. 2014a). As an additional advantage, natural fibers possess a rough and uneven surface texture (that can be enhanced through alkali treatment Zhi et al. 2001; Rout et al. 2001; Venkateshwaran et al. 2013)) resulting in a stronger fiber-matrix bond the weakest part of a composite material. Along with that, no new manufacturing technology other than those already existing for manufacturing of synthetic fiber-based composites is required to manufacture these bio-composites (Bavan and Kumar 2013).

Among the various available natural fibers, for use in structural engineering applications flax, sisal, maize, and banana fibers are mainly explored by the research community owing to their lower cost, easy availability, low processing cost, high strength to weight ratio, and high stiffness (Huang et al. 2016).

3 Applications of Natural Fiber Reinforced Polymer Composites in Structural Engineering

Though currently, bio-composites are primarily used in the automotive, furniture, packaging industry, they have started to gain significant attention from the research community regarding their suitability in structural engineering applications. But as going with the basic engineering principle, before the prototype is manufactured a properly tested and verified design model is required that could ensure successful application leading to savings in cost and time. To fullfill, this requirement bio-composites are currently being investigated by the research community through experimental, analytical, and numerical studies. The research is primarily focused on evaluating their potential as flexure elements in form of sandwich composite beams, solid section beams, and as a strengthening laminate for the repair and retrofitting of the existing reinforced cement concrete flexure elements. In the case of bio-composites, the material properties are a function of internal defects such as voids, delamination, moisture content, etc. hence, for the analytical studies, it is necessary to validate the analysis results with those obtained from experiments (Lim et al. 2003), and for numerical studies (performed using the finite element method (FEM)) in order to correctly predict the structural behavior, it is necessary to have highly accurate input parameters like properties of the constituent materials, boundary conditions, and loading.

3.1 Flexure Elements

Properties such as low density and high specific strength and stiffness, have made the fiber-reinforced polymer (FRP) laminates as an efficient system in civil engineering applications that require a significant reduction in the self-weight of the structure. Also due to their non-corrosive, non-magnetic, and non-abrasive characteristics, they often become a material of choice for the structural designer for use in situations where significant corrosion, chemical, and magnetic resistance is required and traditional materials like steel and concrete are either not durable or not fit for use (like electronics manufacturing units, MRI rooms, chemical processing units, etc.)

FRP composites are generally employed as flexure elements in form of thin-walled pultruded profiles, sandwich beams consisting of a thin outer FRP shell and a soft internal core made up of synthetic material like foam or a lightweight natural material like cork, and as solid sections manufactured using hand layup or pultrusion techniques.

A significant amount of study has been done over the last 50–60 years regarding the structural response of FRP flexure elements reinforced with synthetic fibers of glass, carbon, or a hybrid combination of them. In past, such investigations have led to the availability of various codes, design guidelines, and manuals specifying various design rules and has standardized the manufacturing methods and sizes. However, such a scenario does not exist for those manufactured using the natural fibers and resin, the studies currently reported in this area are so scarce and such randomly distributed in terms of applications that the structural designer would not be able to draw significant outputs in form of design guidelines and calculation models from them. All this has hindered the application of bio-composites in the civil engineering industry.

Following are some of the significant studies carried out in the recent past to study the flexure response of bio-composites that can serve as an initial reference to the problem, enabling the structural designers to bring this renewable, bio-degradable, and environmentally friendly material into some constructive use, instead of letting them end up as landfill, animal food, manure or simply left to rot or set to fire in the fields.

3.1.1 Sandwich Composite Beams

Sandwich composites made of fiber-reinforced polymer laminates and light-weight core are very efficient systems in civil engineering applications that demand a significant reduction in the self-weight of the structure. In a sandwich composite beam, the outer shell made up of FRP laminates resist the bending stresses and the core resists the shear. The core also enhances the insulation, flexure resistance, and stiffness of the beam by increasing the lever arm and moment of inertia of the cross-section, respectively.

To evaluate the potential of all-natural composite monolithic structural panels as load-bearing elements of a low-rise commercial building, Dweib et al. (2004) tested five types of beam elements in a four-point bending test configuration and determined their strength, stiffness, and mode of failure. The beam elements of overall dimensions 1067 × 89 × 203 mm and a nominal wall thickness of 6.4 mm were manufactured from soy oil-based resin reinforced with different fiber combinations like flax mats, woven E-glass fiber, recycled paper-chicken feathers mats, recycled paper-corrugated paper, recycled paper-woven E-glass fiber. Also to manufacture the beams an integral foam core (Elfoam T300) was used as illustrated in Fig. 2. Though the foam core could contribute significantly as thermal and sound insulation, it's a contribution to the strength and stiffness of the beam was ignored.

Test results showed that the composite beams made from recycled paper have strengths and stiffness comparable to that of wood members conventionally used in structural engineering. As predicted, the E-glass fiber beam showed the maximum failure load of 39 kN. An interesting observation to note was that making use of different distribution mediums in the recycled paper did not affect the ultimate strength significantly as all beams failed around an ultimate load of 25 kN. The flax beam failed at a much lower value of around 10 kN. Regarding the mode of

Fig. 2 The cured beam sections with foam core (Dweib et al. 2004)

failure, the flax beam and recycled paper beams with chicken feathers and corrugated paper showed brittle failure which was initiated from the tension fibers and split the beam into two parts. The E-glass beam and the recycled paper beam with the E-glass layer, however, showed ductile failure caused by buckling of the compression fibers with no splitting of the beam. From this observation, the authors concluded that by adding small quantities of glass fibers to the recycled paper beams, the mode of failure could be changed from brittle to ductile.

A similar study was carried out by Sadeghian et al. (2018) to evaluate the flexure behavior of all-natural sandwich composites in civil engineering applications requiring minimum dead load such as cladding, deck, and roofing panels. They conducted a series of experimental parametric studies on sandwich composite beams manufactured by replacing the conventionally used glass fibers and synthetic core with natural flax fibers and cork core, respectively, using the hand lay-up method. The fiber type, type of core, number of lamina (varying from 0 to 2), span length (varying from 150 to 300 mm) were taken as the test variables.

Before subjecting the beams to a four-point bending test, a material characterization of the coupons was done. The flax reinforced composite coupons showed a bilinear stress–strain curve having a secondary slope of around two-thirds of the initial in the strain range 0.002–0.003 mm/mm. For glass-reinforced composite, the stress–strain curve was linear up to failure. A large difference in the tensile strength was observed, with a flax FRP coupon (207.42 MPa) having almost a quarter of the tensile strength of a glass FRP coupon (865.35 MPa). A difference of around 15 GPa was observed in the elastic modulus, with a glass FRP coupon having a higher elastic modulus of around 37.37 GPa as compared to 21.94 GPa of flax FRP coupon.

In the four-point bending test, six different types of failure modes were observed. The shear failure indicated by inclined cracks in the core was observed as the main failure mode in flax FRP sandwich beams with a cork core. In specimens having only a single layer of the lamina and a shorter span, indentations were observed on the skin, however, on increasing the number of flax lamina to two, the failure mode switched to shear failure of the core, again on increasing the span and number of lamina the failure mode change to buckling of the compression laminate at maximum bending moment section. Other failure modes observed were those due to core shear, core buckling, and crushing of core. In the glass FRP sandwich specimens with a honeycomb core, indentation failure was not observed. Also as glass FRP lamina is thicker and stiffer than that with flax fibers, no bucking of the compression laminate was observed. It was observed that the failure pattern turns out to be shifting from shear failure of the core to combined shear and buckling failure of the core as the core thickness increases. A comparative analysis of the test results showed that a two-layered flax FRP sandwich beam with an 11 mm thick cork core can be considered as equivalent to a single-layered glass FRP sandwich beam with a 12 mm honeycomb core, as both possesses an ultimate bending resistance of around 850 N. Similarly, a two-layered flax FRP sandwich beam with 22 mm thick cork has stiffness and transverse shear rigidity equivalent to that of a single layer glass FRP sandwich beam with 12 mm thick honeycomb. The authors also proposed a series of analytical relations (Eqs. 1–6) aimed at computing the flexural stiffness, the maximum mid-span deflection, the shear rigidity of the core, shear strength, and bending stress in laminates of the sandwich composite beams. A perfect bond was assumed to be existing between the FRP and core for ensuring the strain compatibility. The flexural rigidity D of the section as expressed by Eq. (1) was taken as the sum of the individual flexural rigidities of FRP laminates and the core.

$$D = E_f \frac{bt^3}{6} + E_f \frac{btd^2}{2} + E_c \frac{bc^3}{12} = D_1 + D_2 + D_3 \tag{1}$$

where, E_f and E_c correspond to Young's modulus of elasticity of laminate and core, respectively, d is the c/c distance between the confining laminates at the top and bottom of the core, t is the thickness of the laminate, b is the section width, and c is the core thickness.

Previous studies (Allen 1969) have shown that when the d/t ratio is greater than 5.77, and the value of the expression given by, $6\frac{E_f}{E_c} \times \frac{t}{c} \times \frac{d^2}{c}$ is found to be greater than 100, D primarily depends on D_2.

Hence, the flexural rigidity D of the section can be assumed to be as given by Eq. (2)

$$D = D_2 = E_f \frac{btd^2}{2} \tag{2}$$

From the obtained value of D, the maximum mid-span deflection Δ of a sandwich composite beam having a symmetric cross-section be calculated from Eq. (3) as per ASTM D7250 (2016).

$$\Delta = \frac{P\left(2S^3 - 3SL^2 + L^3\right)}{96D} + \frac{P(S - L)}{4U} \tag{3}$$

where, P is the applied service load, S is the effective span of the beam, L is the c/c distance between the loads, and U is the transverse shear rigidity of the section (in terms of force). U can be obtained by simultaneously solving Eq. (3) for a minimum of two known loading cases.

The shear rigidity of the core can be calculated by using Eq. (4) as per ASTM D7250 (2016).

$$G = \frac{U(h - 2t)}{(h - t)^2 b} \tag{4}$$

where, h is the total depth of the section and b, t as defined previously are the width of the section and thickness of the laminate, respectively.

The shear stress, τ can be assumed to be constant over the entire core depth and can be obtained from Eq. (5),

$$\tau = \frac{V}{bd} \tag{5}$$

The bending stress, σ can be obtained from Eq. (6),

$$\sigma = \frac{M}{bdt} \tag{6}$$

where, V and M are the maximum shear force and bending moment in the beam, respectively. Since all the specimens considered in the study satisfied the above-mentioned two conditions, the equations were used to validate the test results and vice versa, and the results were found to be in good agreement.

3.1.2 Solid Rectangular Section Beams

Bavan et al. (2013) carried out a numerical study using a commercially available finite element method simulation software (ANSYS) to evaluate the potential of using a natural fiber-based solid rectangular composite beam for structural engineering applications. The mechanical behavior of a composite beam manufactured from maize stalk fiber and unsaturated polyester, fabricated using the hand lay-up procedure was evaluated. The computer model of the rectangular composite beam section was meshed using SOLID95 elements having isotropic material properties. Uniformly arranged unidirectional fibers varying from 20 to 25% in volume were assumed as 80 mm long cylinders of 1 mm diameter. The Young's modulus was taken as of 8.58 GPa for fibers and of 1.3 GPa for the polymeric resin having a Poisson ratio of 0.3. The linear elastic analysis was performed and a load of 10 N was applied to the top edge.

The simulation results showed that the mechanical properties are in a direct proportion to the fiber content, as the composite with higher fiber content showed a higher ultimate stress level as compared to the one with lower fiber content, hence proving the general assumption of fiber-reinforced composites (also stated in the introduction part of this chapter) that in FRP composites, the load is primarily resisted by the fibers. The Von Mises failure criterion showed a high-stress concentration in the fiber and matrix interphase region. Based on the simulations, the authors stressed on the need for establishing certain assumptions that can clearly define the interface properties so that a perfect bond between the fibers and the matrix can be assured.

3.1.3 Solid Rectangular Section Cantilever Beams

The structural behavior of solid rectangular section cantilever beam elements fabricated from all-natural composite materials was investigated by Lim et al. (2003). The authors analyzed the whole-field deformation of an oil-palm fiber-reinforced composite cantilever beam subjected to static loading, using three-point bending tests and Shadow Moire' method. The results were compared with those obtained from the finite element analysis performed using commercially available finite element analysis (FEA) software. The whole-field measurement technique (*Shadow Moire' method*) was employed to measure the deformations, as the authors believed that in FEA a large amount of strain and displacement data is obtained and hence a more reliable comparison between experimental and FEA results can only be made if the measurement is made over a region of the structure instead of a few selected points, as generally done in the conventionally deployed contact methods.

The physical properties required as inputs in the FEA software were determined from three-point bending tests conducted on five 125 mm long, 160 mm wide, and 4 mm thick specimens. To model the response, the following assumptions were made in the FEA. Firstly, the composite was assumed to show a linear elastic behavior (Dominick 1989). Secondly, both the fiber and matrix phases were assumed to act together so that strain compatibility can be ensured. Thirdly, the laminate was assumed to be elastic and follows the Hook's law. Hence, by using the *rule of mixtures* (Bank, 2006) the stress in the laminate can be calculated by Eq. (7) or Eq. (8)

$$\sigma = \sigma_f a_f + \sigma_m a_m$$
$$\because \sigma_f = E_f \epsilon_f ; \sigma_m = E_m \epsilon_m \tag{7}$$

and $\epsilon_f = \epsilon_m$

$$\therefore \frac{\sigma_f}{E_f} = \frac{\sigma_m}{E_m}$$
$$\therefore \sigma = \sigma_f \left(a_f + \frac{E_m}{E_f} a_m \right) \tag{8}$$

where, σ is the average stress intensity in the laminate; σ_f is the stress intensity in the fibers; σ_m is the stress intensity in the matrix; a_f is the fiber area fraction; a_m is the matrix area fraction; E_f is the elastic modulus of the fiber; E_m is the elastic modulus of the matrix; ϵ_f is the strain developed in the fibers; and ϵ_m is the strain developed in the matrix.

The computer model of the cantilever beam was build using SolidWorks software and analyzed using COSMOS Works FEA software. The model was meshed with a 4 mm size mesh. To obtain well-distributed displacements and strain data, *Shadow Moire' whole-field measurement method* was used. The following section presents a summary of the technique.

The Shadow Moire' Method

The Shadow Moire' method is a method for measuring the surface profile and is frequently employed in the field of metrology with different setup configurations. In this technique, the profile of the surface is mapped in the form of contours called *Moire' fringes*. Figure 3 explains the working principle of the *Shadow Moire' method.*

A grating is placed over the surface and subjected to illumination from a light source. The angle θ_1 is called the angle of incidence, and θ_2 is called the angle of

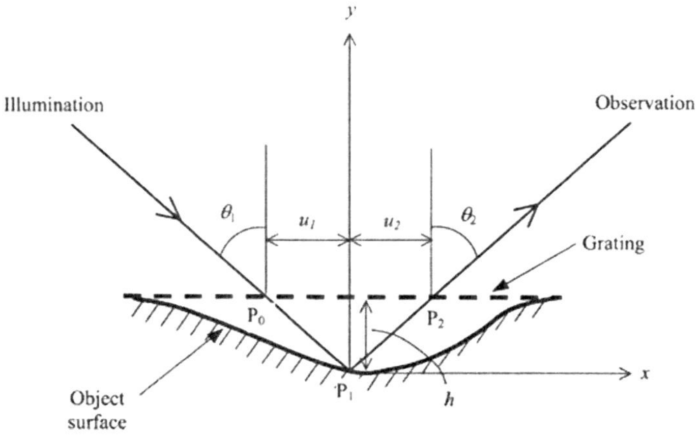

Fig. 3 Working principle of the *Shadow Moire' method* (Lim et al. 2003)

reflection (also called the angle of observation). When θ_1 and θ_2 are equal to 45°, the contour spacing becomes equal to the grating pitch. The length of the surface relative to its shadow u given by the Eq. (9) is approximated by projecting the point P_0 of the grating to a point P_1 on the surface and is observed by re-projecting the point to the grating as point P_2.

$$u = u_1 + u_2 = h(x, y)\tan\theta_1 + h(x, y)\tan\theta_2 \tag{9}$$

where, h is the vertical spacing between the grating and the surface at a point (x, y) given by Eq. (10), and u_1, u_2 as shown in Fig. 3 are the functions of h and θ.

$$h(x, y) = \frac{np}{\tan\theta_1 + \tan\theta_2} \tag{10}$$

where, p is the grating pitch, and n is the fringe order number (n = 0, 1, 2, 3 ...)

If the modulation function φ(x) given by Eq. (11) becomes equal to n, a dark fringe is observed.

$$\varphi(x) = \frac{u}{p} = \frac{h(x, y)\tan\theta_1 + h(x, y)\tan\theta_2}{p} \tag{11}$$

Since, the *Shadow Moire' method* gives only the maxima and minima of the contours, hence to enhance the resolution of the data *the phase-shift method* is used in which the obtained data is processed and the values at each pixel point are found.

Figure 4 shows the experimental setup used for the *Shadow Moire' method* in the study. A 1 mm pitch grating was placed over the cantilever beam, and a standard slide projector was used to project the light at an angle of incidence θ_1 equal to 45°. The observations were recorded through a digital camera place in the normal direction

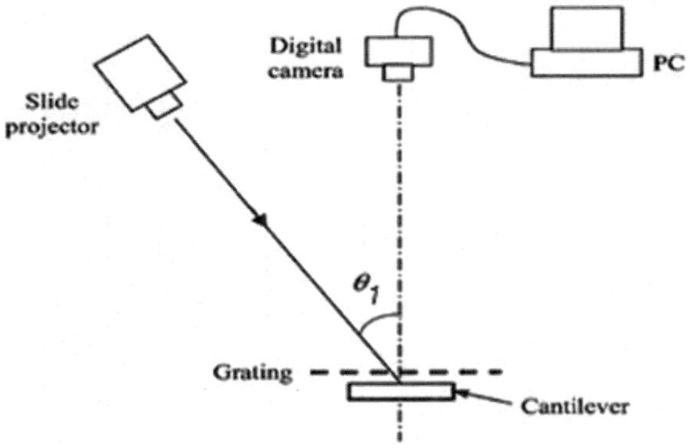

Fig. 4 Experimental set-up for *Shadow Moire' method* (Lim et al. 2003)

of the cantilever and processed using a PC connected to the camera. To deflect the beam a 1.8 N weight was attached to the beam using a string.

The test results showed that in the three-point bending tests the specimens deflected by an average peak deflection of 8.77 mm at the free end and gave an average flexure modulus of 2.41 GPa. The value of modulus of elasticity of the composite beam was validated with those obtained from *Shadow Moire' method* by substituting the deflection profile data into the deflection equation, and only a difference of 4.6% was found. In the FEA the maximum deflection at the free end was obtained as 3.99 mm which differed by 10% from that of *Shadow Moire' method*. On comparison of the results, the authors also observed an error near the fixed end which was primarily due to improper modeling of the fixed end in the FEA. The fixed end was modeled as a welded joint with completely restrained nodes though in the experiment setup it was only clamped by steel plates and not completely restrained.

3.2 Strengthening of RCC Structures

Factors like corrosion of rebars, inadequate design, increase in loads, spalling and delamination of the concrete cover, etc. demands an immediate need for retrofitting of existing concrete structures. Conventionally the concrete structural elements were strengthened using externally bonded steel plates/strips, but inherent problems like corrosion and difficulty in the installation made the engineers replace them with synthetic fiber-reinforced polymer (FRP) composite plate/strips. FRP composites possess a high strength to weight ratio, excellent damping and fatigue resistance, and high resistance to corrosion and electromagnetic radiations. But the problem associated with the non-biodegradability, high manufacturing cost, and non-renewability of synthetic fibers has shifted the focus on replacing these synthetic fibers with naturally occurring fibers of sisal, banana, bamboo, jute maize, etc. (Sen and Reddy 2011).

The following sections summarize the recent developments in the evaluation of flexure and shear responses of the reinforced cement concrete flexure elements strengthened by natural fiber-based FRP laminates.

3.2.1 Flexure Strengthening

To evaluate the potential of sisal fiber-reinforced polymer composites as a possible replacement to the conventionally used externally bonded steel and synthetic fibers reinforced polymer composites for strengthening the flexural concrete elements Sen and Reddy (2011) carried out a numerical study using a commercially available finite element method simulation software (ANSYS) on a plain, and on steel-reinforced cement concrete block externally strengthened by sisal fiber-reinforced polymer (FRP) composite.

The investigations were carried out using a commercially available finite element analysis software, ANSYS. The computer models of the rectangular beam sections having a dimension of 500 × 200 × 200 mm reinforced with longitudinal steel reinforcement (2, 10 mm diameter bars at each face) and shear reinforcement (6 mm diameter stirrups at c/c spacing of 50 mm); and of sections without steel reinforcements were meshed using the SOLID65 elements. The uniaxial Pipe16 elements were used for the rebars, and the SHELL63 element was used for the externally bonded 5 mm thick strengthening plate of sisal FRP composite. Due to small displacements in the concrete structures, geometric nonlinearity was ignored in the analysis, and only material nonlinearity was considered.

The simulation result showed that the load-carrying capacity of reinforced cement concrete beam retrofitted by all around wrapping of sisal FRP got increased by 83.33% over those without any external strengthening. Also, a significant reduction in the deformations was observed in the strengthened beams. The authors concluded that the required strength of the externally bonded composite strips can be achieved by varying the percentage of the sisal reinforcing fibers.

An experimental parametric study was also carried out in this area by Huang (2016) using the flax fibers. The study was aimed to determine the feasibility of environmental-friendly and inexpensive flax FRP composites as an effective alternative to glass/carbon fibers reinforced polymer composites for external strengthening and retrofitting of damaged reinforced cement concrete structures. The authors investigated the flexure behavior of reinforced cement concrete beams externally strengthened with bidirectional natural flax fiber-reinforced polymer composite plates. The specimens were subjected to the four-point flexure test and the thickness of the strengthening plate (4 and 6 layers), percentage of steel reinforcement (0.223 and 0.503%), and the type of section (cracked or uncracked) were taken as the test variables.

Before testing the beam specimens to determine the mechanical properties of the strengthening composite plates along both the warp (0°) and weft (90°) directions, the plates were subjected to a tensile test on the universal testing machine. The test results were compared with those reported in the literature and were found to be in good agreement. The flax fabric layers (varying from 2, 4, 6 layers) were taken as the test variables. The test results showed that as the fiber layers are increased the tensile strength of the flax FRP composite decreases. The maximum strength of 98.8 MPa was reported for composite with only two layers and a minimum of 84.4 MPa for those with six layers. On the contrary, the elastic modulus and the percentage elongation at failure remained approximately the same for all the specimens, i.e., around 10 GPa and 2%, respectively.

For the flexure test, eight 2 m long, 150 mm wide, and 300 mm deep reinforced cement concrete beam specimens were fabricated. The beams had a reinforcement ratio varying from 0.223 to 0.503%, a set of closely spaced links at the ends, and concrete having an average compressive strength of 31.4 MPa. Six of the beam specimens were strengthened with bidirectional flax FRP composite plates of dimensions 1700 mm × 140 mm glued at the soffit using epoxy resin, and two were used as the control beams. Also, to prevent premature debonding of the strengthening strips, three strips of flax FRP composite (700 mm long and 60 mm wide) were glued at the two ends at a c/c spacing of 30 mm around the three faces of the cross-section. Strain gages and transducers were employed to measure the resulting strains and the deflection, respectively.

Two different kinds of failure modes were observed, the control beams failed by the yielding of reinforcement followed by crushing of concrete whereas the strengthened beams failed by yielding of reinforcement followed by rupturing of strengthening plates at the mid-span. In cracked section beams (sections were cracked by applying a preload equal to 80% of the yield load), the cracks appeared earlier and spread at a faster rate however, the failure mode remain the same as in those with an uncracked section. More numbers and wider cracks were observed at failure in the strengthened beams as compared to in control beams. No debonding of the FRP plate was observed in any of the specimens which could be attributed to the presence of holding strips at the two ends. Specimens with maximum reinforcement ratio and strengthened with a composite plate made of maximum flax fiber layers showed the maximum ultimate load-carrying capacity of 175.3 kN which is 1.2 times the capacity of the control beam with equivalent reinforcement ratio. The test results showed that the load-carrying capacity depends much more on the steel reinforcement ratio than on the number of reinforcement layers in the strengthening composite, as can be inferred from the results that in beams strengthened with composite having six layers of reinforcement the ultimate load-carrying capacity increased to almost twice on increasing the reinforcement percentage to 0.503% from 0.223%. On the other hand, on keeping the reinforcement ratio constant and increasing the number of fabric layers from four to six, the capacity got increased by only 1.2 times.

The load-versus-deflection curves of the strengthened beams showed that once the peak response is reached there is a sudden drop in the load-carrying capacity and deflection. This implies that the brittle characteristics of strengthening plates governs the failure of the beams. It was also observed that before yielding of the steel reinforcement, both the control beam and the strengthened beams display a similar load-versus-displacement behavior (with strengthened beams carrying a higher load than the control beams). But after yielding of the longitudinal reinforcement, until the peak load is reached, the strengthened beams showed a more ductile behavior than the control beams.

The energy absorption values were also calculated from the area under the load–deflection curves. The values showed that for lower reinforcement ratios, on varying the number of layers from 4 to 6, the energy absorption capacity increases by up to 54.9%. Whereas for higher reinforcement ratios (for the same variation in the number of layers (4–6)), it increases by only 4.5%. Also, pre-cracking the section decrease the energy absorption capacity of the beam, as among all the strengthened beams the pre-cracked beams gave the smallest energy absorption values.

A comparison of the test results with those of similar dimensions and reinforcements but strengthened with synthetic glass/carbon fiber reinforced composite plates (Rahimi and Hutchinson 2001) showed that the improvement in ultimate load carrying capacity is comparable even though the tensile strength and elastic modulus of natural flax fiber-reinforced composite are much lower than those with synthetic fibers.

3.2.2 Shear Strengthening

To evaluate the potential of high strength natural fiber reinforced polymer composites manufactured from treated and untreated natural fibers of kenaf, jute, and jute rope as possible alternatives to the high-cost carbon fiber reinforced polymer composites that are conventionally used as shear strengthening plates in reinforced cement concrete structures, Alam and Riyami (2018) investigated the shear behavior of reinforced cement concrete beams externally strengthened with natural fibers reinforced polymer composites. To improve the fiber-matrix bond, fibers were subjected to alkali treatment. An empirical relation was derived to obtain an optimum cross-sectional area of the strengthening laminates. The test results were compared with those strengthened with carbon fiber-reinforced composite plates.

The high moisture absorption property and poor interfacial bonding with the polymeric matrix are the main problems that obstructs the use of natural fibers as reinforcements in composite laminates. Previous studies have shown that such problems can be successfully addressed by subjecting the fibers to an alkali treatment before use (El-Shekeil 2012; Sever 2009; Meon et al. 2012; Ray et al. 2001; Keshk et al. 2006). The alkali treatment breaks the hydrogen bonds in the fibers and removes excess lignin and oils from the surface eventually enhancing their adhesion with the polymeric matrix (Mohanty et al. 2001a).

Three different kinds of 400 mm long reinforcements from kenaf fibers, jute fibers, and jute rope were prepared to reinforce a total of six 100 mm wide and 6 mm thick laminates, using the hand-layup process. To improve the fiber-matrix bond, one of the sample set was subjected to alkali treatment using a 6% solution of sodium hydroxide. The volume fraction of fibers in each specimen was kept at around 45%. Eight reinforced cement concrete beams having dimensions as 2300 × 150 × 300 mm, reinforced with 2 nos. of high yield strength, 16 mm diameter rebars, and 6 mm diameter links at c/c spacing of 130 mm were cast using concrete of

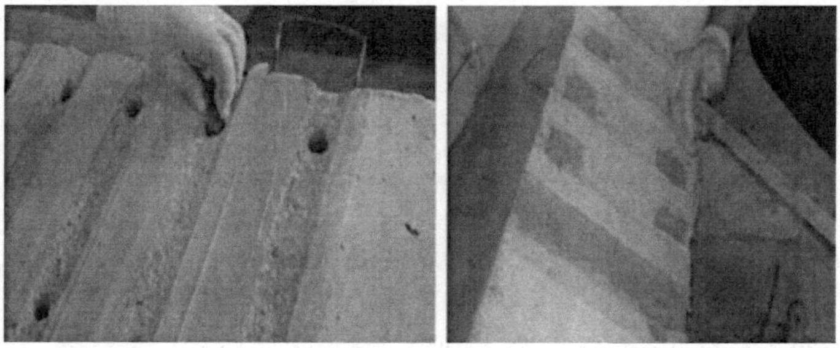

Fig. 5 Shear strengthening of beams using *embedded connector system* (Alam et al. 2017; Alam and Riyami 2018)

30 MPa compressive strength. Three of the beams were strengthened using 300 mm long 35 mm wide and 6 mm thick laminate strips having treated fibers and three with laminate strips having untreated fibers. The strips were placed only in the shear span of the beam (550 mm) with a c/c spacing of 100 mm. For comparison of the test results, one of the specimens was left un-strengthened (to be used as a control beam). To prevent the debonding of the laminates, a technique of *embedded connector system* given by Alam et al. (2017) as shown in Fig. 5 was deployed. In this method 20 mm diameter, 25 mm deep holes were made on the top and bottom of the beam face along the shear span, inside which adhesive was filled, and a 16 mm diameter rebar was inserted to work as a connector that holds the strengthening strips in place. To measure the shear response, the prepared specimens were subjected to a four-point flexure test.

The optimum cross-sectional area of a single natural fiber-reinforced laminate was obtained from the analytical relations (Eqs. 12–21) derived using the design guidelines of EN 1992-1 (2004), and ACI 440.1R, (1999). The first assumption made was that, once the beam develops a shear crack before the shear failure, concrete can no longer resist the shear and all of the shear will get resisted only by the shear link and the strengthening laminate, i.e., Eq. (12).

$$V_d = V_{link} + V_{frp} \tag{12}$$

where, V_d is the shear strength of the beam, V_{link} is the capacity of the links, and V_{frp} is the capacity of strengthening laminates. As per the second assumption, the maximum strain in the laminates was taken as 0.004 as per the ACI 440.1R (1999) recommendations for the design of shear strengthening system of an RCC beam. Thirdly, the inclination of shear crack was assumed to be at 45°.

The maximum shear force in the beam can be calculated using Eq. (13).

$$V_{\max} = \frac{M_{\max}}{l} \tag{13}$$

where, V_{max} is the maximum shear force in the beam, M_{max} is the maximum bending moment in the beam, and l is the shear span.

For safety, as per Eq. (14).

$$V_d = V_{link} + V_{frp} \geq V_{max} \tag{14}$$

From the assumptions considered, the design strain in a single laminate is 0.004 hence as per the Hook's law, Eq. (15).

$$\sigma_{frp} = 0.004 \times E_{frp} \tag{15}$$

where, E_{frp} is Young's modulus of FRP laminate.

The capacity f_{frp} can be calculated using Eq. (16).

$$\begin{aligned} f_{frp} &= \sigma_{frp} \times A_{frp} \\ &= 0.004 \times E_{frp} \times A_{frp} \end{aligned} \tag{16}$$

Since the laminates are generally applied in a 'U' shaped fashion around the beam, i.e. at a particular section two laminate strips are effective. Hence, the capacity of a single strengthening strip, F_{frp} can be calculated using Eq. (17).

$$F_{frp} = 2 \times \left(0.004 \times E_{frp} \times A_{frp}\right) \tag{17}$$

If N_{frp} are the total number of strengthening strips placed at a c/c spacing of S_{frp} along the shear span, the total shear capacity of the strengthening laminates V_{frp} can be calculated using Eq. (18).

$$\begin{aligned} V_{frp} &= N_{frp} \times F_{frp} \\ &= N_{frp} \times \left[2 \times \left(0.004 \times E_{frp} \times A_{frp}\right)\right] \end{aligned} \tag{18}$$

N_{frp} can be calculated using Eq. (19).

$$N_{frp} = \frac{d}{\tan(\theta) \times S_{frp}} \tag{19}$$

where, θ is the angle of inclination of shear crack from the longitudinal axis of the beam (assumed as 45°), d is the effective depth of the longitudinal reinforcement measured from the highly compressed extreme fiber, and S_{frp} is the c/c spacing of the laminates.

Hence, as per Eq. (20),

$$V_{frp} = \left(\frac{d}{\tan(\theta) \times S_{frp}}\right) \times \left[2 \times \left(0.004 \times E_{frp} \times A_{frp}\right)\right] \tag{20}$$

Since,

$$V_d = V_{link} + V_{frp}$$

$$\therefore V_d - V_{link} = \left(\frac{d}{\tan(\theta) \times S_{frp}}\right) \times [2 \times (0.004 \times E_{frp} \times A_{frp})]$$

$$\therefore A_{frp} = \frac{(V_d - V_{link}) \tan(\theta) S_{frp}}{0.008 E_{frp} d}$$

taking $V_d = V_{max}$

$$\therefore A_{frp} = \frac{(V_{max} - V_{link}) \tan(\theta) S_{frp}}{0.008 E_{frp} d} \quad (21)$$

From the tensile tests performed on the laminates, no significant variation in the mechanical properties was observed between those reinforced with treated fibers and those with untreated fibers. Maximum tensile strength was observed in laminates reinforced with treated jute fibers (137 MPa) and minimum for those reinforced with untreated jute rope (110 MPa). Maximum elastic modulus was observed in laminates reinforced with treated kenaf fibers (14.91 GPa) and minimum for those reinforced with untreated jute rope (10.21 GPa).

In the four-point flexure test on beams, the control beam failed by premature shear failure with 38° shear cracks. The beam strengthened by untreated jute rope reinforced polymer laminate failed by fracture of the laminate followed by shear failure with a 45° shear cracks. Due to their high mechanical properties, the natural fiber reinforced polymer laminates served their purpose well as no shear failure was observed in the beams strengthened by untreated jute fiber reinforced polymer laminate, and kenaf fiber-reinforced polymer laminates. Such beams failed in flexure followed by the crushing of concrete. The maximum load-carrying capacity was shown by those with jute fiber-reinforced laminates (213 kN) and lowest by those strengthened with laminates made of jute rope (209 kN). An average 34% increase in the load-carrying capacity relative to the control beam was observed in the beams strengthened with laminates made of untreated fibers. No shear failure was observed in any of the beam specimens strengthened by laminates made up of treated fibers or rope as all failed in flexure followed by crushing of concrete. Such beams also showed ductile behavior before failure. The maximum load-carrying capacity in such specimens was shown by those with jute rope reinforced laminates (204 kN) and lowest by those strengthened with laminates of treated kenaf fibers (156 kN). An average 20% increase in the load-carrying capacity relative to the control beam was observed in the beams strengthened with laminates made of treated fibers.

As with the deflections, among the beams strengthened with laminates made of untreated fibers maximum deflection was observed in beams strengthened with laminates reinforced with jute rope (26% less than control beam) and minimum in those with jute fibers (87% less than control beam). Yielding of main reinforcement leading to a ductile mode of failure was observed in all beams strengthened with untreated laminates. Similar behavior was also observed in the beams strengthened with laminates made of treated fibers or rope. The maximum deflection in such beams was observed in beams strengthened with laminates reinforced with jute rope (25% less than control beam) and minimum in those with jute fibers (68% less than control beam).

On comparing the test results of untreated and treated fibers, the authors concluded that though the alkali treatment helped in enhancing the interfacial matrix-fiber bond leading to a marginal increase in the mechanical properties of the laminate, no significant difference in the structural response of the beams strengthened using the two different types of laminates was observed. Hence, as per the authors when it comes to using the laminates for shear strengthening, untreated fibers can be successfully used, though chemical treatment will enhance the long-term durability of the fibers.

The experiment results were compared with those obtained from calculations made as per ACI 440.1R (1999) design guidelines and were found to be in good agreement. The analytical equations derived to obtain the optimum cross-sectional area of the laminates by assuming the design strain as 0.004 and 45° shear crack, were validated with the experimental results. The validation study showed the ultimate load-carrying capacity obtained from experimental results to be around 12% higher for beams strengthened using untreated fibers reinforced laminates and varying from 2.4 to 8.8% in those with treated fibers.

To see how the natural fiber-reinforced composite strengthened beams behaved relative to those strengthened with conventionally used carbon fibers. The same test was also performed on a beam strengthened with a 300 mm long 35 mm wide and 1.2 mm thick carbon fiber reinforced polymer composite strip. In such a beam debonding of the laminate even after the presence of connectors followed by the shear failure was observed. The beam showed a brittle mode of failure which was sudden without any deflections. The authors (Alam and Riyami 2018) concluded that though the tensile strength and elastic modulus of carbon fibers reinforced polymer laminate are more than that with natural fibers, strain compatibility could not be assumed to remain as established once the shear reinforcement is yielded. On the other hand, since these properties in the case of natural fibers reinforced laminates were very low, they made the laminate relatively very ductile enabling them to maintain the strain compatibility even after yielding of the shear link. Due to the debonding of the laminate, the ultimate failure load of the beam was found to be lower than those strengthened with natural fibers reinforced laminates by an average difference of 14%.

4 Response to Dynamic Loading

Previous studies have shown that the damping characteristic of natural fiber based composites is different from those manufactured from synthetic fibers (Chandra et al. 1999). As the natural frequency of vibration of a reinforced polymer composite depends on several parameters like fiber length, fiber orientation, fiber distribution in the matrix, the fraction of the fibers, and the fiber-matrix interface, it becomes a very challenging problem to find the free vibration properties of structural elements manufactured from such materials. To find the effect of fiber fraction and length on the mechanical and dynamic properties of the synthetic fibers reinforced polymer composite, many research studies have been carried out. Based on such studies the researchers have concluded that the fiber orientation rather than the fiber length and weight fraction affects the mechanical and damping properties of the resulting composite (Suarez et al. 1986). However, such a relation for natural fibers reinforced composites is not well established (Kumar et al. 2014).

The following sections summarize the recent developments made in determining the dynamic response of natural fiber-based composites. The studies are primarily focused on determining the effect of fiber length, fiber fraction, and the effect of enhancement of matrix-fiber interfacial bond through chemical treatment on the free vibration properties of the bio-composite laminate.

4.1 Effects of Fiber Length and Weight Fraction

To investigate how the fiber length and weight fraction affect the mechanical and damping properties of a banana FRP and sisal FRP composites, Kumar et al. (2014) performed a series of experimental parametric studies. The composites were manufactured using the compression molding method and the type of fibers (banana, sisal), the fiber length (3, 4, 5 mm), and the fiber weight fraction (0.30, 0.40, 0.50) were taken as the test variables. Interfacial observations based on microscopy analysis were used to validate the test results. Specimens of size 300 mm × 125 mm × 3 mm were subjected to tensile, flexure, and impact hammer test to measure the mechanical and dynamic properties, respectively.

The tests results showed that in the case of banana fiber-reinforced polymer (FRP) composites on keeping the fiber fraction constant to 50% by weight and increasing the fiber length from 3 to 4 mm, the mechanical properties of composite improves (tensile strength being increased to 30.5 MPa from 28.6 MPa, flexural strength to 47.9 MPa from 46.7 MPa, and impact strength to 215.3 MPa from 187.2 MPa), but on further increasing the fiber length to 5 mm, the entanglement of fibers occurs leading to the deterioration of the mechanical properties (tensile strength reducing to 21.5 MPa, flexural strength to 37.7 MPa, and impact strength to 153.2 MPa). However, due to the random distribution of test data, no relation between the weight fraction of fibers and properties of composites was established. Similarly, the test results for sisal fiber reinforced polymer composites were too randomly distributed to reach a certain conclusion. Such results could be attributed to the random distribution of variable-sized voids in the natural fibers.

To correlate the experiment results with the microstructural properties of the composite, microscopic analysis through a scanning electron microscope (SEM) was conducted. The results showed that in the banana FRP composites for a fiber length of 4 mm and 0.5 weight fraction, there exists a good adhesion between fiber and matrix, as inferred from the appearances of river-like patterns on the fiber-matrix interface. Such a pattern indicates a thick interface and results in a better shear action which leads to a layer by layer fracture of the polymer matrix. Also, these composites showed a more uniform fiber arrangement in the fractured region, which explains their higher impact energy absorption property. On further increasing the fiber length to 5 mm (keeping the weight fraction same to 0.5), agglomeration of fibers due to high cohesive forces was observed that lead to deterioration of the mechanical properties. On the other hand, in sisal FRP composites with a fiber length of 3 mm and 0.5 weight fraction, the SEM images showed hackles and matrix failure in the resin-rich regions. On increasing the fiber length to 5 mm (keeping the weight fraction the same) agglomeration, and improper wetting of the fiber surface by the polymer matrix similar to that in banana FRP composites were observed.

For determining the natural frequency of vibration of the composites, a cantilever plate specimen was subjected to a constant force applied through a piezoelectric impact hammer. The exciting frequency was varied and the force was applied at five different positions. The response was measured using an accelerometer attached to one of the corners of the plate. A frequency response plot was obtained using the Fast Fourier Transform (FFT) analyzer, and by using the half-power bandwidth method the damping ratio ζ was calculated from Eq. (22).

$$\zeta = \frac{\omega_2 - \omega_1}{2\omega_n} \tag{22}$$

where, ζ is the damping ratio, ω_n is the natural frequency of vibration, and ω_1, ω_2 are the forcing frequencies at which the response amplitude becomes equal to $1/\sqrt{2}$ times the resonance amplitude.

The test results showed that the banana FRP composites manufactured using a 0.5 weight fraction of 4 mm long fibers possess the maximum fundamental frequency (around 32 Hz). For sisal FRP composites, the highest natural frequency (31.7 Hz) was found for those reinforced with a 0.5 weight fraction of 5 mm long fibers. From the analysis of the test results, the authors concluded that the fundamental frequency of the composite is a function of the stiffness of the reinforcing fibers and a more rigid fiber will produce a composite with a much higher fundamental frequency than the one with fibers of lesser stiffness. The authors also observed that in the case of banana FRP composites, the fiber length does not significantly influence the natural frequency of vibration.

The damping ratios calculated from the frequency response plots using the half-power bandwidth method were found to be a function of the fiber content and fiber length, though the results were again too arbitrary to derive any rational relationship between the two. From the test results, it can be roughly stated that for the fiber content varying from 30 to 40% by weight and fiber length varying from 3 to 4 mm, the sisal fiber-based composites gave higher values of the damping ratio than those with banana fibers, but as the fiber fraction was increased to 0.5, keeping the length as same, banana fiber-based composites performed better.

From their study for the optimum damping performance of the composite, the authors recommended a combination of 0.5/4 mm for banana fibers based unidirectional composites and of 0.5/3 mm for sisal fibers based unidirectional composites.

4.2 *Effects of the Chemical Treatment*

The natural fiber reinforced polymer composites possess a poor interfacial bond due to the hydrophilic nature of organic fibers (Goriparthi et al. 2012). Cellulose which is the main strengthening component of natural fiber is not able to bond properly with the polymeric matrix due to the presence of hemicellulose and lignin layer. To reduce the amount of hemicellulose and lignin layer, the fibers are pretreated with an alkaline solution, and various researchers (Zhi et al. 2001; Rout et al. 2001; Venkateshwaran et al. 2013) have reported an increase in the mechanical properties like tensile strength, fiber-matrix bond strength, etc. of composites manufactured from such treated fibers. Hence, the effects of such treatments on the free vibration behavior of the resulting composite is also required to be properly understood before recommending them as possible replacements to conventionally used metal and laminated composites load-bearing structural elements.

The effects of chemical treatment done by sodium hydroxide for improving the fiber-matrix bond, and of weight fraction of fibers on the free vibration behavior of a randomly distributed banana and sisal fibers reinforced polyester composites were investigated by Rajesh et al. (2016). The response was studied using an impulse hammer test.

For the test, 270 mm long, 250 mm wide, and 3 mm thick composite laminates were manufactured using the compression molding method. Sisal fibers, banana fibers, and hybrid banana-sisal fibers were used to reinforce the polyester matrix in three different types of specimens. From the cured laminates, two different sized specimens, one for the flexure test (127 $\times$ 13 $\times$ 3 mm) and the other for free vibration tests (200 $\times$ 35 $\times$ 3 mm) were cut.

The test results showed that in composites reinforced with untreated fibers till 0.5 weight fraction, the flexure modulus and fundamental frequency of the resulting composite increase significantly, but as the weight fraction was increased beyond 0.5, the properties decrease. This could be due to the reduction in stiffness of the composite from the agglomeration of fibers as their content increase, resulting in a poor bond between the matrix and fibers. Composites reinforced with 50% sisal fibers showed the highest flexural modulus of 2.61 GPa, whereas those reinforced with a hybrid combination of 50% banana-sisal fibers showed the lowest value of 2.24 GPa owing to partial replacement of sisal fibers with low cellulose content banana fibers. Composites with a 0.35 weight fraction of sisal fibers showed the highest damping ratio of 0.48. On the other hand for the composites manufactured from fibers having a weight fraction of 0.5, and treated with a 1% sodium hydroxide, solution a significant improvement (almost twice) in the mechanical properties was observed. This could be due to the increase in the stiffness of the laminate resulting from the enhanced cellulose-matrix bond. In such composites also, the Sisal fibers performed better than the rest.

5 Summary

From the literature referred, the following observations of importance can be extracted:

- The flax fiber-reinforced polymer composite laminate possesses a bilinear stress–strain curve.
- Flax FRP composite has almost a quarter of the tensile strength (207.42 MPa) of glass FRP composite (865.35 MPa).
- On increasing the fiber length in a laminate, agglomeration of fibers due to high cohesive forces occurs leading to the deterioration of the mechanical properties.
- River-like patterns on the fiber-matrix interface were observed in the SEM images of the banana FRP composite, indicating a thick interface that results in a layer by layer fracture of the polymer matrix.
- Composite beams consisting of recycled paper as reinforcement have strengths and stiffness comparable to that of wood members used in structural engineering.
- On adding small quantities of glass fibers to the composite beams consisting of recycled paper as reinforcement, the mode of failure changes from brittle to ductile.
- The composite beams consisting of flax fibers, and those consisting of recycled paper with chicken feathers and corrugated paper as reinforcements possess brittle failure that initiates from the tension fibers and splits the beam into two parts.
- The shear failure indicated by inclined cracks in the core is the main failure mode in flax FRP sandwich beams with a cork core.
- In sandwich beams, the failure pattern shifts from the shear failure of the core to the combined shear and bucking failure of the core on increasing the core thickness.
- A two-layered flax FRP composite sandwich beam with an 11 mm thick cork core can be considered as equivalent to a single-layered glass FRP sandwich beam with a 12 mm honeycomb core.
- The value of modulus of elasticity of the composite beam obtained from experimental investigation and that from the *Shadow Moire' method* differs by approximately 4.6%.
- Beams strengthened using treated natural fibers based laminates and those from untreated fiber base laminates shows no significant difference in their structural response.
- The load-carrying capacity of reinforced cement concrete beam retrofitted with sisal FRP composites wrapped all around gets increased by 83.33% over those without any external strengthening.

- No shear failure is observed in the beams strengthened by untreated jute fiber reinforced polymer laminate, and kenaf fiber-reinforced polymer laminates. Such beams fail in flexure followed by the crushing of concrete.
- The load-carrying capacity of the beam increases by an average of 34% and the deflection reduces by 26% on strengthening with laminates made from untreated fibers.
- The load-carrying capacity of the beam increases by an average of 20% and the deflection reduces by 25% on strengthening with laminates made from fibers treated with 1% sodium hydroxide solution.
- Debonding of the carbon fibers reinforced polymer laminate can led to a relative decrease in the ultimate load-carrying capacity of the strengthened beam by around 14%, as compared to the beams strengthened with natural fibers reinforced laminates in which no debonding of the laminate occurs.
- The fundamental natural frequency of the composite is a function of the stiffness of the reinforcing fibers.
- In the case of banana FRP composites, the fiber length does not significantly influence the natural frequency of vibration.
- For the optimum damping performance of the unidirectional composite laminate, a combination of 0.5 weight fraction and 4 mm fiber length for banana fibers and 0.5 weight fraction and 3 mm fiber length for sisal fibers is recommended.
- Composites with a 35% weight fraction of sisal fibers possess the highest damping ratio of 0.48.
- Composites manufactured from fibers treated with a 1% sodium hydroxide solution shows a significant improvement (almost twice) in the mechanical properties due to a better cellulose-matrix bond.

6 Research Gaps and Future Research Directions

- Natural fibers and bio-resins can be highly anisotropic due to the presence of a large number of internal defects and imperfections like voids, micro-cracks, etc. Due to such ambiguity regarding the anisotropy of the natural fibers and bio-resin, the applicability of analytical equations that are based on the isotropic nature of the fiber and the matrix is questionable. Hence, to check the suitability of all such previously derived analytical equations of synthetic fiber-based composites to NFRP composites, the equations need to be validated from experimental results.
- In the previous studies, NFRP composites are manufactured either by hand layup technique or by compression molding method. The feasibility of using advanced fabrication techniques like pultrusion to make NFRP profiles needs to be investigated.
- Structural members made from NFRP open/closed section profiles need to be investigated for their short-term behavior; and for their long-term behavior by considering the effects of creep, shrinkage, and fatigue.

- As seen from previous studies, agglomeration of fibers due to the high thickness of natural fibers decrease the mechanical properties of the resulting composite significantly and restricts the fiber fraction to only 0.3–0.5 by weight. Restricting the fiber content to such a low amount will make the resulting laminate highly inefficient. Hence, a detailed parametric study focusing on optimizing the fiber content, fiber orientation, and fiber distribution in the matrix need to be carried out for different combinations of natural fibers and bio-resins.
- In the previous studies, the material characterization of NFRP composites is carried out at normal temperatures only. For using NFRP composites in structures that are or might be subjected to elevated temperatures during their service life, the effect of elevated temperatures on the mechanical and physical properties of NFRP composites needs to be investigated.
- A full-scale fire testing of structural components built from NFRP laminates needs to be carried out. Such tests will help in understanding their high-temperature response, and based on the results from such tests, different fire protection, and insulation techniques /guidelines can be developed.
- Previous studies have shown that alkali treatment of fibers leads to only a marginal increase in the properties of the resulting laminate. Also, no significant difference in the structural behavior of RCC beams strengthened with treated fiber-based laminates and those with untreated fiber-based laminates was observed. Hence, more advanced techniques for removing the lignin, oils, and other constituents present on the fiber surface that hinders the proper formation of a cellulose-matrix bond needs to be explored.
- Unlike synthetic fibers, natural fibers are highly susceptible to moisture changes. This could lead to continuous expansion and contraction of fibers, eventually leading to the weakening of the fiber-matrix bond. Hence, more advanced techniques that can reduce the susceptibly of natural fibers to water needs to be explored.
- Due to the high biodegradable nature of natural fibers, a detailed life cycle assessment of NFRP composites is required.
- Among the various available natural fibers, flax, sisal, and kenaf are the only fibers that are mainly explored by the research community. Studies on composites manufactured from other natural fibers like those obtained from jute, hemp, bamboo, wheat stalk, rice stalk, etc. (that are locally available in abundance in various developing countries like India) are scarce.
- Like, polylactic acid (PLA) which is a biodegradable thermoplastic polymer having lactic acid as the primary constitutional unit and is generally obtained through bacteria aided fermentation of corn. Other farm and dairy wastes can be explored for manufacturing bio-resins or fillers.
- Previous studies have shown that as compared to synthetic fibers, the mechanical properties of natural fibers are very low. Hence instead of completely replacing the synthetic fibers, a partially biodegradable hybrid system consisting of both natural as well as synthetic fibers can be manufactured.

Acknowledgements Financial support provided by the Council of Scientific & Industrial Research (CSIR), New Delhi through Scheme no. (22(0781)/19/EMR-II) sanctioned to the second author is highly appreciated for helping in the execution of the project.

References

American Concrete Institute (ACI) (1999) Guide for the design and construction of structural concrete reinforced with fiber-reinforced polymer bars. ACI 440.1R-15. Farmington Hills, MI

Alam MA, Riyami KA (2018) Shear strengthening of reinforced concrete beam using natural fibre reinforced polymer laminates. Constr Build Mater 162:683–696

Alam MA, Sami A, Mustapha KN (2017) Embedded connectors to eliminate debonding of steel plate for optimal shear strengthening of RC beam. Arab J Sci Eng 42(9):4053–4068

Allen HG (1969) Analysis and design of structural sandwich panels. Pergamon Press, Oxford

ASTM International (2016) Standard practice for determining sandwich beam flexural and shear stiffness. ASTM D7250. West Conshohocken, PA, USA

Bank LC (2006) Composites for construction: structural design with FRP materials. Wiley, New Jersey

Bavan DS, Kumar GCM (2013) Finite element analysis of a natural fiber (Maize) composite beam. J Eng

Cevallos OA, Olivito RS, Codispoti R, Ombres L (2015) Flax and polyparahenylene benzobisxazole cementitious composites for the strengthening of masonry elements subjected to eccentric loading. Compos B 71:82–95

Cevallos OA, Olivito RS (2015) Effects of fabric parameters on the tensile behavior of sustainable cementitious composites. Compos B 69:256–266

Chandra R, Singh SP, Gupta K (1999) Damping studies in fiber-reinforced composites—a review. Compos Struct 46:41–51

Daniel IM, Ishai O (2007) Engineering mechanics of composite materials. Oxford University Press, New York

Dominick RV (1989) Designing with reinforced composites. Carl Haneer Verlag, Germany

Dweib MA, Hu B, O'Donnell A, Shenton HW, Wool RP (2004) All natural composite sandwich beams for structural applications. Compos Struct 63:147–157

El-Shekeil YA (2012) Influence of chemical treatment on the tensile properties of kenaf fiber reinforced thermoplastic polyurethane composite. eXPRESS Polymer Lett 6(12):1032–1040

European Standard (CEN) (2004) Design of concrete structures—part 1-1: general rules and rules for buildings. Eurocode 2: EN 1992-1. Brussels

Goriparthi BK, Suman K, Rao NM (2012) Effect of fiber surface treatments on mechanical and abrasive wear performance of polylactide/jute composites. Compos A 43:1800–1808

Huang L, Yan B, Yan L, Xu Q, Tan H, Kasal B (2016) Reinforced concrete beams strengthened with externally bonded natural flax FRP plates. Compos B 91:669–578

John MJ, Uomas S (2008) Biofibres and biocomposites. Carbohyd Polym 71(3):343–364

Keshk S, Suwinarti W, Sameshima K (2006) Physicochemical characterization of different treatment sequences on kenaf bast fiber. Carbohyd Polym 65(2):202–206

Kumar KS, Siva I, Jeyaraj P, Jappes JTW, Amico SC, Rajini N (2014) Synergy of fiber length and content on free vibration and damping behavior of natural fiber reinforced polyester composite beams. Mater Des 56:379–386

Lim JH, Ratnam MM, Khalil HPSA (2003) An experimental and finite element analysis of the static deformation of natural fiber-reinforced composite beam. Polym Testing 22:169–177

Mallick PK (2007) Fiber-reinforced composites: materials, manufacturing, and design. CRC Press, Boca Raton

Meon MS, Othman MF, Husain H, Remeli MF, Syawal MSM (2012) Improving tensile properties of kenaf fibers treated with sodium hydroxide. Procedia Eng 41:1587–1592
Mohanty AK, Misra M, Drzal LT (2001) Surface modifications of natural fibers and performance of the resulting biocomposites: an overview. Compos Interfaces 8(5):313–343
Mohanty AK, Misra M, Hinrichsen G (2001) Biofibres, biodegradable polymers and biocomposites: an overview. Macromol Mater Eng 276–77:1–24
Morales AP, Güemes A, Lopez AF (2017) Bamboo-polylactic acid (PLA) composite material for structural applications. Materials 10:1286
Mukhopadhyay M (2004) Mechanics of composite materials and structures. University Press, Hyderabad
Rahimi H, Hutchinson A (2001) Concrete beams strengthened with externally bonded FRP plates. J Compos Constr 5(1):44–56
Rajesh M, Pitchaimani J (2016) Dynamic mechanical analysis and free vibration behavior of intra-ply woven natural fiber hybrid polymer composite. J Reinf Plast Compos 35:228–242
Rajesh M, Pachamanca J, Rajinib N (2016) Free vibration characteristics of banana/sisal natural fibers reinforced hybrid polymer composite beam. Procedia Eng 144:1055–1059
Rajini N, Jappes JTW, Rajakarunakaran S, Jeyaraj P (2012) Mechanical and free vibration properties of montmorillonite clay dispersed with naturally woven coconut sheath composite. J Reinf Plast Compos 31:1364–1376
Ray D, Sarkar BK, Rana AK, Bose NR (2001) Effect of alkali treated jute fibres on composite properties. Bull Mater Sci 24(2):129–135
Rout J, Misra M, Tripathy S, Nayak S, Mohanty A (2001) The influence of fiber treatment on the performance of coir-polyester composites. Compos Sci Technol 61:1303–1310
Sadeghian P, Hristozov D, Wroblewski L (2018) Experimental and analytical behavior of sandwich composite beams: comparison of natural and synthetic materials. J Sandwich Struct Mater 20(3):287–307
Sen T, Reddy HNJ (2011) Finite element simulation of retrofitting of RCC beam using sisal fibre composite (natural fibre). In: 3rd international conference on electronics computer technology, Kanyakumari, India, pp 29–33
Sever K (2009) The improvement of mechanical properties of jute fiber/LDPE composites by fiber surface treatment. J Reinf Plast Compos 29(13):1921–1929
Singh SB (2014) Analysis and design of FRP reinforced concrete structures. McGraw Hill Education(India) Private Limited, New Delhi, India
Suarez SA, Gibson RF, Sun CT, Chaturvedi SK (1986) Influence of fiber length and fiber orientation on damping and stiffness of polymer composite materials. Exp Mech 26:175–184
Suhad DS, Mohaiman JS, Leman Z, Sultan MTH, Ishak MR, Cardona F (2015) Physical, mechanical, and morphological properties of woven kenaf/polymer composites produced using a vacuum infusion technique. Int J Polymer Sci 894565
Susheel K, Kaith BS, Inderjeet K (2009) Pretreatments of natural fibers and their application as reinforcing material in polymer composites-a review. Polym Eng Sci 49:1253–1272
Tong FS, Siew CCSC, Doh SI, Gimbun J (2017) Natural fiber composites as potential external strengthening material—a review. Indian J Sci Technol 10(02)
Venkateshwaran N, Elayaperumal A, Arunsundaranayagam D (2013) Fiber surface treatment and its effect on mechanical and visco-elastic behavior of banana/epoxy composite. Mater Des 47:151–159
Wallenberger FT, Weston N (2003) Natural fibers, plastics and composites. Kluwer Academic Publishers
Wool RP, Sun XS (2005) Bio-based polymers and composites. Elsevier, Amsterdam
Yan L, Chouw N (2013) Crashworthiness characteristics of flax fibre reinforced epoxy tubes for energy absorption application. Mater Des 51:629–640
Yan L, Chouw N, Jayaraman K (2014) Effect of triggering and polyurethane foam filler on axial crushing of natural flax/epoxy composite tubes. Mater Des 2014(56):528–541

Yan L, Chouw N, Jayaraman K (2014) Flax fibre and its composites—a review. Compos B 56:296–317
Zhi RM, Qiu ZM, Yuan L, Cheng YG, Min ZH (2001) The effect of fiber treatment on the mechanical properties of unidirectional sisal reinforced epoxy composites. Compos Sci Technol 61:1437–1447

Structural Response of FRP Truss Bridge

Himanshu Chawla and S. B. Singh

Abstract This chapter focus on the current state of the steel truss bridges in construction and development of alternative truss bridge using FRP. A model of FRP truss bridge is fabricated at structural testing center of BITS Pilani. In order to overcome the deficiencies of steel bridges, a FRP truss bridge is proposed. The experimental response of the FRP truss bridge is predicted under static loading. Further, truss bridge is modelled using ABAQUS and it is observed that experimental and numerical responses are in good agreement with each other. From the numerical study, it is noted that without stiffening of the top chord, truss deflect in the lateral direction as well as top chord buckles before overall failure of the bridge. From this study, it is concluded that connection of the top chord with horizontal strut is more efficient than vertical strut. Top chords and struts of FRP truss bridges made by CFRP and the floor beams made by GFRP box section, are economical and efficient for heavy loading.

1 Introduction

Trusses are one of the most widely adopted structural systems used in different applications such as bridges, roofs, cranes, aircrafts, and even robots. Trusses have triangular arrangement of members such as angles, channel sections, etc. Triangles provides strength to the structure through tension and compression. Truss structures are economical and efficient load carrying structural systems. Some examples of truss structures are shown in Fig. 1.

Truss bridges are most suitable for pedestrians, roads and railway bridges. These bridges compete against plate girders for shorter spans, against box girders

H. Chawla (✉)
Civil Engineering Department, TIET-Patiala, Patiala, India
e-mail: himanshu.chawla@thapar.edu

S. B. Singh
Civil Engineering Department, BITS Pilani, Pilani 333031, India
e-mail: sbsingh@pilani.bits-pilani.ac.in

S. B. Singh et al. (eds.), *Emerging Trends of Advanced Composite Materials in Structural Applications*, Composites Science and Technology,
https://doi.org/10.1007/978-981-16-1688-4_15

(a) Truss bridge (b) Effile tower

Fig. 1 Applications of truss structures

for medium spans and cable-stayed bridges for long spans. Since beginning steel members have been used in truss bridges. Steel bridges have lot of disadvantages such as they are highly susceptible to corrosion, heavy, have high transportation and maintenance costs. Due to various pitfalls, there is a need to replace steel with fiber reinforced polymer (FRP) in construction.

During the Second World War, glass fiber reinforced polymer (GFRP) was first used to build aircrafts. Later in the late 1950s, carbon fiber was discovered and widely used in aviation, automobiles and sports industries. Since three decades, FRP has replaced the steel in bridges such as carbon fiber reinforced polymer (CFRP) tendons, CFRP reinforcement bars in deck slabs and GFRP railings (Bakis et al. 2002). Like in bridges, FRP can replace steel members in truss bridges to overcome deficiencies of steel truss bridges. Some of the issues related to the earlier trends in construction of steel truss bridges (Acton 2013) are as follows:

- Maintenance cost such as protection from rust, which is due to marine (salt spray) exposures and industrial environment (Fig. 2).
- Steel truss bridges are very heavy, therefore transportation cost is high and assembly of elements is not easy.
- Repairing cost is too high.
- Regular painting is required for aesthetic point of view.

With these deficiencies, there is a need for development of new material to be used in truss bridges. The various advantages of FRP (Raasch 1998) in construction are as follows:

- FRP resists a broad range of chemicals and is unaffected by moisture or immersion in water.
- It has high strength-to-weight ratio and stiffness-to-weight ratio.
- FRP is a good insulator with low thermal conductivity.
- Nonconductive and high dielectric capability.

(a) Honore Mercier Bridge, Canada

(b) Devil's Elbow Bridge Missouri, United States

Fig. 2 Corrosion of steel truss bridge

2 Salient Features of FRP Truss Bridge

- Angle sections and gusset plate were fabricated from hand-layup process, while the box section was made by pultrusion process.
- All the structural elements of FRP bridge were connected with steel bolts and nuts.
- Deck slab is supported on FRP box sections (floor beams), while box sections are connected to truss through gusset plate, while plane trusses are connected with each other by top struts, which resists the out-of-plane bending of truss (see Fig. 3).
- Stacking sequence of structural elements is given in Fig. 4.
- Top chords and diagonal members are angle section with stacking sequence of the laminae is shown in Fig. 4a.
- Use of random oriented chopped strands mat (CSM) is for uniform distribution of strength in all directions.

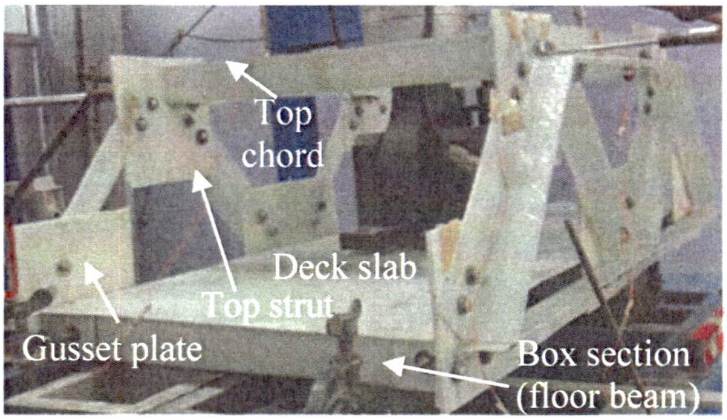

Fig. 3 FRP truss bridge

Fig. 4 Stacking sequence of truss bridge elements

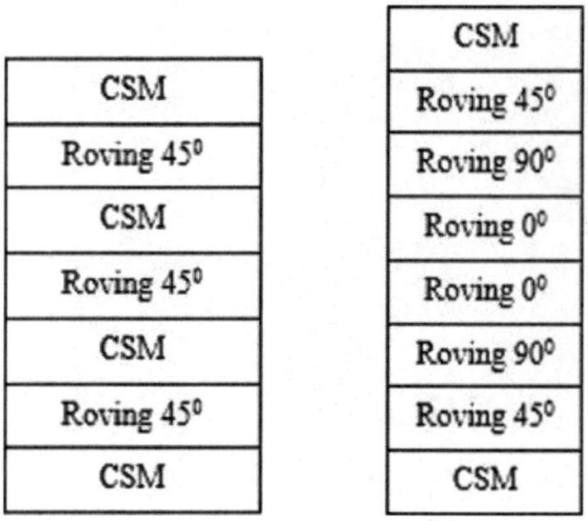

(a) Angle section

(b) Box section & gusset plate

3 Fabrication Process

The structural elements of FRP truss bridge were made by hand layup process and pultrusion technique. FRP angle sections and plates were made by handlayup process. In order to make angles, first channel section was made by orientations of the lamina shown in Fig. 4. Then channels were cut from middle along the length of the channel to get two angle sections. Handlayup process for fabrication of channel is described below:

- Firstly, wooden moulds of channel section was made. A mixture of resin and hardener was prepared. Amount of hardener used was 8% of weight of epoxy. The paste of resin and hardener is spread evenly over the mould using putty knife.
- A layer of fibers was wrapped around the mould above the layer of resin. Again, paste is spread over the fabrics and the layer is compressed using roller to squeezed out the extra resin from fiber and to obtain uniformity. Second layer of fibers was again placed on this layer and the same procedure was repeated, till all the plies placed.
- Similarly, another FRP channel section was made. A polythene sheet was wrapped on each mould separately, for avoiding the contact of layers of both moulds.
- In order to compress the layers of web of channel section, both moulds were placed back-to-back and the ends of moulds were connected together by bolts.
- Fiber layers in the flanges of the channel section were compressed by placing wooden plates over the flanges of moulds. Heavy weights were kept over the plates.

Moulds and plates were removed after 24 h. Later, specimens were cured under open atmosphere for 7 days, then it was further used in fabrication of truss bridge (Fig. 5).

Fig. 5 Fabrication process of FRP channels/angles using handlayup process

(a) Spreading of resin over fiber

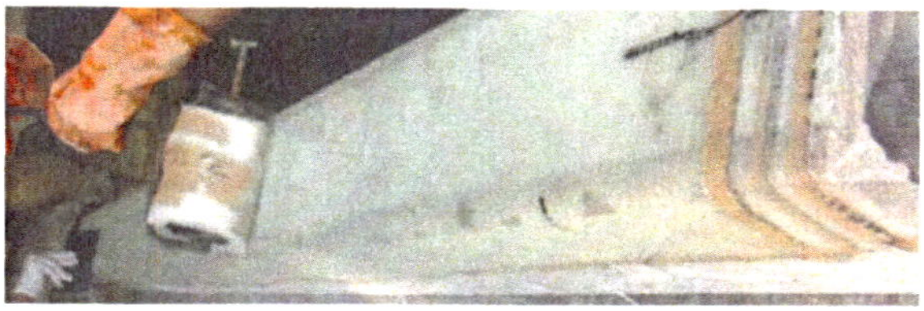

(b) Compaction the layers of fiber over mould

(c) Joining the two moulds together

Similarly, FRP gusset plate and deck slab was fabricated by handlayup process. These elements were fabricated by compressing the FRP layers with matrix between two flat plates. Box section was fabricated by pultrusion technique and was supplied by Agni fiber pultrusion private limited. At a big scale, angle section can also be made by pultrusion technique.

4 Experimental Study

The performance of the FRP bridge was predicted by test under static loading (Harik 1999). Load was applied on the middle girder of the bridge and simply supports were provided under the end girders of the bridge as shown in Fig. 3. Deflection of the top strut of bridge was measured using linear variable displacement transducer (LVDT). Bridge was loaded till the buckling of top chord. Predicted load–deflection response is presented in the Fig. 6.

Fig. 6 Load versus deflection response of the truss bridge under loading

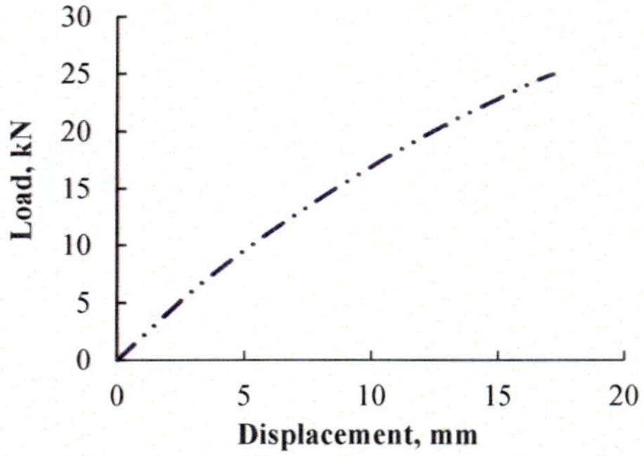

5 Numerical Modelling

FRP truss bridge is modelled in the finite element software such as ABAQUS 6.13 (ABAQUS 2013). Beams and stiffening elements are modeled with linear shell element with reduced integration, i.e., S4R. Load is applied through two bearing plates similar to experimental test. Bearing plates are modeled with shell discrete rigid object (i.e., undeformable object). Bearing plates are restrained to displace in transverse direction and allowed to translate in vertical direction, while restrained to rotate about vertical axis. Interaction between bearing plate and the beam is given 'Surface-to-surface interaction'. The contact property in tangential direction (i.e., parallel to interaction) is given penalty with friction coefficient 0.3, while in normal direction (i.e., perpendicular to the interaction), contact is assigned by hard contact. Structural elements of the bridge were connected using fasteners. Truss bridge modelled in ABAQUS is shown in Fig. 7. First, linear buckling analysis is performed to determine the elastic critical buckling load using 'Buckle' step under Linear perturbation module. Later, post-buckling response of the truss is determined for the first mode shape. Imperfection of 0.1% of depth of floor beam is imparted in this analysis.

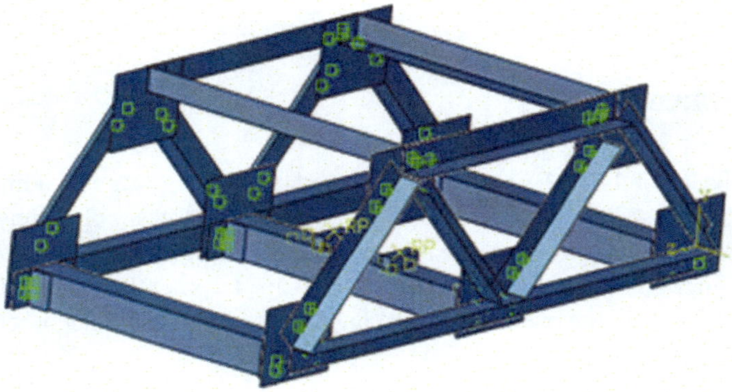

Fig. 7 FRP truss bridge modelled in ABAQUS

6 Results and Discussions

The load versus deflection responses of bridge obtained from experimental investigation and using ABAQUS is presented in Fig. 8. In this figure, 'EXPT' denotes experimental response and 'WVH' denotes ABAQUS response without vertical and horizontal stiffening elements. The results obtained from ABAQUS are in good agreement with experimental results.

Figure 9 shows that top chord buckled and trusses deflected in lateral direction during loading. The stresses in the truss determined from numerical study reveals that buckling of top chord is due to the compressive stress. Buckling of the top chord can be resisted by three methods: first, stiffening of top chord with struts; second, increasing the thickness of the chord and third, by replacing the low stiff layers of

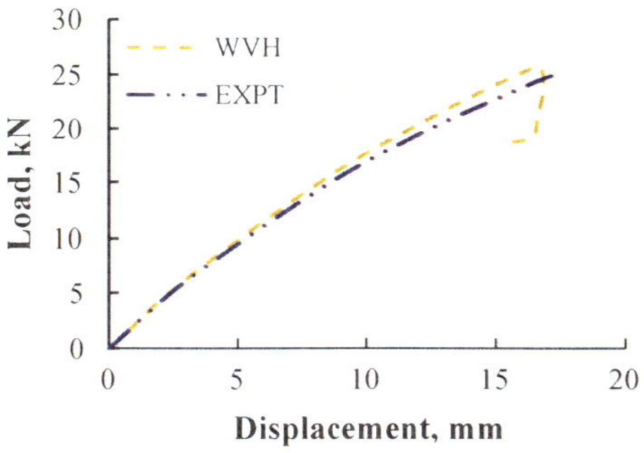

Fig. 8 Comparison of load vs deflection response of truss bridge obtained from experimental test and ABAQUS

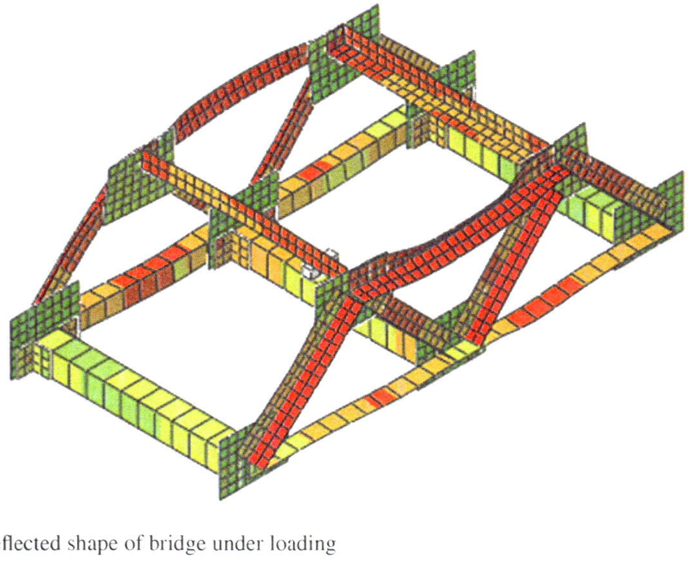

Fig. 9 Deflected shape of bridge under loading

chord with high stiff layers such as carbon fibers. Effect of stiffening of the top chord on the flexural strength of the bridge is explained in the following sections.

(a) **Addition of horizontal strut**
In order to restrain the buckling of top chord and lateral deflection of the truss, top chords of both trusses are connected together with a horizontal strut as shown in Fig. 10. Lateral buckling of the truss was restrained but the chord buckled locally. The load versus deflection response of the truss with and without horizontal strut is shown in Fig. 11. It is observed that flexural strength of the truss bridge increased with addition of horizontal strut.

(b) **Addition of vertical member**
In another bridge, a vertical member is provided under the top chord of both truss. One end of the vertical member is connected with top chord and other end is connected with gusset plate. The deflected shape of the truss obtained from

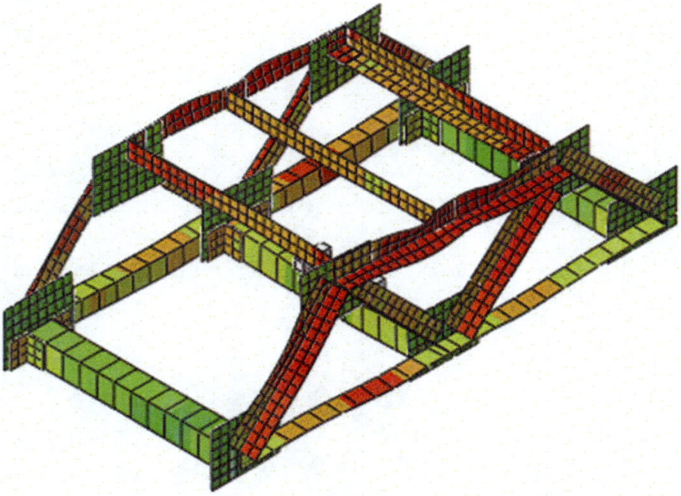

Fig. 10 Deflected shape of bridge with horizontal strut under loading

Fig. 11 Load versus deflection responses of truss bridge with addition of horizontal strut

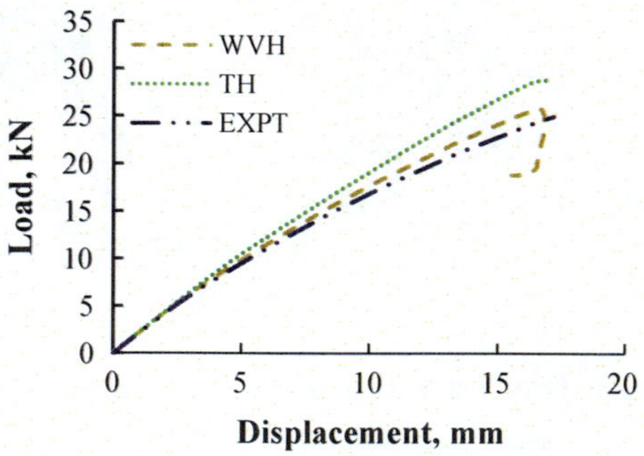

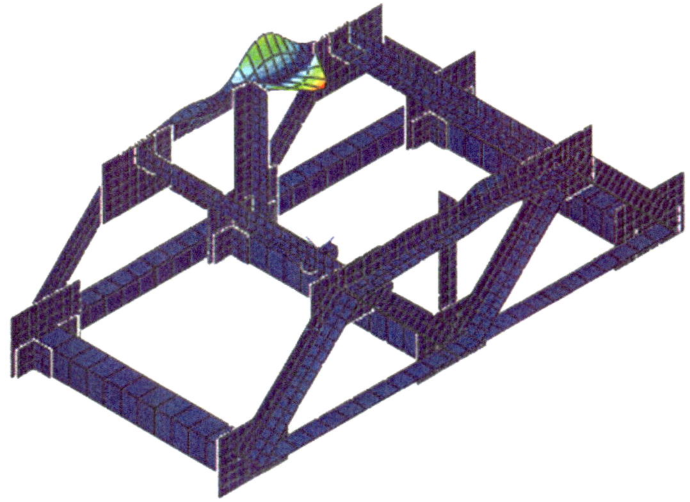

Fig. 12 Deflected shape of bridge with vertical member under top chord

ABAQUS is shown in Fig. 12. It is observed that lateral buckling of the top chord is restrained but trusses buckled laterally. The comparison of responses of the bridge with and without vertical members is shown in Fig. 13. In this figure, 'TV' denotes truss with vertical member. It is noticed that there is little increase in the stiffness of the bridge with addition of vertical member.

(c) **Addition of horizontal strut and vertical member**

In this section, the effect of addition of horizontal and vertical struts on the flexural response of the bridge is studied. The flexural strength of bridge obtained is more than bridge stiffened with horizontal or vertical members only. Response of bridge (THV) is nearly equals to response of bridge with horizontal strut (TH) (see Fig. 14) while lateral displacement of the bridge is restrained (see Fig. 15).

(d) **Effect of addition of fabrics**

Fig. 13 Flexural response of the bridge with vertical member under top chord

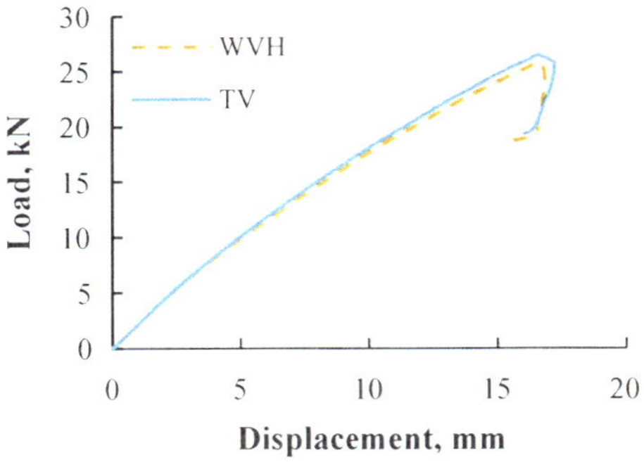

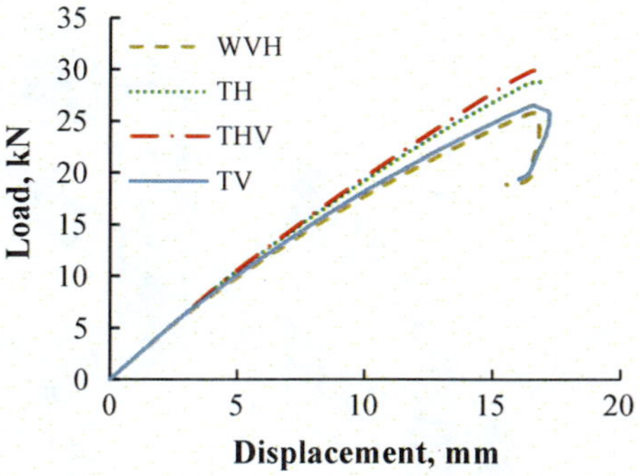

Fig. 14 Flexural response of the bridge with vertical member under top chord and horizontal strut

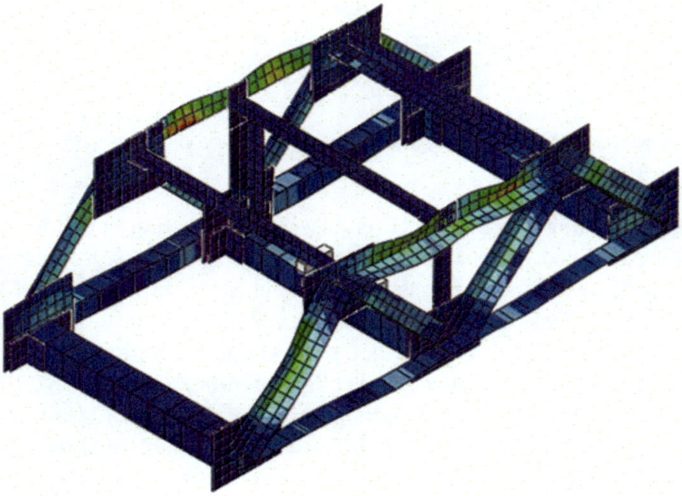

Fig. 15 Deflected shape of bridge with vertical member under top chord and horizontal strut

In this section, effect of addition of number of layers in top chord on the flexural strength is determined. Number of uni-directional glass fabric layers in 0° (longitudinal axis) is increased to enhance the buckling load of the bridge. Table 1 presents the percentage increase in the ultimate load carrying capacity with increase in the thickness and addition of layers. In the specimen $(0/90/0/0/0/0/0/0)_{2s}$ the failure mode changed from buckling of top chord to buckling of bottom chord.

(e) **Effect of addition of high stiff layer**

The ultimate load carrying capacity of the bridge is predicted by increasing the high stiffness layers (i.e., carbon layers) in the top chord. Increase in the load carrying capacity with addition of high stiffness layers is presented in Table 2. It is noted that addition of two 0° layers in the specimen $(0/90)_{2s}$ increase in the ultimate load is

Table 1 Effect of addition of fabrics in top chord on buckling load of bridge

Stacking sequence	Thickness (mm)	Ultimate load (kN)	% increase in load carrying capacity
$(0/90)_{2s}$	2.54	24.32	–
$(0/90/0/0)_{2s}$	3.81	26.78	9.18
$(0/90/0/0/0/0)_{2s}$	5.08	29.12	16.48
$(0/90/0/0/0/0/0/0)_{2s}$	6.35	32.56	25.31
$(0/90/0/0/0/0/0/0/0/0)_{2s}$	7.62	35.03	30.57

Table 2 Effect of addition of carbon fabrics in top chord on buckling load of bridge

Stacking sequence	Thickness (mm)	Ultimate load (kN)	% increase in load carrying capacity
$(0/90)_{2s}$	2.54	24.32	-
$(0/90/0/0)_{2s}$	3.81	28.56	17.43
$(0/90/0/0/0/0)_{2s}$	5.08	31.86	31.00
$(0/90/0/0/0/0/0/0)_{2s}$	6.35	34.81	43.13
$(0/90/0/0/0/0/0/0/0/0)_{2s}$	7.62	37.78	55.34

increased for specimen is 4.24 kN, while the addition of two layers in the specimens $(0/90/0/0/0/0)_{2s}$ and $(0/90/0/0/0/0/0/0)_{2s}$ is aprox. 3.3 kN.

7 Concluding Remarks

In this study, experimental and analytical responses of the FRP truss bridge fabricated at structural testing center of BITS Pilani is predicted. Flexural strength is determined with addition of horizontal, vertical struts and fiber layers in top chord. Based on this study following conclusions can be made:

- FRP truss bridges are suitable for light loadings or light weight vehicles such as foot-over bridges, walkways, etc. FRP truss bridges can replace traditional steel bridges.
- Increasing the thickness of struts increases the load carrying capacity of FRP truss bridges.
- Connection of top chord by horizontal strut are more effective than connection by vertical strut.
- Chords and struts of FRP truss bridges made by CFRP, while the floor beams made by GFRP box section, is economical and efficient for heavy loading.

References

Acton A (2013) Issues in structural and materials engineering. Scholarly Editions, USA

ABAQUS Standard User's Manual (2013) Version 6.13, Pawtucket RI: Hibbit, Kalsson & Sorensen Inc., ABAQUS

Bakis CE, Bank LC, Brown VL, Cosenza E, Davalos JF, Lesko JJ, Machida A, Rizkalla SH, Triantafillou TC (2002) Fiber-reinforced polymer composites for construction—state-of-the-art review. J Compos Constr 6(2):73–87

Harik, Static testing on FRP bridge deck panels. In: Proceedings of 44th International SAMPE symposium and exhibition, vol 2. Society for the Advancement of Material and Process Engineering, Covina, California, pp 1643–1654

Raasch JE (1998) All-composite construction system provides flexible low-cost shelter. Compos Sci Technol 4(3):56–58

A Review of Natural Fiber Composites for Structural, Infrastructural and Ballistic Applications

P. Siva Sankar and S. B. Singh

Abstract Natural fiber composites are becoming suitable alternative material to replace synthetic fibers like carbon, Kevlar and E-glass. Natural fibers are having the properties of biodegradability, renewability, and low density. Nowadays, researchers are showing interest in natural fibers to draw natural fiber-reinforced composites (NFRC). NFRC is generally used in the construction sector for non loading members and researchers are evaluating the applications of NFRC in load-carrying members of the structural and infrastructural areas. The hybridization of natural fibers and synthetic fibers have good results over the plain natural fiber composites. The main problem with natural fibers is that properties are significantly influenced by the source of fiber, type of fiber, and fiber treatment. This chapter presents the review of natural fiber composites applications in structural, infrastructural, ballistic, and fire resistance wings.

Keywords NFRC · Plant fiber · Hybridization · Fire resistance

1 Introduction

Synthetic fibers are commonly used in different engineered structures due to their high specific stiffness and high specific strength. The fibers like Kevlar, glass and carbon are mainly used in the industries of aircraft and aerospace. Synthetic fibers are reinforced with the polymer type of matrix to provide cushioning for fibers to protect them from the impact loads. The cushioning action of matrix also plays a crucial role in seismic attacks by absorbing the rigid structures vibration energy. There are many benefits with the advanced synthetic composites, even though the public commonly criticizes few issues:

P. Siva Sankar (✉) · S. B. Singh
Department of Civil Engineering, Birla Institute of Technology and Science, Pilani 333031, India
e-mail: p20190444@pilani.bits-pilani.ac.in

S. B. Singh
e-mail: sbsingh@pilani.bits-pilani.ac.in

S. B. Singh et al. (eds.), *Emerging Trends of Advanced Composite Materials in Structural Applications*, Composites Science and Technology,
https://doi.org/10.1007/978-981-16-1688-4_16

1. Disposal of synthetic fibers is creating environmental issues; still, the recycling process of the synthetic fiber composites is not developed correctly.
2. The structures made by the synthetic fibers are sometimes having over strength and it is not required to maintain that much strength; this problem mainly occurs when the carbon fiber composites are used.
3. Using synthetic fibers in the construction increases the material cost.

Therefore, an alternative material to be identified to overcome the effects of synthetic fibers. For the past few decades, researchers are working on composites that are available naturally (Lau et al. 2018). Natural fibers like hemp, flax and jute have gained the researcher's interest by due to their specific properties like recyclability, renewability, low energy production and short growth cycle time (Bambach 2020). The main advantage of using natural fibers is that the composites are lighter in weight and bio-degradable. The cost of manufacturing composites is also significantly cheaper compared to advanced synthetic fibers. By considering the various advantages of natural fibers, a lot of research work is carried out in the past. Most of the natural fibers are environmentally friendly but not all the fibers.The best natural composite is made with the fiber which is available from the short renewable sources of the plant and easily biodegradable under the suitable conditions (Dittenber and Ganagarao 2012). Natural fibers are preferable to replace the E-glass fibers when the load applied to the member is medium or low. These fibers are reinforced with the matrix in different forms like woven, typically short and random orientation. Among all these orientations, the natural fibers which are placed in the woven form influence the mechanical and dynamic properties of the composite (Ramana and Ramprasad 2017).

Natural fibers are hybridized by another natural type of fiber or with synthetic fiber. This type of hybridization enhances the mechanical properties of the composites. Hybridized composites are mainly used for the structural and infrastructural point of view based on their load carrying capacity, energy absorption properties, housing panels and stability (Safri et al. 2018). People are getting more interest in the applications of natural fiber composites in structural and infrastructural elements where lower cost, environmentally friendly and moderate strength are required. The hybridized natural fiber composites develop load-bearing features like beams, roof, housing panels, pedestrian bridges, and water tanks in structural and infrastructural applications. Natural fibers are also used in non-loading members like wall cladding, insulation panels, and architectural elements. Natural fiber composites are also used for ballistic applications to safeguard humans against the impact and shock loads (Ticoalu et al. 2010).

1.1 Types of Natural Fibers

Nowadays, researchers are mainly focused on the natural fibers due to an environmental point of view and the increasing cost of synthetic fibers (Khan et al. 2018).

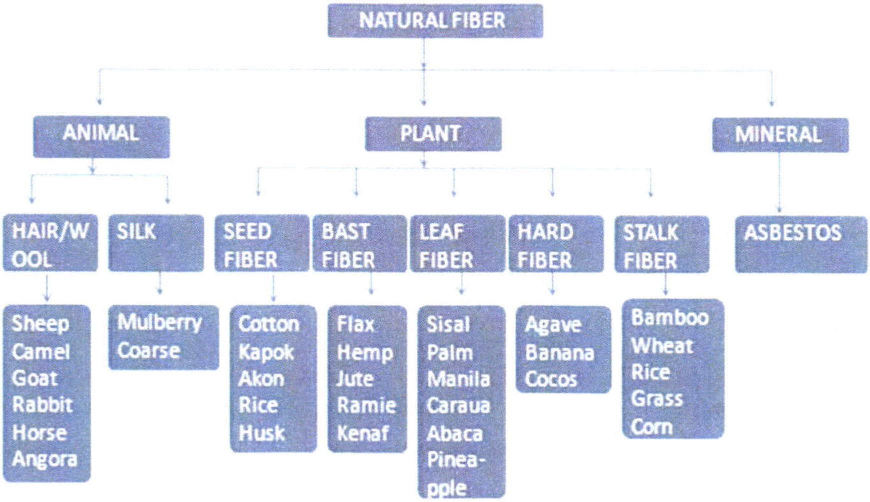

Fig. 1 Sources of natural fibers (Kumar et al. 2019)

Natural fibers are generally available naturally, and the sources of the natural fibers are mainly two categories, i.e., plant-based fibers and animal-based fibers. Plant-based fibers are derived from the plant's different parts like bast, leaves, fruits, and stalks. As same, the animal-based fibers are also derived from the body parts of the animal like fur, shell, bone, etc. (Nayak et al. 2020). Natural fibers are generally harvested once a year and few crops like hemp, jute and kenaf can be produced two times for a year. Natural fibers are having a high rate of assimilation of CO_2 and reducing the carbon content from the atmosphere. Figure 1 shows the different types of natural fibers based on the source of plant and animal, and Fig. 2 shows the life cycle of the natural fiber composites (Kumar et al. 2019).

1.2 Chemical Composition of Fibers

Cellulose is the major component in the natural fibers that develop the stability and strength of the cell as well as fiber. The amount of percentage of cellulose in the fibers influences the mechanical properties, economics of fiber production and usage of fibers in various types of applications. Cellulose contents in the fiber are used to identify the potential application of that fiber. If the fiber's cellulose content is very high, it is used for textiles and if the hemicellulose content is very low, then it will be used for making the ethanol.

Hemicellulose is slightly crosslinked when compared to cellulose and it is the combination of multi-polysaccharide polymers with the degree of polymerization and orientation is less than the cellulose. Generally, the filler part between the cellulose and lignin is hemicellulose, consisting of mannose, glucose and galactose. Lignin

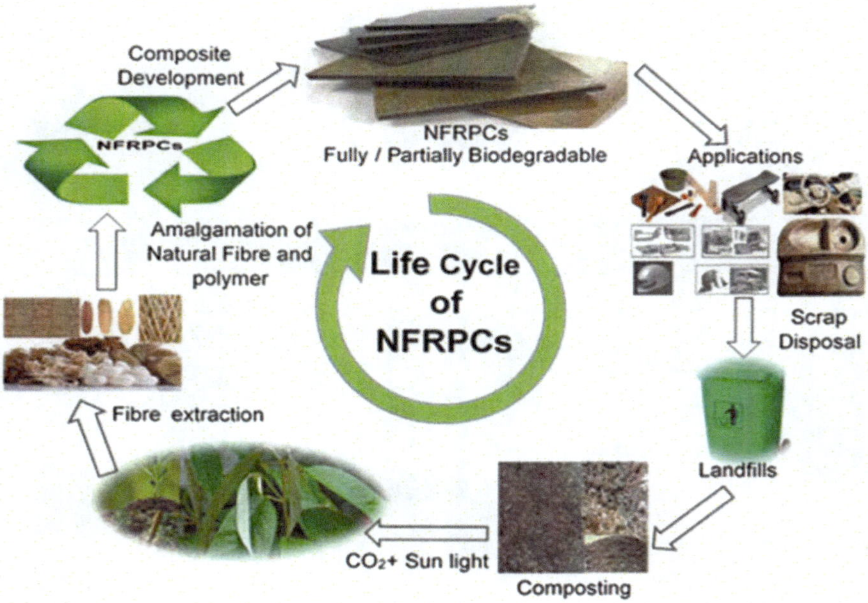

Fig. 2 Life cycle of Natural Fiber Reinforced Composites (NFRP) composites (Khan et al. 2018)

is a highly crosslinked molecular complex and acts like glue between fibrils, which form the cell wall.

Lignin provides the strength and stiffness of the cell wall and it protects carbohydrates from physical and chemical attacks. Natural fibers, which are having a high content of lignin, appears like finer and flexible (Reddy and Yang 2005). The strength of the fibers also depends upon the microfibrillar angle and aspect ratio. The higher strength of the fiber achievable when the microfibrillar angle of the fiber is low. Addition of lignin to the composites beneficial to the fibers up to a specific limit. Kraft lignin is added to the natural hemp fiber and it increases the impact, flexural and tensile strength of the composite. Simultaneously, if the added amount of lignin is excessive, it shows adverse effects on the composite (Wood et al. 2011). Cellulose, hemicellulose and lignin amount values vary from plant to plant and fiber type. Table 1 represents the chemical properties of the different kinds of natural fibers (Khan et al. 2018).

1.3 Mechanical Properties of Natural Fibers

Generally, the type of fiber is based on its source: animal, plant, or mineral. Cellulose is the main constituent of the fibers and protein is the main constituent in animal fibers. Due to the higher strength and stiffness, the plant fibers are used over the animal fibers.

Table 1 Chemical properties of natural fibers (Khan et al., 2018)

Fiber type	Chemical composition percentage (%)					
	Cellulose	Hemicellulose	Lignin	Pectin	Waxes	moisture
Flax	64.1	16.7	2	1.8	1.5	10
Hemp	55–80.2	12–22.4	2.6–13	0.9–3	0.2	6.5
Jute	64.4	12	0.2	11.8	0.5	10
Remie	68.9	13.1	0.6	1.9	0.3	10
Kenaf	37–49	18–24	15–21	8.9	0.5	–
Sisal	65	12	9.9	0.8	0.3	10
Cabuya	68–77	4–8	13	–	2	–
Abacca	56–63	15–17	7–10	–	3	–
Coir	19.9–36.7	11.9–15.4	32.7–53.3	4.7–7.0	–	0.2–0.5
Banana	48–60	10.2–15.9	14.4–21.6	2.1–4.1	3–5	2–3
Betelnut	35–64.8	29–33.1	13–26	9.2–15.4	0.5–0.7	–
Rice	28–48	23–28	14	–	20	–
Wheat	29–51	26–32	16–21	–	7	–
Oat	31–48	27–38	16–19	–	7.5	–
Sea grass	57	38	5	10	–	–
Bagasse	28.3–55	20–36.3	21.2–24	–	0.9	–
Bamboo	48.2–73.8	12.5–73.3	10.2–21.4	0.37	–	11.7

The top performance is mainly secured with the high cellulose percentage and the alignment of microfibrils in fiber direction, which is occurring in bast fibers(flax, hemp, jute, and kenaf). While comparing the plant fibers with the synthetic fibers, they have lesser strength. Still, the specific modulus and elongation at break have signified the importance of natural fibers to replace the synthetic fibers.

By considering the mechanical properties of different types of bast fibers, jute, flax, and hemp fibers will give better results compared to other natural products (Kabir et al. 2012) (Table 2).

1.4 Chemical Treatment of Natural Fibers

The main drawback of the natural fibers composite is that the combination of the hydrophilic nature of the fiber and hydrophobic nature of the matrix. Due to these different phases of the mixture, a weakening bond may be developed at the surface. To improve the significance of the fiber's interfacial bonding, surface is to be treated with different chemicals. By conducting the chemical treatment on the fiber, the hydrophilic nature would reduce and increasing the compatibility with the matrix.

Table 2 Mechanical properties of natural fibers (Khan et al. 2018)

Fiber type	Physical and mechanical properties of natural fibers and natural fibers						
	Density (g/cm^3)	Tensile strength (MPa)	Specific strength (MPa)	Youngs modulus (GPa)	Specific modulus (GPa)	Elongation (%)	Specific gravity
Flax	1.40	340–1600	535–1000	25–81	16.7–54	1.1–3.3	1.5
Hemp	1.48	550–900	372–608	70	–	0.8–3	1.5
Remie	1.50	200–1000	147–625	41–130	27–81	1.5–4	4
Kenaf	1.20	223–1191	641	11–60	10–42.9	1.6–4.3	1.3
Jute	1.46	385–850	269–548	9–31	6.9–20.7	1.4–2.1	1.3–1.5
Sisal	1.33	400–700	366–441	8.5–40	6.5–30.8	1.9–15	1.3
Abacca	1.80	980	–	–	–	–	–
Coir	1.25	170–230	146	3–7	2.5–5	14–30	1.2–1.4
Banana	1.35	711–789	444	4–32.7	3.6–27.3	2.4–3.5	1.1–1.2
Betelnut	0.2–0.4	120–166	–	1.3–2.6	1.0–1.9	22–24	1.3–1.4
Wheat	1.45	–	–	–	–	–	–
Elephant grass	0.817	185	–	7.4	–	3	–
Bamboo	1.10	500–575	454	27–40	50–67.9	1.9–3.2	0.4–0.8
Bagasse	1.50	170–350	–	5.1–6.2	3.6–4.1	6.3–7.9	1.4–1.5

Chemical treatment also improves the mechanical properties of the composites significantly.

Commonly used chemical treatments are alkaline, silane, acetylation, permanganate, isocyanate, and benzoylation. Out of these chemical treatments, alkaline treatment is the best chemical treatment for the natural fibers due to its inherent advantages. The alkaline concentration changes the densely packed crystalline to the amorphous phase in the fiber, and this phase will allow the penetration of chemicals (Ghelli and Minak 2011). The alkaline will reduce the amount of wax, oils, and lignin, which covered the outer part of the fiber cell wall. Fibers are immersed in NaOH for a certain time to influence the composition of natural fibers and increase the fiber's mechanical properties. For a flax-epoxy composite, the tensile properties have increased up to 30% by conducting alkaline treatment. In the case of sisal-polyester composite, treatment with 4%NaOH has given maximum tensile strength properties. As in the case of hemp fiber, 8% NaOH increased thermal stability. In the case of jute-vinyl ester the fiber treated with the 5%NaOH has increased the flexural strength by 20% and tensile strength by 19% (Reddy and Yang 2005). At the same time, the excess usage of alkaline will lead to the damage of the fiber (Ismail et al. 2019).

2 Applications of NFRP Composites in Structural Member

At present, the researchers are showing more attention towards natural fiber-reinforced composites' applications in a structural and infrastructural applications where the materials need to be lower in weight and moderate strength is required. Natural fibers composites are used in automotive body parts, wind blades, and non-structural elements from the past decades. While considering the structural and infrastructure market, the natural fibers are used in making load-carrying members such as beams, water tanks, columns, bridge decks for pedestrians and panels for multi-purpose (Lau et al. 2018).

2.1 *Mechanical Properties of the Different Types of Synthetic/Natural Hybrid Composites*

Hybrid composite is generally a combination of two or more fillers into a matrix. Fillers are mainly characterized based on their origin, i.e., either natural or synthetic fibers. Natural fibers have some specific advantages compared to synthetic fibers in the aspects of low weight, renewable and sustainability. Synthetic fibers will give better strength when compared to natural fibers. Many researchers have found that natural fibers are hydrophilic and synthetic fibers are hydrophobic. Both the type of fibers are having certain limitations when compared to each other. Researchers identified that to increase the performance of the composite, the hybridization of natural fibers and synthetic fibers is needed. And the mechanical properties of these hybrid composites are mainly based upon the selection of matrix, surface modifications, the orientation of the fiber and interfacial adhesion of different types of lamina (Mochane et al. 2019).

2.2 *Performance of Synthetic-Natural Fibers Hybrid Composites*

To enhance the performance of the resultant composite, the hybridization of synthetic and natural fibers is required. Generally, natural type of fibers are attracted towards moisture and it will affect the composite performance. To reduce the moisture absorption, surface modification was required by the chemical treatment. These chemicals sometimes harsh in nature and damage the fiber properties. When the hybridization takes place between the synthetic and natural fibers, then it will reduce the moisture absorption, maintain the balance between the cost of the materials and it gives a chance to change the surface modification chemical. The fiber fraction will influence the performance of the composite. The addition of carbon content increases the tensile properties of the composite and strength of the composite.

Akil et al. (2014) have investigated the synthetic-natural hybrid composite's performance by considering the glass fiber as synthetic and jute fiber as a natural type of fiber. These fibers are added to the polyester resin to manufacture the composite. The fibers are having the uni-directional. The composites are manufactured by considering a high fiber to matrix ratio of 70:30 and from the volume point of view, the ratio was considered as 50:50. Tests are conducted on four types of laminates by changing the lamina's fiber orientation and stacking sequence. They have conducted the moisture absorption test and the tensile test to determine the properties of the composite. Based on the results, Akil et al. (2014) identified that the hybrid composite has superior mechanical properties over the nonhybrid composite made with jute fibers only. The hybridization also reduced the moisture absorption of the composite. This type of hybridization makes the composite perform better and balances the cost of the materials.

Ramana et al. (2017) investigated on the jute-carbon hybrid epoxy composites. To reinforce the fibers, the thermosetting resin epoxy has been used. Four plates have been manufactured by fixing the epoxy percentage as 55% and changing the carbon and jute laminas. These composites include plain jute composite, plain carbon composite and jute carbon composite. By conducting the tests the tensile strength, flexural strength, and impact strength have been examined. Based on the results, the tensile strength of jute/carbon composite is almost equal to the carbon epoxy and is more than the jute epoxy composite. While considering the impact strength, the jute/carbon hybrid performs similar to the jute-epoxy composite and much more than the carbon-epoxy composite. The flexural strength of the composite made with jute/carbon has served better than the othercomposites.

Kureemun et al. (2018) have conducted the tests on carbon- flax hybrid composite. These fibers combined with the thermosetting resin, i.e. epoxy for fabricating the composites. By considering the different volume fractions of the carbon and flax, the laminates are made. For every laminate, they have considered four laminas by changing the fiber orientation and stacking sequence. The fiber volume fractions of plain carbon epoxy and plain flax epoxy are 66 and 29%. From the results with the 8% of carbon by volume the strength and stiffness are increased more than 50 and 30% increse in specific strength compared to aluminum. Overall the hybrid composite has increased the mechanical properties over the non-hybrid flax composite.

2.3 *Performance of Natural-Natural Fibers Hybrid Composites:*

The hybrid composites can be manufactured with even natural-natural type of fibers. The composites made with different natural fibers have exhibited significant advantages in various fields like automotive parts, agricultural, biomedical, residential, and building applications. These types of hybrid composites have reduced the cost of the materials, and the resultant composite weight is also reduced (Mochane et al. 2019).

The composite has been manufactured by considering the jute, kenaf fiber and hemp fiber. The epoxy resin is used to reinforce the fibers. The composites are fabricated by the woven type of mats. By changing the stacking sequence different type of composites has been produced. To identify the mechanical properties, various tests have been conducted, such as tensile test, flexural test, and water absorption test. Based on results, it was observed that the hybridization of the natural fibers reduced the water absorption capacity of the fibers. If the fibers are immersed longer, then it will effect the strength of the composite. Sometimes, the fibers' delamination and cracking in the matrix also increase the water absorption capacity and influence the composite's strength parameters. The hybridization between the different types of natural fibers has increased the tensile strength and flexural strength of the composite due to the load sharing between the different types of fibers (Maslinda et al. 2017).

The composite has been made by considering the fibers of sisal, jute and curaua fiber. The epoxy resin is used to reinforce the fibers. The fiber fraction is 30 and 70% of epoxy are considered for manufacturing the composite. The fibers are treated by the process of alkalization to increase the mechanical properties of the composite. The hybridization process produced the better results of the composites. The improvement in tensile strength of the jute/sisal, jute/curua and jute/glass over the untreated pure jute composite is 68%, 77% and 90%, respectively. The flexural strength has increased in all the composites except jute/sisal composite due to the chemical treatment. Even the impact strength of all the composites has increased due to the chemical treatment except in the case of jute/curua hybrid composite. Based on the results, it has been observed that the usage of sisal and curua fibers will replace the glass fibers in composites. These types of natural hybrid composites can be used for the low load-bearing structures (Cavalcanti et al. 2019).

2.4 Impact Characteristics of Natural Fiber Composites

Most of the engineering structures are often exposed to impact load by the foreign objects in the maintenance of the composite, repairing, manufacturing, and service conditions. When compared to the metallic structures, the laminated composites are more vulnerable to damage created by impact loading. In manufacturing or repairing the composite, sometimes the tools may fall on composite and make impact load on that. In aircraft landing, the impact load is added on the cover material by the debris projectile from the runway due to high velocity contact between runway and tires. Impact loading is generally in two forms like low velocity and high velocity. Due to the low velocity of the impact, the fibers face matrix cracking failures, internal damage, and fibers delamination. The resistance of the impact loading mainly depends on the fiber's stacking sequence and type of fiber. To examine the impact resistance of fiber, the low velocity and high velocity tests are used. In low-velocity test, the impactor is released between 5 and 15 m/s while in the high-velocity impact test the impact is 2500 m/s. The low-velocity tests damage the composite by applying the load still the composite is able to function.While in case of a high-velocity test, the composite is completely damaged due to the penetration of the impactor. The composite manufactured by the glass-carbon fiber has performed better than the metallic bumper, which is also reduced by 33% in the weight point of view. Further research has been extended to hybrid composites by replacing the synthetic fibers with the natural type of fibers. The hybridization of natural-synthetic fibers will reduce the weight of the composite (Safri et al. 2018).

Generally, for structural design, the strength of roof and impact strength are the important aspects. The coir fiber-based composites are used for automotive applications because of their higher impact strength and it is the beneficial to the automotive industry. Usually, the coir composites are low in tensile strength, but the composite's impact strength is comparatively better than other composites. The cotton fiber composites reinforced in epoxy resin also exhibit better results in past research (Ticoalu et al. 2010).

Natural-Natural type of hybridization of the composites also improves the impact strength of the composite. The impact strength of the pure epoxy is 4 kJ/m^2. The hybridization of the composite manufactured by the Jute/Hemp/Flax/Epoxy has the maximum impact strength of 10.19 kJ/m^2. While in the case of Jute/Epoxy and Jute/Hemp/Epoxy has a full impact strength of 7.68 kJ/m^2 and 6.93 kJ/m^2, respectively. These variations are developed due to the interfacial bonding between the types of natural fibers added with epoxy resin (Chaudary et al. 2018).

The hybridization of synthetic-natural fibers alters the impact strength of the composite. A Charpy test has been conducted on the flax fiber composite and the result value is only 38.4 kJ/m^2. Whereas if we consider the carbon composite, then the composite impact strength is 820% greater than the flax fiber composite. The hybridization of the flax and carbon fiber composite has improved the impact strength of about 400% over the flax fiber composite. This much of strength to the composite is incorporated by adding 12% of carbon fiber (Flynn et al. 2016).

In the process of hybridization among the different types of fibers, the stacking sequence also affects the composite's impact strength. The composite prepared by the alternative layers of synthetic and natural fibers will have greater impact strength (Arpita et al. 2017).

2.5 *Advanced Composite Beams and Columns*

The steel used in the beams and columns is viable to attract the moisture, leading to the steel's corrosion. Then the deterioration of the member will take place due to the corrosion of the steel. And considering the density point of view, the members constructed by using the steel and concrete leads to an increase in the member's weight. The advanced composite material is one of the best options to replace the steel plates or steel jackets that are used in construction. For a long span of a building, the composite columns and beams are suitable. Figure 3 shows the different types of composites beams and columns (Fan and Fu 2017).

Fig. 3 Natural fiber composites for beams and columns in construction (Fan and Fu 2017)

The beam is the primary component in any type of building, bridge or any engineered structure. The structural member which is in the mode of bending or in the mode of flexural then that member is treated as a beam. Based upon the design requirements, the shape of the beam is either in rectangular or square mostly. The flexural strain parameters, stress, modulus of elasticity, and deflection of the beams have been found out by using the three-point or four-point bending tests (Ticoalu et al. 2010).

Using the natural fibers to manufacture the beams, a composite sandwich is mainly used among the other types. By incorporation of different types of layers, the composite beam is fabricated. The core material is sandwiched between the same type of material at top and bottom, which have more strength than the inner material. Dweib et al. (2004) investigated the sandwich beam made with cellulose fibers and reinforced with the acrylate epoxidized soy-bean oil incorporated by foam care. Vacuum Assisted Resin Transfer Molding (VARTM) method is used to manufacture the beams. Flax beams are cast by wrapping the foam core with sufficient reinforcement material. The beams are casted by various fiber combinations like flax mats, woven glass fiber, recycled paper with chicken feather mats, recycled paper with corrugated paper and recycled paper with woven glass fiber. Four-point bending test is conducted on the beams to evaluate the mode of failure, strength and stiffness. From the failure loads, it has been observed that the E- glass beam is having the ultimate load of 39kN and flax beam is having the lowest ultimate load of 10kN. The ultimate loads of recycled paper with chicken feathers and corrugated is almost equal, and the values are 24.2 kN and 25.8 kN respectively. Figure 4 shows the different fracture surfaces of the recycled paper beam. Normally wood members are generally used as beams in buildings from the past. The strength of the wood beams mostly will fall within the range of 9 kN to 26 Kn,whereas the recycled paper beam with chicken feathers and corrugated cardboard has more strength than wood beams.

Sandwich composites are lightweight composites and having applications in marine fields, aerospace, wind energy and civil engineering structures. These composites are manufactured by two stiff thin skins sandwiched by the thick lightweight core material. Waddar et al. (2019) conducted the sandwich beam composite's buckling and free behavior under the axial compressive loads. Sisal

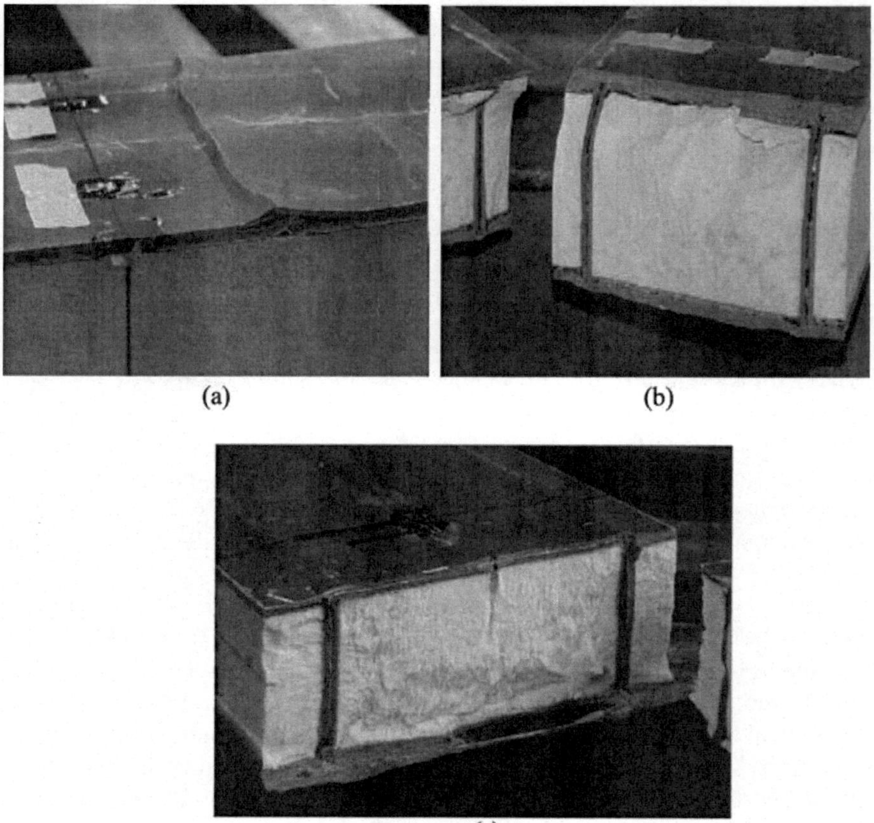

(a) (b)

(c)

Fig. 4 Fracture surfaces of the recycled paper beams **a** recycled paper with chicken feathers. **b** Recycled paper with corrugated paper. **c** Recycled paper with woven E-glass fiber (Dweib et al. 2004)

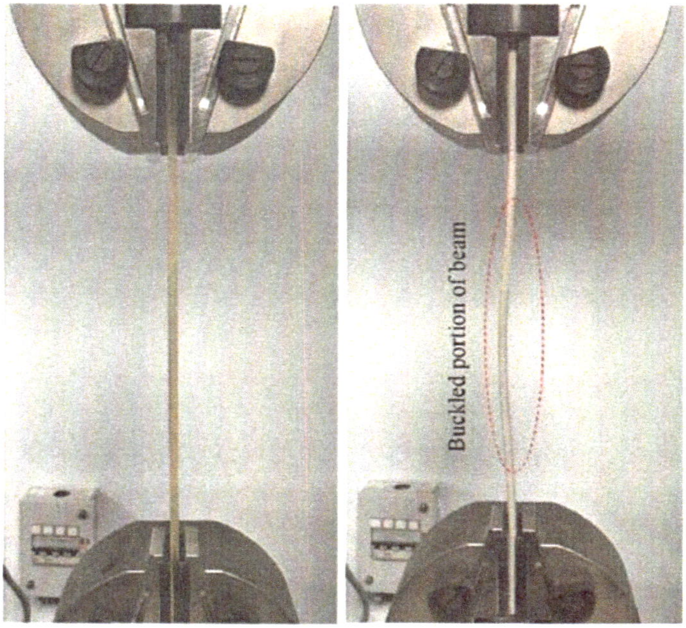

(a) (b)

Fig. 5 Syntactic foam sandwich beams **a** before buckling, **b** after buckling (Waddar et al. 2019)

is considered the syntactic foam which fills the top and bottom skin of the sandwich and core part of the sandwich is filled through the epoxy resin.To increase the stiffness of the foam, the cenospheres is used and the treated cenospheres is used to improve the modulus and stiffness of the foam. After demolding of composite, the resultant thickness of the manufactured sandwich is 4 mm. The beams are tested against the axial compressive load using the universal testing machine. The beams exhibited the global buckling modal shape without skin wrinkling and skin delamination. And due to the increase of the filler in the core, the buckling and natural frequencies are also raised. During the buckling test the prepared sandwich composites show the global buckling mode and maximum displacement at the middle of the composite, as shown in Fig. 5. The buckling load is increased in the untreated and treated composites compared to neat epoxy composites in the range of 7.85–25.44% and 19.92–38.98% (Waddar et al. 2019).

2.6 Ballistic Applications

Day by day, conflicts are rising worldwide and safeguarding the man's power of a particular country is a critical issue. The humans should be safeguarded against the shock and impact loads by the well-developed protection system known as personal protective armor. Generally, a multilayered ballistic armor system(MBAS) is used as a protection system which includes a ceramic front plate, flexible composite laminate, lightweight and ductile material as shown in Fig. 6 due to various stages of its layers MBAS protects the user from ballistic loads. The ballistic load is an impact caused by low mass projectile with high velocity triggered by some propelling force. These types of impact loads can also lead to fatal conditions to humans (Nayak et al. 2020). Composite laminate which is used in MBAS is synthetic fiber-reinforced composite laminate mostly. MBAS should be light in weight because it provides mobility to the user. To make the composite lighter in weight, various alternative materials have been identified, and natural fibers are alternative materials to synthetic fibers (Medvedovski 2010).

Tirillo et al. (2017) investigated the response of the basalt/ carbon fiber reinforced with epoxy resin subjected to high-velocity impact. The different stacking sequences of the fiber mats sandwich-type composite and intercalated type composite have been manufactured. To evaluate the difference between the carbon and basalt, the plain type of carbon/epoxy and basalt/ epoxy composites were also fabricated. For all the composites, the plate's thickness is 4 mm, and fiber volume fraction is considered in the range of 58–64%. With the speed of 165–475 m/s, a spherical tempered steel projectile was considered to conduct the impact test.

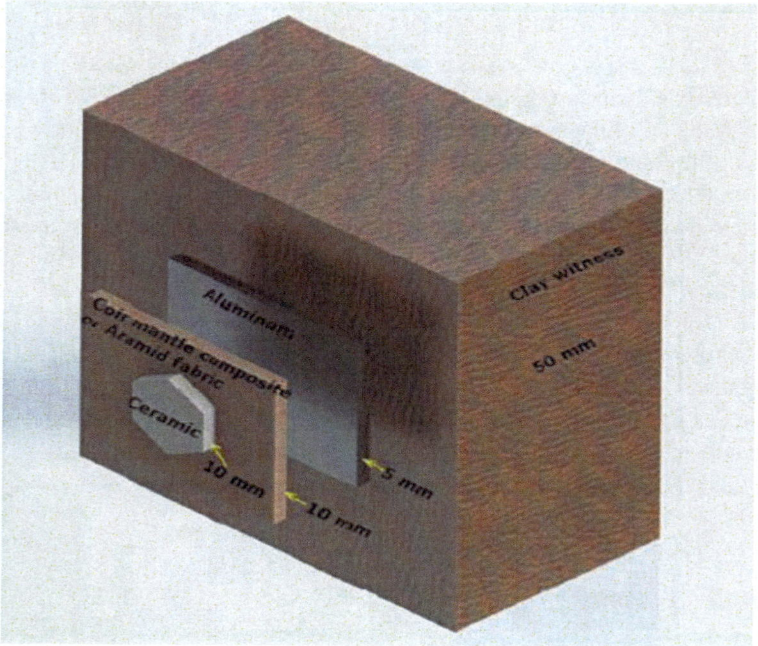

Fig. 6 Typical MBAS model for testing (Medvedovski 2019)

The results show that the hybrid type of composites have performed better than all the configurations of carbon/epoxy composites when the ballistic load is applied. Hybrid composites are failed because of the fibers' debonding and in the case of the carbon/epoxy composites, the failure in the form of fiber splitting and brittleness of the fiber. Intercalated composites are manufactured with the less basalt fiber content even though it exhibits better performance over the sandwiched type due to the different stacking sequence of the fibers.

Architecture
Up to nineteenth century almost all the buildings are modeled with natural fibers that are extracted from the bamboo. Most builders use the vernacular architecture these days by considering steel, concrete, glass, carbon, and bricks. When the non-vernacular architecture occurs in buildings and energy consumption is about 40% CO_2 emissions and 15% of it is needed to produce the materials required. By considering the natural fibers to replace non-renewable materials, the construction cost of buildings will be cheaper.

In most developing countries, the walls' internal and external panels are constructed by the use of natural fibers. The hemp fibers are mixed with concrete to casting the side walling of the buildings in England, as shown in Fig. 7. The main advantages of using natural fibers in vernacular architecture are wall cladding in the deck and outdoor furniture construction. The main aspect of architecture is the used materials should be easily convertible to the required shape. By using the natural type of fibers the resultant composites can be converted easily into any required shape (Steffens et al. 2017).

2.7 *Retrofitting of RCC Structures*

The reinforced concrete structures sometimes not be durable for a long time due to the members' deterioration. This deterioration is taking place due to inappropriate reinforcement, fatigue failure and corrosion under aggressive weather conditions (Padanattil et al. 2017).To overcome this condition, the deteriorated members should be retrofitted with some other external material. The wrapping of hybrid sisal glass fiber reinforced composite polymer (HSGFRP) around the concrete cylinder has greatly influenced the specimen's results. Three types of models are considered for testing the strength properties of the cylinder. In the first model, one layer of HSGFRP wrapped around the cylinder, in second model two layers of HSGFRP is placed, and in the third model three layers of HSGFRP is wrapped around the cylinder as shown in Fig. 8

The energy absorbed in the plain concrete specimen when tested against the load is 64.86 Nm whereas in the case of one layer wrapped cylinder it is 1032.25 Nm; in two layers wrapped cylinder it is 1415.64 Nm, and in the case of three layers wrapped cylinder it is 1886.25 Nm. The axial strain has improved by almost 44% in the case of HSGFRP. This improvement of axial strain also improves the ductility of

Fig. 7 Hemp line concrete walling(Steffens et al. 2017)

(a) plain concrete cylinder (b) cylinder wrapped with HSGFRP

Fig. 8 Retrofitting the cylinder specimens (Padanattil et al. 2017)

the member. The wrapping of HSGFRP is also increasing the seismic resistance of the reinforced members. To enhance composites' results, the fibers are required to be treated chemically with the process of alkalization (Padanattil et al., 2017).

2.8 Acoustic Applications

Berradi and Lannace (2015) identified that in recent years, natural fibers became the alternative for synthetic fibers to use in the sound-absorbing panels. Natural fibers have good thermal insulation properties, and these fibers are not harmful for health while manufacturing. The materials' high porosity content is required to absorb the sound and pass it to the matrix to the dissipation of sound. Authors considered different types of fibers like kenaf, jute, wood, cane, cork, sheep wool, and cardboard to prepare the composites and tested against the sound absorption capacity and flow resistance.

Among all the composites, sheep wool and coconut have higher values and are recognized as promising materials for absorbing the sound. The noise reduction coefficient is maximum for sheep wool of about 0.7 and a minimum for wood mineralized of 0.20. All-natural fibers are good while absorbing the medium or high frequencies of sound. But whereas in the low frequencies, the absorbing capacity is low, this drawback can be balanced by using the fiber's thickness and incorporating the air gaps in the composite.

2.9 Fire Resistance of Natural Fiber Composites

Natural fibers are considered to replace the synthetic type of fibers for use in the structural members. Flammability is one of the main limits of applications of natural fibers. Generally, bio fibers are flammable. Thus the heat released from the natural fibers is much more than the synthetic fibers (Chapple and Anandjiwala, 2010). The strength and stiffness of the fiber-reinforced polymers can be reduced at high range of temperatures. The glass fibers have a high thermal capacitance and if the fire once has started on the glass fiber due to its stored thermal energy, the reignition also occurs on the surface of the fiber. These fires will get ignited at one and it leads to difficulty to the firefighters.

In case of natural fibers, the fiber's performance is based on the chemical composition of the fiber. Natural fibers with the cellulose content in high ranges are highly flammable, and the fibers having the hemicellulose content are lesser flammable. The formation of char is better with more lignin content. If the fibers have the silica/ash, they also perform better against the fire. Apart from the fiber's chemical composition, the microstructure that represents the lower polymerization and higher crystallinity improve the fire performance (Dittenber and Ganagarao 2012).

Based on the thermogravimetric analysis, the thermal degradation of the fibers can be evaluated. Plant fibers like sisal and jute have an equal value of thermal degradation of 340 °C and the flax fiber has the value of 345 °C. Cellulose decompositions is occurred between 260 and 350 °C and in the process of decomposition, it will release tars, noncombustible gases and some char. As in the case of hemicellulose, the fiber decomposition falls in the range of 200 and 260 °C and releases more combustible gases and less tar when compared to cellulose.

Composite flammability mainly depends upon the type of matrix, and along with that, it also depends upon the type of fiber and the bonding between the fiber and matrix. Sometimes, the core materials present in the composite also affect the composite's fire properties. The composite fire resistance can be improved by adding a fire retardant or a thermal barrier coating. These additions can easily be added to the thermoset resins, but thermoplastic resins still need to be done (Burgueno et al. 2005).

3 Concluding Remarks

Natural fibers used in composites as a reinforcing material have been explored throughout the world due to its inherent advantages and many applications. Most of the natural fibers are produced from agricultural waste and it adds some contribution to the national GDP of agro-based economics. The remarks from the present investigation are summarized as follows:

- Natural fibers are becoming promising materials to replace the E-glass fibers for secondary level structural members.
- The moisture absorption capacity of the natural fibers can be reduced by the treatment of the fibers with different types of chemicals and among all those alkalinizations has resulted better properties.
- Flax fibers have a superior energy absorption capacity. Due to this, flax fibers have better performance over the jute and hemp fibers in case of ballistic loads.
- Composites are added with the flame retarders to improve the flammability and thermal stability.
- Compared to glass fiber, the natural fiber-reinforced composites have comparable structural properties in the view of strength and stiffness.
- NFRPCs produce better mechanical properties when the fiber weight fraction is around 30–40%.
- With the 5 wt% alkali treatment, the NFRPs will achieve better tensile and flexural properties.
- The addition of lignin content to the natural fibers also improves the impact properties of the composite.
- Hybridization of natural-synthetic fiber is influencing the composite properties in a greater way. With 8% of carbon by volume in the composite, about 50% more strength and stiffness are achievable.

- The addition of filler or nanofiller also reduces the few limitations of the composite. By using the flame retarders, the flammability and thermal stability of the composite is increased.
- The composites' strength and modulus will reduce if the composites are immersed in water for a longer duration.
- Significant research has been carried out to investigate biopolymers, fire retarders, and chemical treatment of the fibers.

4 Research Gaps and Future Research Needs

Even though many of the factors affect the usage of natural fibers in structural load members still, those fibers are attracted by many researchers by their renewability, biodegradability, and cost competitiveness (Lau et al. 2018). Further, to improve the properties of natural fiber composites, a lot of research is required in the future according to the following aspects:

- In the previous studies, thermosetting resins and thermoplastic resins are added to the natural fibers to manufacture the composites. The addition of these types of resins made composites partially biodegradable. To manufacture composites as fully biodegradable, further research needs to be carried using biodegradable resins for manufacturing the natural fiber composites.
- The main drawback of the natural fibers is that they have a hydrophilic nature, absorbing more moisture. Many of the researchers have identified different chemical treatments to reduce the hydrophilic nature of the fiber. Much research has determined that alkanization is the effective chemical treatment compared to all, but surplus amounts of alkaline will damage fiber texture. Further research is needed to identify alternative chemical treatments by considering different chemicals to improve natural fiber properties.
- Many of the researchers have investigated the properties of natural fiber composites in normal temperature conditions only. But if the natural fiber composites are using in industries and ballistic applications, they are subjected to elevated temperatures. Further research has to be investigated inflammability property to identify the different fire retarders and methods to improve natural fibers fire resistance at high temperatures.
- The past research has concluded that natural fiber composites mechanical properties are low compared to synthetic fibers. To enhance natural fiber composites properties competitively, hybridization is needed among the natural fibers and synthetic fibers to manufacture the composites. Hence, further research has been needed to identify the hybridization methodology while incorporating the natural fibers into synthetic fibers.
- Previous studies have shown that bast fibers like flax, hemp, and jute have better ballistic resistance than other natural fibers. Further research must be needed by considering the fibers from the stalk, seed, and hard fiber in ballistic applications.

- In the previous studies, a lot of research has been done by considering the natural fibers to make the composites. Nowadays, composites filled with natural fillers like cellulose nanocrystals, nano fibrillated cellulose are attracting researchers. These type of fillers improves the mechanical and thermal properties. Further research has to be carried out by identifying different kinds of natural fillers and the methodology of adding filler material into the polymers.

Acknowledgements Financial support provided by the Council of Scientific & Industrial Research (CSIR), New Delhi through Scheme no. (22(0781)/19/EMR-II) sanctioned to the corresponding author is highly appreciated for helping in the execution of the project.

References

Akil HM, Santulli C, Sarasini F, Tirillo J, Valente T (2014) Environmental effects on the mechanical behaviour of pultruded jute/glass fibre-reinforced polyester hybrid composites. Compos Sci Technol 94:62–70

Arpita GR, Sanjay MR, Senthamaraikannam P, Barile C, Yogesha B (2017) Hybridization effect of sisal/glass/epoxy/filler based woven fabric reinforced composites. Express Technol 41:577–584

Bambach MR (2020) Direct comparison of the structural compression characteristics of natural and synthetic fiber-epoxy composites: flax, jute, hemp, glass and carbon fibers. Fibers 62:01–14

Berradi U, Lannace G (2015) Acoustic characterization of natural fibers for sound absorption applications. Build Environ 94:840–852

Burgueno R, Quagliata MJ, Mehta GM, Mohanty AK, Misra M, Drzal LT (2005) Sustainable cellular biocomposites from natural fibers and unsaturated polyester resin for housing panel applications. J Polymer Environ 13:139–149

Cavalcanti DKK, Banea MD, Neto JSS, Lima RAA, Da Silva LFM, Carbas RJC (2019) Mechanical characterization of intralaminar natural fibre-reinforced hybrid composites. Compos Part B: Eng 175:107149

Chapple S, Anandjiwala R (2010) Flammability of natural fiber-reinforced composites and strategies for fire retardancy: a review. J Thermoplast Compos Mater 23:871–893

Chaudary V, Bajpai PK, Maheswari S (2018) Studies on mechanical and morphological characterization of developed jute/hemp/flax reinforced hybrid composites for structural applications. J Nat Fibers 15:80–97

Dittenber DB, Ganagarao HVS (2012) Critical review of recent publications on use of natural composites in infrastructure. Compos A Appl Sci Manuf 43:1419–1429

Dweib MA, Hu B, Donnel A, Shenton HW, Wool RP (2004) All natural composite sandwich beams for structural applications, composite structures. Compos Struct 63:147–157

Fan M, Fu F (2017) Introduction: a perspective-natural fibre composites in construction. In: Advanced high strength natural fibre composites in construction, pp 1–20

Flynn J, Amiri A, Chad U (2016) Hybridized carbon and flax fiber composites for tailored performance. Mater Des 102:21–29

Ghelli D, Minak G (2011) Low velocity impact and compression after impact tests on thin carbon/epoxy laminates. Compos Part B: Eng 42:2067–2079

Ismail MF, Sultan MTH, Hamdan A, Shah AUM, Jawaid M (2019) Low velocity impact behaviour and post-impact characteristics of kenaf/glass hybrid composites with various weight ratios. J Mater Res Technol 8:2662–2673

Kabir MM, Wang H, Lau KT, Cardona F (2012) Chemical treatments on plant-based natural fibre reinforced polymer composites: an overview. Compos Part B: Eng 43:2883–2892

Khan MZR, Srivasthava SK, Gupta MK (2018) Tensile and flexural properties of natural fiber reinforced polymer composites: a review. J Reinf Plast Compos 37:1435–1455
Kumar R, Ul Haq MI, Raina A, Anand A (2019) Industrial applications of natural fibre-reinforced polymer composites–challenges and opportunities. Int J Sustain Eng 12:212–220
Kureemun U, Ravandi M, Tran LQN, Teo WS, Tay TE, Lee HP (2018) Effects of hybridization and hybrid fibre dispersion on the mechanical properties of woven flax-carbon epoxy at low carbon fibre volume fractions. Compos Part B: Eng 134:28–38
Lau K, Hung P, Zhu MH, Hui D (2018) Properties of natural fiber composites for structural engineering applications. Compos Part B: Eng 136:222–233
Maslinda AB, Abdul Majid MS, Rizuan MJM, Afendi M, Gibson AG (2017) Effect of water absorption on the mechanical properties of hybrid interwoven cellulosic-cellulosic fibre reinforced epoxy composites. Composites Struct 167:227–237
Medvedovski E (2010) Ballistic performance of armour ceramics: Influence of design and structure. Part 2. Ceram Int 36:2117–2127
Mochane MJ, Mokhena TC, Mokhothu TH, Mtibe A, Sadiku ER, Ray SS, Ibrahim ID, Daramola OO (2019) Recent progress on natural fiber hybrid composites for advanced applications: a review. Express Polymers Lett 13:159–198
Nayak SY, Sultan MTH, Shenoy SB, Kini CR, Samant R, Shah AUM, Amunthakkanan P (2020) Potential of natural fibers in composites for ballistic applications—a review. J Nat Fibers 1–11
Padanattil A, Karingamanna J, Mini KM (2017) Novel hybrid composites based on glass and sisal fiber for retrofitting of reinforced concrete structures. Constr Build Mater 133:146–173
Ramana MV, Ramprasad S (2017) Experimental investigation on jute/carbon fibre reinforced epoxy based hybrid composites. Mater Today: Proc 4:8654–8664
Reddy N, Yang Y (2005) Biofibers from agricultural byproducts for industrial applications. Trends Biotechnol 23:22–27
Safri SNA, Sultan MTH, Jawaid M, Jayakrishna K (2018) impact behaviour of hybrid composites for structural applications: a review. Compos Part B: Eng 133:112–121
Steffens F, Steffens H, Oliveira FR (2017) Applications of natural fibers on architecture. Procedia Eng 200:317–324
Ticoalu A, Aravinthan T, and Cardona F (2010) A review of current development in natural fiber composites for structural and infrastructure applications. In: Southern Region Engineering conference, pp 113–117
Tirillo J, Ferrante L, Sarasini F, Lampini L, Barbero E, Sanchez Saez S, Valente T, Gaudenzi P (2017) High velocity impact behaviour of hybrid basalt-carbon/epoxy composites. Composite Struct 168:305-312
Waddar S, Pitchamani J, Doddamani M, Barbero E (2019) Buckling and vibration beviour of syntatic foam core sandwich beam with natural fiber composite facings under axial compressive loads. Compos Part B:Eng 175:107133
Wood BM, Coles SR, Maggs S, Meredith J, Kirwan K (2011) Use of lignin as a compatibiliser in hemp/epoxy composites. Compos Sci Technol 71:1804–1810